China's Long-Term Low-Carbon Development
Strategies and Pathways

Institute of Climate Change and Sustainable
Development of Tsinghua University et al.

China's Long-Term Low-Carbon Development Strategies and Pathways

Comprehensive Report

Institute of Climate Change and Sustainable
Development of Tsinghua University et al.
Beijing, China

ISBN 978-981-16-2526-8   ISBN 978-981-16-2524-4   (eBook)
https://doi.org/10.1007/978-981-16-2524-4

Jointly published with China Environment Publishing Group Co., Ltd.
The print edition is not for sale in China (Mainland). Customers from China (Mainland) please order the
print book from: China Environment Publishing Group Co., Ltd.

This Springer imprint is published by the registered company Springer Nature Singapore Pte Ltd.
The registered company address is: 152 Beach Road, #21-01/04 Gateway East, Singapore 189721,
Singapore

Lead authors:

Jiankun He, Zheng Li, Xiliang Zhang

Co-authors:

Hailin Wang, Wenjuan Dong, Shiyan Chang, Xunmin Ou, Siyue Guo, Zhiyu Tian, Alun Gu, Fei Teng, Xiu Yang, Can Wang, Bin Ouyang, Li Zhou, Xuecheng Wang, Jianhui Cong, Ershun Du, Siyuan Chen, Mingtao Yao, Zhiyi Yuan, Quan Bai, Qiang Zhang, Xiaofan Zhao

List of the Research Projects of China's Long-Term Low-Carbon Development Strategies and Pathways

Leading Institute: Institute of Climate Change and Sustainable Development of Tsinghua University

Project Advisor: Xie Zhenhua

Principal Investigators of the Project: He Jiankun, Li Zheng

List of Projects, Institutions and Principal Investigators:

1. Midterm and long-term goals, strategies, and pathways for China's socioeconomic development: State Information Center, Zhu Baoliang
2. Low-carbon strategies and pathways in the coordinated development of the eastern, central, and western China's economy and the process of urbanization: Institute of Ecological Civilization of Chinese Academy of Social Sciences, Pan Jiahua
3. Impacts of international trade and industrial relocation on China's low-carbon development in the context of globalization and policy recommendations: Chinese Academy of International Trade and Economic Cooperation of the Ministry of Commerce, Gu Xueming
4. Midterm and long-term strategies and pathways for China's energy system transformation: Energy Research Institute of the National Development and Reform Commission, Wang Zhongying
5. Optimized composition and technology roadmap of power source and grid, Department of Energy and Power Engineering of Tsinghua University, Li Zheng
6. China's midterm and long-term energy conservation potential, goals, policies, and cost–benefit analysis: Energy Research Institute of the National Development and Reform Commission, Dai Yande
7. China's midterm and long-term emission reduction technology evaluation, cost-benefit analysis, and technology roadmaps: School of Environment of Tsinghua University, Wang Can
8. China's industrial sector transformation and upgrading and low-carbon emission strategies and pathways: Energy Research Institute of the National Development and Reform Commission, Bai Quan

9. Low-carbon emission strategies and pathways of China's building sector: School of Architecture of Tsinghua University, Jiang Yi

10. Low-carbon emission strategies and pathways of China's transportation sector: China Academy of Transportation Sciences, Ouyang Bin

11. Study on investment strategies of China's midterm and long-term energy infrastructure transformation and development: Energy Research Institute of the National Development and Reform Commission, Kang Yanbing

12. Measures and pathways for China's consumption pattern transformation and low-carbon society construction: Department of Communications and Education of the Ministry of Ecology and Environment, Jia Feng

13. Strategies, measures and pathways for China's non-energy-related carbon dioxide and other greenhouse gases and in agriculture and forestry and land use: Institute of Energy, Environment, and Economy of Tsinghua University, Teng Fei

14. Coordinated measures and effect analysis of China's greenhouse gas emissions reduction and environmental governance: School of Environment of Tsinghua University, He Kebin

15. Policy guarantee system for China's realization of low-carbon development: Institutes of Science and Development of Chinese Academy of Sciences, Wang Yi

16. Recommendations for promoting global climate governance and international cooperation: National Center for Climate Change Strategy and International Cooperation, Xu Huaqing

17. China's midterm and long-term low-carbon emission scenario analysis and pathways: Institute of Energy, Environment, and Economy of Tsinghua University, Zhang Xiliang

18. Comprehensive report on China's low-carbon development transformation strategies and pathways: Institute of Climate Change and Sustainable Development of Tsinghua University, He Jiankun

Project management office: Ma Linwei, Kong Geng, Hong Yi

Foreword

2020 is an unusual year in which the COVID-19 pandemic has raged through the globe, infecting tens of millions of people and killing hundreds of thousands. The pandemic has not only wreaked havoc on public health systems, economic activities, and people's lives, but also has greatly affected and will continue to reshape the world's political, economic, and trade patterns.

Just like the coronavirus pandemic, climate change is also a major and urgent global challenge facing humankind. The difference lies in that the coronavirus pandemic emerged as a sudden and pressing crisis affects human health and lives, whereas climate change is a longer-term and deeper challenge threatening humans' survival and development. We can observe that during the past decades, as greenhouse gas (GHG) concentrations continue to climb, climate change and more frequent extreme climate events are increasingly threatening humankind's survival and health, endangering terrestrial and marine ecosystems and harming biodiversity. The Intergovernmental Panel on Climate Change (IPCC) Fifth Assessment Report elaborated on eight catastrophic risks brought about by climate change and pointed out that climate change is no longer a future challenge, but is already an immediate threat. United Nations Secretary-General António Guterres has said that the world is at a life-and-death juncture and has called for all countries in the world to join efforts in combating climate change, which is the most significant and urgent problem facing humanity today.

In 2020, in the face of grave crises, especially the coronavirus pandemic and climate change, people have begun to rethink about the relationship between humans and nature, and are increasingly aware that people and the environment are an interdependent community with a shared future. People should make more effort to respect, adapt to, and protect nature, should attach more importance to the harmonious coexistence of people and nature, and should coordinate present and future with a view to planning ahead for global challenges. This means that we must fundamentally change traditional production modes, lifestyles, and consumption patterns, promote transformation and innovation, and proceed on a green, low-carbon, and circular development path. Climate, development, and environment should not be isolated or addressed alone. Instead, climate actions, together with economic, social, environmental, health, employment, stability, security, and other issues, should be regarded as a large-scale system to achieve coordinated development. By taking the path of sustainable development, people of the world must deal with climate change, protect the environment, reverse the trend of biodiversity loss, and ensure the long-term health and safety of humankind.

After the outbreak of COVID-19 in 2020, it has become the universal consensus of the international community to achieve economic recovery through green and low-carbon development. On Earth Day on April 22, Secretary-General Guterres proposed a global initiative of green, high-quality recovery and urged countries to stress climate actions in economic recovery measures. As of October 2020, over 100 countries have already committed to being carbon neutral in 2050. The European Union (EU) unveiled its Green Deal at the end of 2019, and committed Europe to achieve carbon neutrality by 2050. Meanwhile, the EU has issued a roadmap for policies and measures in seven fields, namely, energy, industry, building, transport, food, ecology, and environmental protection, to facilitate a green recovery. Germany, which has the new Presidency of the Council of the EU, has proposed a recovery plan from the coronavirus pandemic in which there is strong support for green growth and climate change mitigation is listed as one of the three priorities. Even the US House of Representatives, which announced on a Climate Crisis Action Plan in June 2020, has proposed efforts to limit global warming to 1.5°C by the end of this century. The plan points out that addressing climate change should be regarded as a national priority, aiming to reduce GHG emissions by 88% below 2010 levels in 2050 and to reach net-zero carbon dioxide (CO_2) emissions by 2050. The plan also details planned future measures in the areas of economy, employment, infrastructure, public health, and investment. It has been recognized and supported by Joe Biden, who has just won the 2020 presidential election.

China has more recently attached great importance to addressing climate change and has been adhering to green, circular, and low-carbon development as important strategic measures for promoting high-quality sustainable development. China has integrated tackling climate change into overall socioeconomic development. Since the 12th Five-Year Plan (2011–2015), China has been promoting low-carbon development based on the systematic and binding goal of reducing per capita GDP carbon intensity. In 2015, China set Nationally Determined Contributions (NDCs) of carbon emissions peaking around 2030 and peaking at the earliest time possible. Various policies and measures have been adopted, including adjusting the industrial structure, conserving energy and resources, improving utilization efficiency of energy and resources, optimizing energy structure, developing non-fossil fuels, developing a circular economy, increasing forests as a carbon sink, establishing and operating a carbon market, and South–South cooperation, to facilitate green and low-carbon transformation of the entire society. In 2019, per capita GDP CO_2 emissions in China dropped by 48% from 2005 levels, equivalent to a reduction of about 5.62 billion tons of CO_2 emissions (or 11.92 million tons of sulfur dioxide or 11.3 million tons of nitrogen oxides). GDP growth during 2005 to 2019 more than quadrupled, and 95% of the poor were lifted out of poverty. The proportion of the tertiary sector rose from 41.3% to 53.9% while coal consumption dropped from 72.4% to 57.7%, and non-fossil fuels accounted for 15.3% in primary energy, up from 7.4%. Average life expectancy increased from 72.9 years to 77.3 years. All the data show that policies and actions for addressing climate change will not hinder economic development, but will achieve the co-benefits of improving the quality of economic growth, cultivating new industries and markets, boosting employment, improving people's livelihood, protecting the environment, and enhancing people's health.

However, we should also note that, alongside the potential there are huge challenges in China's low-carbon transformation. First, the manufacturing sector, with its high energy and material consumption and low added-value rate, is still in the middle and low end of the international industrial value chain, posing daunting tasks of economic structural adjustment and industrial upgrading. Second, coal consumption still accounts for a high proportion of energy use (over 50%). CO_2 emissions per unit of energy consumption are about 30% larger than the world average, making the task of energy structural optimization formidable. Third, energy consumption per unit of GDP is still high at 1.5 times the world average and 2–3 times that of developed countries. Establishment of a green and low-carbon economic system is arduous work.

Addressing climate change meets the domestic demand for sustainable development. Looking to the future, as China's endeavor to achieve a modern socialist economy with Chinese characteristics enters a new era, the transformation to green and low-carbon development is a fundamental way to solve the problems of imbalanced and insufficient development, coordinate efforts for promoting economic development, improving people's livelihoods, eliminating poverty, and preventing or controlling pollution, and to realize the goals of building a moderately prosperous society in all respects. These goals would realize socialist modernization by 2035 and develop China into a prosperous, strong, democratic, culturally advanced, harmonious, and beautiful country by 2050.

Addressing climate change is the common mission of humankind. President Xi has pointed out clearly that China has been actively fostering international cooperation in its response to climate change, and has become an important participant, contributor, and torchbearer in the global endeavor for ecological civilization. From a global viewpoint, green and low-carbon development has become the irreversible trend in the economic, energy, technology, and governance systems of all countries, and is the fundamental path to dealing with humankind's common crises. China's efforts to promote green and low-carbon transformation around the world and to build a human community with a shared future under the guidance of Xi's thoughts on ecological civilization demonstrates tremendous political courage and a sense of mission of China as a great developing country.

Therefore, we need to, during the period of the 14th and 15th Five-Year Plans and even longer into the future, maintain strategic resolve, insist on a green and low-carbon development philosophy, promote ecological conservation, continue to adopt active policies and actions to respond to climate change, fulfill all of the NDCs, and strive to do even better than the NDCs. Faced with an increasingly difficult international situation at this historic juncture, it is particularly meaningful to conduct in-depth research on such topics as how China could plan low-carbon development strategies, pathways, and measures in a blueprint of socialist modernization in the new era, how China should assume international responsibilities and obligations suited to its national conditions and capabilities based on the principles in the Paris Agreement, and finally, how China should promote and lead the process of global climate governance.

To address the abovementioned topics, since the beginning of 2019, the Institute of Climate Change and Sustainable Development (ICCSD) of Tsinghua University has been cooperating with dozens of research institutes to undertake a research project, *China's Long-Term Low-Carbon Development Strategies and Pathways*, supported by the Special Fund for Climate Change and Global Green Development of Tsinghua University Education Foundation and the Energy Foundation. The research institutes with which the institute has been cooperating include the National Center for Climate Change Strategy and International Cooperation, the State Information Center, the Institute for Urban and Environmental Studies of the Chinese Academy of Social Sciences (recently renamed the Institute for Ecological Civilization), the Institute of Science and Development of the Chinese Academy of Sciences, the Energy Research Institute of the National Development and Reform Commission, the Department of Communications and Education of the Ministry of Ecology and Environment, the Academy of International Trade and Economic Cooperation of the Ministry of Commerce, the Academy of Transportation Sciences of the Ministry of Transportation, and the Department of Energy and Power Engineering, Institute of Energy, Environment, and Economy, School of Environment and School of Architecture, Tsinghua University. Based on China's national conditions, the team, after comprehensively considering socioeconomic, policy, energy, and other macro-development trends and needs, has proposed suggestions on China's low-carbon development strategies, pathways, technologies, and policies until 2050 in order to realize the goals of building a great socialist modern country, a beautiful China, and to achieve the global warming goals set in the Paris Agreement alongside global sustainable development.

This comprehensive report builds on, extends, and enriches the reports for the 18 sub-projects in this overall project. We hope that our research findings can serve as reference for all sectors of society in their research efforts, offer support for the formulation and implementation of national low-carbon development strategies and policies, and help to tell China's stories well to the world in the general trend of global green, low-carbon development.

July 2020

Xie Zhenhua
Beijing, China

Preface

Global climate governance has entered the post-Paris Agreement era in which full implementation is placed at the forefront of climate action priorities. The Paris Agreement aims to keep global temperature rise well-below two degrees Celsius while active pursuing efforts to limit it within 1.5 degrees Celsius, compared to pre-industrial levels. It introduced a new bottom-up mechanism where countries submit their targets and action plans in their nationally determined contributions (NDCs). However, the current emission reduction pledges still fall far in achieving the goal of limiting temperature rise below two degrees. Therefore, the Paris Agreement requests all binding parties to communicate and strengthen their 2030 NDC ambitions and actions, and submit their mid-century, low greenhouse gas emission development strategies by 2020.

The 19th National Congress of the Communist Party of China put forth the goals, fundamental strategies and key action plans towards a socialist modernization in the new era. Aimed at building China into a modern socialist country that is prosperous, strong, democratic, culturally advanced, harmonious, and beautiful by the middle of the century. While making new and greater contributions to global ecological safety and human progress, and the building of a community with a shared future for humankind. Thus, China needs to align its overall strategies both at the national and international level, implement its national strategy on climate change mitigation and adaptation under the guidance of President Xi's Thought on Socialism with Chinese Characteristics for a New Era, conform and lead the low-carbon transition of energy and economy internationally. By the middle of this century, China should ensure the attainment of the second centenary goal of building a modern socialist country, it must also strive toward the long-term global target set forth by the Paris Agreement for addressing climate change, and embrace a deep decarbonization pathway consistent with the global target of capping temperature rise within two degrees Celsius. Thus, it is also imperative to study and devise a mid and long-term low-carbon emission development strategy in China now, and incorporate into the overall plan of socialist development in the new era.

Countries around the world are either formulating or have already submitted their mid- and long-term low-carbon development strategies. Some developed regions such as the EU, and the UK, and developing regions such as Chile, Ethiopia and

Fiji have already committed to net-zero carbon emissions by 2050. Thus far, 121 countries have set out mid-century carbon neutrality goals, and 114 countries have declared intentions to update their 2030 NDCs. The flourishing of alliances and advocacy organizations for climate action from various industries, enterprises and NGOs, along with the blooming number of new international rules, industry norms and social codes of conduct for low-carbon transition have also encouraged countries to take more ambitious actions against climate change.

In implementing its national climate change strategy, China has incorporated energy-saving, emission and carbon reduction goals into its Five-Year Plan for National Economic and Social Development, and carried out institutional safeguards and policy measures, scoring enormous achievements that received internationally recognition. In addition, China has submitted an ambitious NDC and action plan under the Paris Agreement, which should be executed step-by-step in the next two five-year plans, coupled with enforced measures and actions to ensure the achievement of the NDC goals. China will adhere to a new set of development principles in the new era. It seeks to transform development drivers through innovation, reshape development method through green initiatives, create and form a green, low-carbon circular industrial system with a clean, low-carbon, safe and efficient energy system, step up institutional capacity-building for ecological civilization, enable coordinated governance to facilitate high-quality economic growth between the economy, energy, environment, and climate change for an all-win scenario. All of the above efforts have laid solid institutional and policy groundwork to fulfill the 2030 voluntary reduction commitments and the long-term goal of deep decarbonization.

China upholds Xi Jinping Thought on Ecological Civilization and the philosophy of building a community with a shared future for humankind, actively participating in and encouraging international cooperation on climate change. Working together to build an equitable, balanced, and win-win mechanism of global climate change governance, to become an active participant, contributor and torchbearer in building global ecological civilization. The experiences and case study of China's low-carbon transition embodying Chinese wisdom and approach to contribute to the global pursuit of a climate-friendly low-carbon path of economic development.

The coronavirus has taken a heavy toll on the world economy, triggering greater changes in global political, economic and technological competition, and prompting an overhaul of the global governance system. This overshadows international cooperation on climate change with great uncertainty. Now, there is a groundswell of support internationally to a post-COVID green economic recovery, swifter low-carbon transition and joint actions against the furthering the effect of climate change, which is emerging as a global consensus. This will produce a profound impact on countries' policies and pathways to economic recovery and development in the post-COVID era. In researching and developing a long-term low-carbon development strategy for mid-century, China is closely following and leading the global trends of energy and economic low-carbon transition, contributing to the ecological safety of the planet, and the survival and development of all humankind that reflects its domestic capability and international influence. Moreover, it helps foster China's own economic, trade and technological competitive edge to secure sustainable development and enhance its global competitiveness and influence.

To adapt to the new realities of domestic and international climate change efforts, the Institute of Climate Change and Sustainable Development of Tsinghua University, with the support from relevant authorities, took the leadership in bringing together a team of more than a dozen domestic research institutions for comprehensive studies in multiple fields and disciplines on China's long-term low-carbon development strategies and pathways, covering 18 sub-projects. This comprehensive report builds on, extends, and enriches the reports of the 18 sub-projects. The comprehensive report consists of twelve chapters, each of which has a list of authors as shown below, and finally, He Jiankun was responsible for the compilation of the draft and short version report.

Chapter 1: Dong Wenjuan, He Jiankun
Chapter 2: Chang Shiyan, Wang Hailin, He Jiankun
Chapter 3: Ou Xunmin, Tian Zhiyu, Guo Siyue, Bai Quan, Yuan Zhiyi, Wang Xuecheng, Ouyang Bin
Chapter 4: Dong Wenjuan, Chen Siyuan, Du Ershun, Li Zheng
Chapter 5: Zhou Li, Wang Hailin, Zhang Qiang, Zhang Xiliang, He Jiankun
Chapter 6: Gu Alun, Teng Fei
Chapter 7: Ou Xunmin, Chang Shiyan, Cong Jianhui, Wang Can, Yuan Zhiyi
Chapter 8: Chang Shiyan, Ou Xunmin, Yao Mingtao, Tian Zhiyu, Guo Siyue, Dong Wenjuan, Du Ershun
Chapter 9: Wang Hailin, Zhao Xiaofan, He Jiankun
Chapter 10: He Jiankun, Wang Hailin, Zhao Xiaofan
Chapter 11: Wang Hailin, Zhao Xiaofan, He Jiankun
Chapter 12: He Jiankun, Li Zheng, Zhang Xiliang, Yang Xiu

The research on China's long-term low-carbon emission strategy and pathways aims to provide holistic evaluation on the strategic linchpins, policies and pathways of implementation for the energy and economic deep decarbonization that would echo of China's two-stage approach for socialist modernization: the country's efforts to honor and enhance its NDC commitments during the first phase, and achieve the target of limiting global temperature increase to well-below 2 degrees Celsius and ideally 1.5 degrees Celsius during the second phase. We hope that our research findings can serve as reference for all sectors of society in their research efforts, and offer support for the decision-making of government authorities.

We would like to express our deep gratitude to the Special Fund for Climate Change and Global Green Development of Tsinghua University Education Foundation and Energy Foundation for their generous support to the research project.

Beijing, China

He Jiankun
Li Zheng
Zhang Xiliang

Contents

Abbreviations and Units of Measure

Abbreviations

AC	Alternating Current
AWE	Alkaline Water Electrolysers
BCA	Border Carbon Adjustment
BECCS	Biomass Energy Carbon Capture and Storage Systems
BF-BOF	Blast Furnace – Basic Oxygen Furnace
BIGCC	Biomass Gasification Combined Cycle
BM	Bio-Methanol
BRI	Belt and Road Initiative
CAES	Compressed Air Energy Storage
CBD	Convention on Biological Diversity
CBDR	Common But Differentiated Responsibilities
CCHP	Combined Cooling, Heating and Power
CCS	Carbon Capture and Storage
CCUS	Carbon Capture, Utilization and Storage
CDM	Clean Development Mechanism
CDR	Carbon dioxide Reduction
CER	Certified Emission Reduction
CFC	Chlorofluorocarbon
CH_4	Methane
CHP	Combined Heat and Power
CM	Chemical Methanol
CNNC	China National Nuclear Corporation
CO_2	Carbon dioxide
CO_2e	Carbon dioxide equivalent
COP	Conference of the Parties (within the United Nations Framework Convention on Climate Change)
COVID-19	Coronavirus Disease 2019
CPC	Communist Party of China
CSP	Concentrating Solar Power

DACCS	Direct Air Capture with Carbon Storage
DC	Direct Current
DRI	Direct Reduced Iron
EAF	Electric Arc Furnace
EIB	European Investment Bank
EPA	US Environmental Protection Agency
EU	European Union
EV	Electric Vehicle
FCV	Fuel Cell Vehicle
FYP	National Five-Year Plan of for Social and Economic Development, China
GDP	Gross Domestic Product
GHG	Greenhouse Gas
GTAP	Global Trade Analysis Project
GWP	Global Warming Potential
HCFC	Hydrochlorofluorocarbon
HFC	Hydrofluorocarbon
HTGR	High Temperature Gas-Cooled Test Reactor
HVAC	High Voltage Alternating Current
HVDC	High Voltage Direct Current
IAEA	International Atomic Energy Agency
IATA	International Air Transport Association
ICCSD	Institute of Climate Change and Sustainable Development of Tsinghua University
ICE	Internal Combustion Engine
IDP	Internally Displaced Person
IEA	International Energy Agency
INDC	Intended Nationally Determined Contribution (within the United Nations Framework Convention on Climate Change)
INET	Institute of Nuclear and New Energy Technology, Tsinghua University
IoT	Internet of Things
IPCC	Intergovernmental Panel on Climate Change
IT	Information Technology
JAEA	Japan Atomic Power Agency
LEILAC	Low Emissions Intensity Lime and Cement
LNG	Liquefied Natural Gas
LPG	Liquefied Petroleum Gas
LULUCF	Land Use, Land-Use Change and Forestry
MIIT	Ministry of Industry and Information Technology (China)
MRV	Measurement, Reporting and Verification
MTO	Methanol to Olefins
MTP	Methanol to Propylene
NbS	Nature-based Solutions
NCM	Nickel, Cobalt and Manganese (Materials of Lithium Battery)

NDC	Nationally Determined Contributions to the Paris Agreement (within the United Nations Framework Convention on Climate Change)
NEV	New Energy Vehicle
NGO	Non-Governmental Organization
NO	Nitrogen Monoxide
N_2O	Nitrous Oxide
NO_2	Nitrogen Dioxide
NO_x	Nitrogen Oxides
ODP	Ozone Depletion Potential
OECD	Organization for Economic Co-operation and Development
PFC	Perfluorocarbons
PHEV	Plug-in Hybrid Electric Vehicles
$PM_{2.5}$	Particulate matter with an aerodynamic diameter equal to or less than 2.5 μm
PPP	Public-Private Partnership
PtG	Power-to-Gas
PV	Photovoltaics
PWR	Pressurized Water Reactor
R&D	Research and Development
REACH	Registration, Evaluation, Authorisation and Restriction of Chemicals (a regulation of the European Union)
RMB	Renminbi
SAI	Stratospheric Aerosol Injection
SCWR	Supercritical Water Cooling Reactor
SDG	Sustainable Development Goals
SF_6	Sulphur hexafluoride
SMR	Steam Methane Reforming
SNAP	Significant New Alternatives Policy (a program of the US EPA to identify and evaluate substitutes for ozone-depleting substances)
SOC	Soil Organic Carbon
SO_2	Sulphur dioxide
SRFs & CRFs	Slow Release Fertilizers & Controlled Release Fertilizer
SRM	Solar Radiation Management
SR15	IPCC Special Report on Global Warming of 1.5°C
SSAB	Swedish Steel AB
TMR	Total Mixed Rations
UK	United Kingdom of Great Britain and Northern Ireland
UN	United Nations
UNCED	United Nations Conference on Environment and Development (1992)
UNECE	United Nations Economic Commission for Europe
UNEP	United Nations Environment Programme
UNFCCC	United Nations Framework Convention on Climate Change
US	United States of America

USD	US Dollar
VSG	Virtual Synchronous Generation
WMO	World Meteorological Organization
WTO	World Trade Organization

Units of Measure

a	Year
cm	Centimeter
EJ	Exajoule (10^{18} joules)
gce	Gram of coal equivalent
Gt	Gigaton (10^9 metric tons)
GW	Gigawatts
ha	Hectare
kg	Kilogram
kgce	Kilogram of coal equivalent
km^2	Kilo square meter
kW	Kilowatt (1,000 watts)
kWh	Kilowatt-hour
mm	Millimeter
m^2	Square meter
m^3	Cubic meter
Mt	million metric tons
MPa	Million Pascal
Mtce	Million metric tons of coal equivalent
MW	megawatts (a million (10^6) watts)
Nm^3	Normal cubic meter
ppm	Parts per million
PWh	Petawatt-hour (10^{15} watt-hours)
t	Metric ton
tce	Metric tons of coal equivalent
Tg	Teragram (10^{12} gram)
Wh	Watt-hour
yr	Year
μg	Microgram (one millionth of a gram)

List of Figures

List of Tables

Chapter 1
Introduction

In 2015, 195 countries and regions signed on to the Paris Agreement, which laid the institutional framework for global cooperation on climate change after 2020. The Paris Agreement becomes effective when 55 Parties representing at least 55% of global emissions join the Agreement, a criteria met on October 5, 2016. Officially taking effect on Nov 4 2016,[1] the Agreement established a long-term temperature goal to combat climate change: keeping the global temperature rise (hereafter increase) this century well below 2°C above pre-industrial levels and pursuing efforts to limit the temperature increase even further to 1.5°C. To achieve this temperature goal, the Parties aim to reach global peaking of greenhouse gas emissions (GHGs) as soon as possible and achieve net zero emission of greenhouse gases in the second half of this century. By 2020, parties are expected to communicate or update their 2030 NDCs as well as their mid-century, long-term low greenhouse gas emission development strategies to the UNFCCC secretariat. As of the end of 2019, 195 parties have signed on to the Agreement; 187 parties have ratified or accepted the Agreement; 184 parties have submitted their first NDCs and two have updated their NDCs. In addition, 14 other parties have submitted their long-term low-emission development strategies.[2]

China's 19th Party Congress laid out the overarching goal and masterplan of achieving socialist modernization by 2050. Targets and strategies for addressing climate change that align with the country's blueprint for development should be studied and formulated. China, like all nations that have signed the Paris Agreement, is obliged to submit a mid-century, long-term low greenhouse gas emission development strategy. The research on China's Long-term Low-carbon Development Strategies and Pathways emerged out of this context. The project centers around two areas: (1) given the condition of limiting global temperature increase within 2°C and

---

[1] The Paris Agreement was open for signature from April 22, 2016, to April 21, 2017 and, like the Kyoto Protocol, would only enter into force when 55 parties accounting for at least 55 percent of global greenhouse gas emissions ratified it.

[2] https://unfccc.int/process/the-paris-agreement/long-term-strategies.

© China Environment Publishing Group Co., Ltd. 2022
Institute of Climate Change and Sustainable Development of Tsinghua University et al.,
China's Long-Term Low-Carbon Development Strategies and Pathways,
https://doi.org/10.1007/978-981-16-2524-4_1

developing China into a great modern socialist country, the pathways of China's low-carbon transition to 2050 from aspects of funding, technology, policy required; and (2) examining the feasibility for China to deliver carbon neutrality by 2050 under the 1.5°C target, and evaluating the corresponding technical pathways, costs, obstacles and socioeconomic impact.

1.1 Global Climate Change Impact and Response

1.1.1 The Global Impact of Climate Change

Climate change refers to a rise in the average global temperature, extreme weather and other climate anomalies caused by increasing CO_2 and other greenhouse gases in the atmosphere caused by human activities, such as the burning of fossil fuels and deforestation, as well as natural changes in the environment. The United Nations Framework Convention on Climate Change (UNFCCC) defines climate change as "a change of climate which is attributed directly or indirectly to human activity that alters the composition of the global atmosphere and which is in addition to natural climate variability observed over comparable time periods." The Intergovernmental Panel on Climate Change (IPCC) believes that human activities in the industrial age have contributed to climate change in two main ways: first, the use of fossil fuels has led to increasing atmospheric concentrations of CO_2, aerosols and other pollutants, resulting in an increase in the levels of heat-trapping greenhouse gases in the atmosphere as well as pollution; and second, changes in land use and land cover affect global climate via reflections between the land surface and the atmosphere. These reflections modify near-surface energy, moisture, and momentum fluxes via changes in albedo and surface roughness, among others. According to the IPCC Fifth Assessment Report, "it is extremely likely (95 percent confidence) that human influence on climate caused more than half of the observed increase in global average surface temperature from 1951 to 2010" [1].

According to the World Meteorological Organization (WMO) bulletin, the physical signs and socio-economic impacts of climate change are accelerating as record greenhouse gas concentrations drive global temperatures towards increasingly dangerous levels. Atmospheric CO_2 levels have risen from 357 ppm in 1994 to 405 ppm in 2017. The warming trend in the global climate system continues. Global mean warming reached 1°C above pre-industrial levels in 2018, and the previous five years from 2014 to 2018 were the warmest years in the modern record of meteorological observation. In addition, other obvious signs of climate change, including sea level rise, warming ocean water, ocean acidification, melting glaciers and ice sheets continue to emerge while extreme weather and climate events such as tropical storms, floods, heavy precipitation, heat waves, droughts, cold waves and snow storms have wrought havoc across continents (Fig. 1.1) [2].

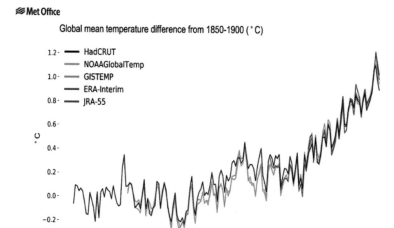

Fig. 1.1 Global mean surface temperature changes since 1850 (*Source* World Meteorological Organization (WMO), 2019. WMO statement on the state of global climate in 2018)
(*Note* HadCRUT—UK Met Office Hadley Centre for Climate Science and Services and the Climatic Research Unit of the University of East Anglia; NOAAGlobalTemp—US National Oceanic and Atmospheric Administration Global Surface Temperature Dataset; GISTEMP—NASA Goddard Institute for Space Studies (GISS) surface temperature analysis for the globe; ERA-Interim-European Centre for Medium-Range Weather Forecasts; JRA-55—Japan Meteorological Agency's the Japanese 55-year Reanalysis)

In the meantime, the impact of climate change is increasingly visible. Natural disasters, mostly linked with extreme weather and climate events, hit 62 million people in 2018. Floods affected the largest number of people—over 35 million. It was estimated that the number of undernourished population increased to 821 million in 2017, partly because of severe droughts linked to El Niño between 2015 and 2016. Climate change has also triggered population displacement. As of September 2018, out of the 17.7 million Internally Displaced Persons (IDPs) tracked by the International Organization for Migration, over two million were displaced due to disasters linked to weather and climate events. Climate and air quality have many interactions, which are exacerbated by climate change. According to the World Health Organization, between 2000 and 2016, the number of people exposed to heat waves grew by around 125 million, while the duration of the average heatwave is 0.37 days longer compared to the period 1986–2008. Other impacts include the loss of "blue carbon" ecosystems found along the world's coasts and ocean in habitats such as mangrove forests, seagrass meadows, and salt marshes [2].

The *IPCC Special Report on Global Warming of 1.5°C (SR15)* released in 2018 pointed out that climate change is no longer a future challenge, but an immediate threat. The report emphasized the need for urgent climate actions. Global warming has already reached 1°C above pre-industrial levels and might reach 1.5°C as soon as 2030. At this rate, the warming might reach or even surpass 2°C around 2065, which would mean that the global temperature increase target set by the Paris Agreement for the end of the century would be exceeded four decades in advance. The report also indicated that there is a big difference in climate impacts between 1.5 and 2°C. If global warming rises from 1.5 to 2°C, more devastating consequences, such as loss of habitats, melting of ice sheets, sea level rise, will occur. This would further threaten the survival and development of human beings, and inflict more damage on the global economy. IPCC points out that in order to hold global warming within 1.5°C, rapid and far-reaching transformation is required in land, energy, industry, buildings, transport, cities, and other relevant areas. By 2030, global net human-caused CO_2 emissions would need to reduce by around 45% from 2010 levels, reaching net zero around 2050. Meanwhile, a significant reduction in non-CO_2 GHG emissions is also required [3].

In 2019, the IPCC issued the *Special Report on Climate Change and Land*, which underscored the threat climate change posed to food security, how it could be a driver of desertification and land degradation, and called on countries to use land in a sustainable manner. Land is of vital importance in the climate systems across the world. Agriculture, forestry and other types of land use account for 23% of human greenhouse gas emissions. In the meantime, the terrestrial biosphere absorbs one-third of carbon emissions from fossil fuel and industry. The land has long been overtaxed by human activities, while climate change exacerbated the situation. At 2°C of global warming, the risks from permafrost degradation and food supply insta-bilities are projected to be very high. At 1.5°C, the risks from dryland water scarcity, wildfire damage, permafrost degradation and food supply instabilities are all on the rise. Land must remain productive to maintain food security as the population grows and the negative impacts of climate change on vegetation is expected to increase. This has posed limits to the role that land can play in addressing climate change. In addition, it also takes time for trees and soil to effectively store carbon efficiently [4].

In the subsequent *Special Report on the Ocean and Cryosphere in a Changing Climate*, IPCC highlighted the urgency of timely, resolute and determined and coor-dinated actions to address unprecedented and enduring changes in the ocean and cryosphere. The ocean and cryosphere (the frozen parts of the planet) have served as a crucial buffer for life on earth against global warming, with the ocean absorbing more than 90% of the excess heat trapped on earth by greenhouse gases, and soaking up a quarter of human CO_2 emissions. The report assessed and encapsulated a series of scientific discoveries related to climate change. For instance, the sea level rose globally by 15 cm during the whole of the twentieth Century, and it is currently rising at double that speed; if greenhouse gas emissions are not curtailed, by 2100, the rate of sea level rise will be over 10 times that in the twentieth century; high greenhouse gas emissions will lead to the loss of over one-third of glacier ice on average by 2100, threatening people who rely on glaciers as a water resource, etc. With the increase in ocean surface temperature and acidification, marine life and marine ecosystems are facing tremendous challenges. Even if the global warming is held at 1.5°C, it is estimated that as much as 90% of warm water coral reefs that exist today will disappear [5].

1.1.2 The Impact of Climate Change on China

China is one of the regions in the world that are vulnerable to climate change. The country has experienced a faster rate of warming than the global average, which has produced a significant impact on its water resources, agriculture and ecosystems, etc. In 2018, warming persisted in the climate system ofChina. The mean surface temperature showed a clear upward trend; extreme weather and climate events increased with greater intensity; the cryosphere melted faster; and climate risk was on the rise. The climate change in China in the past 50 years has been sparked by a combination of rising temperatures due to human activity and natural climatic fluctuations. Meanwhile, China saw higher rates of warming between the 1970s and the turn of the century, and rapid urbanization may also have had varying degrees of impact on local temperature rises [6].

Atmosphere: In line with global surface temperature warming trends, not only has China witnessed rising surface temperatures, temperatures recorded across the country have been rising faster than the global average. The past two decades have been the warmest since the turn of the century, and 2018 was an exceptionally warm year. From 1951 to 2018, China's annual average temperature increased by 0.24°C every decade, and the rate of increase was remarkably higher than the global average. Consistent with the global and China's surface temperature warming trends, the temperature in the upper-troposphere has also shown a notable increase since 1961. In the meantime, annual precipitation averaged across the country has increased slightly, with apparent regional and seasonal differences, and extreme precipitation events has also increased. From a long-term perspective, manual observations and satellite imagery have shown a decreasing trend of total cloud cover over the country since 1961. During the same period, consistent with global changes in solar radiation, the average amount of sunlight reaching the Earth's surface in China has gone down steadily. A significant reduction in sunshine duration of 1.4% every decade on average was also reported. The average terrestrial wind speed has also decreased. Extreme heat events have seen a sharp uptick since the mid-1990s [6].

Ocean: Climate along the coast is dominated by monsoon winds. Changes in the East Asian monsoon, subtropical highs and Arctic Oscillation (atmospheric circulation pattern over the mid-to-high latitudes) have generated profound impact on the variability of parameters such as offshore surface temperature and sea level. China's coastal sea surface temperatures have shown significant increase since 1980, rising by 0.23°C on average every decade between 1980 and 2018, and remained at a historically high level for four consecutive years between 2015 and 2018. During the same period, the sea levels along the coast saw fluctuations in an upward trend at an average rate of 3.3 mm/year from 1980 to 2017, higher than the global average. China's coastal areas, featured by developed local economies, dense populations and fragile ecosystems, are especially vulnerable to the impacts of climate change. In the context of global warming, warmer seas, higher sea levels and increasing intensity of extreme weather and climate events (such as typhoons, tidal waves, and storm surges) have all had a sizable impact on China's marine environment and the socioeconomic conditions of the coastal communities [7]. Due to various challenges such as data collection through ocean observations, climate model uncertainties and water pollution, current research on the impact of climate change on China's offshore waters is still lacking severely.

Cryosphere: China is the largest country at mid- and low- latitudes cryosphere region. A great deal of research has found that the cryosphere in China has been shrinking in the past few decades, and this trend has markedly accelerated in the past ten years and more. The second Glacier Inventory of China confirmed that there were 48,571 glaciers with an area of 51,800 km^2 in the country, where glaciers have exhibited a general trend of retreat. The total area has reduced by 18% and lost some 243.7 km^2 in mass on average each year. Over the past few decades, permafrost temperatures, principally on the Qinghai-Tibetan Plateau, have risen on the whole. The specific symptoms include: decrease in size and thickness of frozen ground; rise in the lower altitudinal limit of permafrost; shortening of period in which the ground is frozen; increase in maximum thaw depths while larger areas of permafrost are experiencing melt, and the disappearance of discontinuous permafrost zones. China has a snow cover totalling to 9 × 10^6 km^2, of which 5 × 10^4 km^2 is stable. They are scattered in the accumulation zones of alpine glaciers in western China. In the five decades between 1958/1959 and 2007/2008, the number of snow-covered days and the maximum snow depth in spring and autumn slowly dropped, while such days and depth rose in winter [8].

Extreme weather and climate events: From 1961 to 2018, the interdecadal changes in the frequency of extreme heat events in China are apparent. The frequency of extreme daily precipitation events has increased. Since the late 1990s, the average intensity of typhoons landing in China has risen dramatically. A litany of extreme weather and climate events such as heat waves, cold waves, drought, heavy precipitation, floods, typhoons, sandstorms have become commonplace and resulted in extensive impacts. Between 1961 and 2018, China's climate risk index was generally on the rise, with major changes in stages and upward fluctuations since the late 1970s. Climate change is and will continue to affect natural ecosystems and socioeconomic development, with a rise in the frequency and intensity of extreme climate events such as heat waves and heavy precipitation and growing instability in the cryosphere. As of today, climate change now constitutes a grave challenge to a wide spectrum of fields including food security, water resources, ecological environment, energy, development of major projects, human health, and socioeconomic development. With the ongoing pace of socioeconomic development and growing urbanization, the risks of high temperature, floods, and droughts facing China is expected to be aggravated. It is imperative that serious attention be paid to climate security [6].

1.2 Theoretical Framework

The key to tackling climate change lies in the reduction of greenhouse gas emissions from human activities. CO_2 emissions from the burning of fossil energy accounts for two-thirds of the global greenhouse gas emissions. The fact that fossil fuel use is concentrated makes the implementation of emission reduction measures easier. The ever-increasing demand for energy to support economic development and the need to cut CO_2 emissions are in fundamental conflict. Low-carbon development has become the only viable path in dealing with climate change under the framework of sustainable development. It refers to a model of economic growth where relatively high carbon productivity, high level of economic and social development and quality of life are achieved with less consumption of natural resources, emissions, and pollution. At its core lies the development of low-carbon technologies, the improvement of energy efficiency and energy structure, the establishment of a low-carbon economic development and social consumption model, the complete decoupling of long-term economic and social development from GHGs.

The Paris Agreement sets out the goal of holding global warming to well below 2°C and pursuing efforts to limit it within 1.5°C. In order to meet this target, global emissions of greenhouse gases must reduced sharply, and net-zero emissions should be achieved by the second half or even the middle of this century. The world must continue to ramp up emission reduction efforts. All countries, especially the developing nations, are also faced with pressing challenges in sustainable development such as economic growth, improvement of people's livelihood, promotion of social progress, and the protection of regional ecological environment. The response to climate change must be aligned with domestic sustainable development priorities. Meanwhile, cooperation among countries and joint actions must be strengthened to create win–win for all parties.

1.2.1 Climate Change and Sustainable Development

Climate change and sustainable development have always been intertwined in global diplomacy and governance. The United Nations 2030 Agenda for Sustainable Development (hereafter the Agenda) adopted in 2015 emphasized that the economic, social and environmental damages are interrelated and inseparable. The 2030 Agenda called attention to the fact that climate change has put at risk the well-being of humanity and the planet's life support system. Therefore, the Agenda calls for urgent actions to combat climate change and its impacts (SDG13), and ensure climate actions and climate restoration is at the core of the sustainable development transition. Eleven targets in the agenda explicitly related to addressing climate mitigation, adaptation, and resilience efforts, while a total of 28 secondary targets are specific climate actions related. In recent developments, the two agenda of sustainable developmetn adn mitigating climate change are inclined to merge into one discourse. From the perspective of the goals set out in the Agenda, sustainable development is presented as a solution to the humanity's development problem, and climate change response is the pathway to solving the global ecological crisis. The ultimate goal of these two agendas is to achieve global sustainable development [9]. Globally speaking, since Paris Agreement took effect, the actions taken by countries in response to climate change have also became the key driver for achieving other Sustainable Development Goals [9].

There is a complex relationship between climate change and sustainable development. From a systematic point of view, it is a relationship that consists of both trade-offs and synergies. Within this context, there are also differences in the impacts of climate adaptation and mitigation efforts on sustainable development. In general, limiting warming to 1.5 and 2°C makes it easier to achieve a number of the Sustainable Development Goals, such as those related to poverty, hunger, health, water, sanitation, cities and ecosystems (SDGs 1, 2, 3, 6, 11, 14, 15) has been made easier. The avoided climate change impacts on poverty reduction and inequality eradication would be greater if global warming were Limiting to 1.5°C warming shows greater potential, such as in poverty alleviation and inequality eradication. By targeting 1.5°C, we could reduce the number of people under poverty line due to climate change by 62 million more than the 2°C target. Under the 1.5°C scenario, there are multiple synergies and trade-offs between mitigation measures and Sustainable Development Goals, but there are more synergies than trade-offs. Further breaking matters down, strong synergistic effects of mitigation and sustainable development can be found in the areas of health, energy, responsible consumption and production and ocean sustainability (SDGs 3, 7, 12, 14); and trade-offs or negative effects rest between strict mitigation measures and SDGs in the areas of poverty and hunger reduction, water, sanitation and energy (SDGs 1, 2, 6, 7). In contrast, the impact of adaptation measures on poverty alleviation, reduction of general inequalities, agriculture, health, and SDGs is projected to be mostly positive [3].

1.2.2 Promote the Coordinated Governance of Climate, Energy and Environment

Climate change is a critical environmental challenge facing humanity. In the meantime, other environmental problems are also threatening human health and the Earth's ecosystems on which humanity depends. In almost all countries, the most important driver of climate policy is not only to avoid the long-term effects of climate change, but also to achieve the SDGs in the short term. The most pressing of which is combating domestic air pollution [1]. The very sources and impacts of air pollution and climate change are closely linked (Fig. 1.2). Therefore, seeking policies or action plans that synergize mitigating climate change and reducing air pollution enable countries to better integrate short-term Sustainable Development Goals with the long-term global climate change mitigation strategy, encouraging them to take more ambitious emission reduction actions. In practice, the close interrelationship

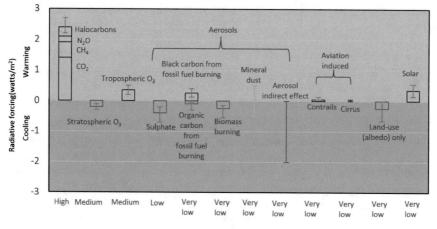

Fig. 1.2 The global mean radiative forcing of the climate system for the year 2000, relative to 1750 (*Source* IPCC. 1990. First Assessment Report. Intergovernmental Panel on Climate Change)

between environment and climate issues, effective policy measures for one cause may benefit the other. Coordinated policies and actions, i.e. coordinated governance, proves to be an effective win–win approach [3].

Many recent studies have suggested that actions to mitigate the impacts of climate forcing agents can have an immediate positive effect on economic and social welfare by improving public health and ecosystem sustainability. For instance, a model-based analysis found that, compared with the 2°C scenario, stricter short-term emission reduction measures under the 1.5°C scenario can reduce the number of premature deaths due to air pollution by 110 to 190 million [10]. Similarly, a comprehensive assessment of air pollution and climate change mitigation in the Asia–Pacific region concluded that by implementing 25 measures to improve air quality, CO_2 emissions can be cut by nearly 20% by 2050 [11]. According to a study by the United Nations Economic Commission for Europe, the average economic value of the health benefits arising from reduced air pollution ranges from $58 to $380 per ton of CO_2 worldwide, and the health benefits are even more significant in developing countries. In East Asia, the benefits are expected to be 10–70 times the marginal abatement cost by 2030 [12]. In 2011, the United Nations Environment Program and the World Meteorological Organization conducted a comprehensive assessment of control measures for short-lived climate pollutants, and discovered that these measures can slow the rate of global warming by half while avoiding 4 to 32 billion U.S. dollars annually of crop yield losses (estimated using global market prices) as well as 2 to 10 trillion U.S. dollars each year spent on health damages [13]. The mitigation of climate change has also created synergies in improving energy efficiency and the development of renewable energy. The synergies in these areas are equivalent to 50% to 350% of project investment. Research shows that in certain forestation projects aimed at mitigating

climate change, the co-benefits can make up between 53 to 92% of total benefits
[14].

1.2.3   Pathways and Technology Options to Mitigating Climate Change

Under the pathways for keeping global temperature rise to within 2°C and/or
1.5°C, the reduction of greenhouse gas emissions across all sectors, including
industry, building, transport, energy, agriculture, forestry and other land use, is
essential, where technology will play a key role. Across all industries, it is imperative to establish a non-fossil based energy supply system. Energy consumption
by the end-use sectors needs to be revolutionized; industrial production needs to
shift toward low-carbon technology, eco-design, flexible manufacturing systems
and circular economy; carbon capture, storage and removal technologies should
be applied on a large scale; innovations in energy saving and carbon emission
reduction should be coordinated including materials, intelligent manufacturing, and
information technology. In addition, revolutionary technological breakthroughs are
essential.

1. **Conserving energy and improving energy efficiency:** China has made
    improvements in energy efficiency mainly through technological advancement.
    Energy conservation and energy efficiency improvement should be the long-
    term priorities for low-carbon development. China is one of the countries with
    the fastest improvement in energy efficiency aroundthe world. From 2005 to
    2018, energy consumption per unit of GDP dropped by 41.5%, the cumulative
    energy savings reached 2.12 billion tonnes of coal equivalent (tce). Presently,
    China has put in place an energy-saving policy system with Chinese character-
    istics, which has evolved from restricting energy intensity alone to coordinated
    control of both energy intensity and total consumption. However, daunting chal-
    lenges remain in further improving energy conservation and energy efficiency.
    In the next stage, efforts need to shift from single pieces of technology and equip-
    ment to system optimization, from structural adjustments to quality enhance-
    ment and content upgrade, and from curbing the growth in energy consumption
    to transforming consumption and reducing demand.
2. **Low-carbon transition of the energy mix:** Limiting global temperature rise
    within and/or 1.5°C requires a non-fossil fuel -based energy system. Coal is
    rapidly exiting the energy system, and oil consumption is expected to peak
    before 2035. In the past decade, the cost of renewable energy technology has
    been effectively reduced, wind and solar power are set to reach grid parity
    between 2020 and 2030. The biggest challenge appears on the grid side. When
    renewables, which are associated with such issues as supply volitility and uncer-
    tainty, become the dominating source of energy, the mechanisms for balance

management and overall planning in the power system will have to undergo qualitative changes. Therefore, it is crucial to address the issues of power system stability and optimized grid operations across time series under the paradigm of high penetration of power electronics throughout all sections of the power system, i.e. source, grid and load.

3. **Revolutionizing energy consumption in end-use sectors:** Electrification is the most important hallmark of the transformation of energy consumption in the end-use sectors. Both pathways of 2°C and 1.5°C targets call for deeper electrification of the end-use sectors. The rate of electrification society-wide should be more than doubled by 2050. Besides, on average two-thirds of the energy efficiency potential for end use remains untapped globally. The combination of energy efficiency technologies with digitalization, intelligent solutions and additive manufacturing will usher in new potential.

4. **Revolutionary advanced technological breakthroughs:** By 2050, under a deep decarbonization scenario that is compatible with the target of holding temperature rise within 2°C, the marginal abatement cost is expected to show a rapid, nonlinear upward trend. Technological breakthroughs are essential for such break through. It is vital to intensify the R&D and proliferation of technologies that drive deeper reductions in CO_2 emissions, such as large-scale energy storage and smart grid technologies with high levels of renewables connected to the grid, negative emissions technologies such as Biomass Energy Carbon Capture and Storage systems (BECCS), technologies for the production, storage and utilization of hydrogen energy—a clean zero-carbon secondary source of energy and an important industrial raw material, and zero carbon manufacturing technologies for raw materials such as chemicals, steel, cement, etc.; in the meantime, the R&D and pilotingof deep emissions reduction technologies for non-CO_2 greenhouse gases such as methane are required to achieve deep emissions reduction of all GHGs [15].

1.2.4 Nature-Based Solutions

The role of natural ecosystems in mitigating and adapting to climate change is getting increasing attention, nature-based solutions (NbS) have been put forward in this context (see Fig. 1.3). It refers to the conservation, restoration and sustainable management of ecosystem for mitigating climate change, and the use of ecosystem and its services to help humans and wildlife adapt to the impacts and challenges brought by climate change [16]. Research on the global carbon cycle shows that terrestrial ecosystems have absorbed roughly one-third of anthropogenic CO_2 emissions since the industrial revolution [4]. NbS can also provide a buffer against the impacts and long-term risks of climate change.[17], [18]. Therefore, efforts to halt the loss or degradation of ecosystems, other misuses of land and ocean, and the protection, restoration, and sustainable management of global ecosystems can ensure that nature continues to provide these essential services to humanity [19].

Fig. 1.3 NbS as an umbrella term for ecosystem-related approaches (*Source* Cohen-Shacham E, Walters G, Janzen C, Maginnis S (eds) 2016 Nature-based solutions to address global societal challenges. IUCN, Gland, Switzerland, 11 pp)

The Paris Agreement recognizes the importance of natural ecosystems in mitigating and adapting to climate change, as well as their broader social values. Throughout the agreement, in the Preamble and Articles 4, 5, 6, and 7, concepts associated with "Nature-based Solutions" are referred to either directly or indirectly. In the Nationally Determined Contributions provided by the nations signed on the Paris Agreement, at least 66% of the NDCs contain some form of NbS to help achieve climate mitigation and/or adaptation goals. At present, NbS mostly aim to achieve climate change mitigation goals through the management, conservation, and restoration of terrestrial forest ecosystems, while the role of grasslands, drylands, wetlands, coastal and marine ecosystems in climate change mitigation has yet to garner sufficient attention. In this sense, enormous potential remains for the application of NbS in future climate change mitigation and adaptation programs [20]. According to an analysis of 16 dynamic global vegetation models provided by the Global Carbon Project, the increase in atmospheric CO_2 concentrations and climate change resulted in an annual mean net carbon storage of ecosystems in China of 85~310 TgC yr^{-1} [21] from 2000 to 2017. In spite of large uncertainties, the methods based on field surveys and atmospheric retrieval all point to China's terrestrial ecosystems being a major carbon sink (96~435 TgC yr^{-1}) than had, offsetted 6~29% of CO_2 emissions from fossil fuels between 2000 and 2009 in the country [21].

1.3 China's Policies on Low-Carbon Transition

1.3.1 Green Development and the Construction of Ecological Civilization

As a developing country, China faces daunting challenges on the triple fronts of climate change, environment and development. As of the end of 2018, China's per capita GDP is approximately US$ 9,780, while its per capita disposable income was only US$ 4,266 [22]. In the past few decades, China has relied on a path adopted by western countries in their early stages of development, extensive growth based on resources input, thus formed an industrial structure dominated by heavy industries. After the reform and opening-up in 1978, China followed the example of Japan and other East Asian countries and pursued an export-led growth strategy, which sought to create a net export surplus to make up for insufficient domestic demand. This "extensive growth" and "export-oriented strategy" have directly led to vast fossil fuel consumption and severe environmental pollution [23]. Meanwhile, the soaring energy consumption makes China the world's largest energy consumer and carbon emitter. China's primary energy consumption in 2018 amounted to 3.27 billion tons of oil equivalent (4.68 billion tce), which is 23.6% of the world total. CO_2 emissions in China reached 9.43 billion tonnes in the same year, or 27.8% of the world's total [24].

The Chinese government has always had grand ambitions for green development. As early as the 1990s, sustainable development became part of the top agendas officially in government documents. In 1992, shortly after the United Nations Conference on Environment and Development, the Chinese government released the *Ten Major Measures to Enhance the Environment and Development*, which, for the first time, proposed the implementation of sustainable development strategy in China, and formulated China's Agenda for 21st Century, the world's first national-level strategy for sustainable development. At the 17th National Congress of the CPC in 2007, the philosophy of "ecological civilization" was brought forward. The 18th CPC National Congress in 2012 elevated the building of ecological civilization into China's guiding principle and governance philosophy. Ecological civilization is also a key building block of the 13th Five-Year Plan adopted in 2016.

In 2015, the Chinese government issued the *Opinions of the Central Committee of the Communist Party of China and the State Council on Accelerating the Construction of Ecological Civilization*. The document evolved the notion of ecological civilization from a governance philosophy into a concrete strategy for action. The main goals for the development of ecological civilization were established, and the philosophy of ecological civilization was integrated into all spheres of China's economic, political, cultural, and social development, including the blueprints for the country's new model of urbanization, industrialization and agricultural modernization. More importantly, the document proposed the "formation of a full-fledged system of ecological civilization", which entailed the improvement of legislation, standards, the system of property rights of natural resource assets and usage regulation, the monitoring and supervision of the ecological environment, the strict observance of the red lines of resources, environment and ecology, the betterment of economic policies, the roll-out of market-based mechanisms, the enhancement of mechanism of compensation for ecological protection, government performance assessment as well as the accountability system. In September 2015, the Chinese government introduced the Integrated Reform Plan for Promoting Ecological Progress, setting clear target and timeframe for the reform. By then, ecological civilization had become the overarching policy to guild China's economic transformation.

The Chinese government has embraced ecological civilization as a central policy objective, and elevated it into one of the strategic pillars of the country's long-term development. China pursues a coordinated relationship among economic development, social progress and environmental protection under the steering of ecological civilization in a bid to strive for the coordinated and sustainable development. The country's efforts in promoting green development have also been recognized internationally. The UNEP's 27th council meeting in 2013 adopted a decision promoting the philosophy of ecological civilization, and in 2015, UNEP published a report entitled *Green is Gold: The Strategy and Actions of China's Ecological Civilization*. The country's remarkable achievements and successful practices in building an ecological civilization and protecting the environment will contribute China's experience and wisdom to a global community faced with the common threat of an environmental crisis.

Resource conservation, environmental protection, and CO_2 emission reduction are the key pillars of China's efforts in building ecological civilization. In the 11th Five-Year Plan (FYP), the country has set binding targets for reducing energy intensity and for the share of new and renewable energy in total energy consumption. The targets were further tailored to the provinces and municipalities and became an important metric for evaluating government performance at all levels. The 12th FYP added the target of CO_2 intensity reduction per capita of GDP, and share of non-fossil fuels in primary energy consumption mix. The 13th FYP put a ceiling for total energy consumption, strengthened policies on energy conservation and carbon reduction, and introduced a range of fiscal, tax and financial incentives. China's 2030 NDC commitments under the Paris Agreement will also be incorporated and implemented in the 14th and 15th FYPs. The NDC targets are an integral part of building ecological civilization during the 14th and 15th FYP periods.

1.3.2 Xi Jinping's Thought on Ecological Civilization and New Development Philosophies for the New Era

The report of the 19th National Congress of the CPC laid out the goals, basic strategies and blueprint of socialist modernization with Chinese characteristics in the new era. It listed climate change as a major non-conventional security threat, and stated that China has been actively fostering international cooperation in response to climate change, and has become an important participant, contributor, and torchbearer in the global endeavor for ecological civilization. It was highlighted in the report that China would pursue environmentally friendly growth and international cooperation on climate change, protect the planet that humanity depends on, and contribute to global ecological safety. In regard to the construction of ecological civilization, China aims to facilitate green, circular and low-carbon development, foster a new model of modernization with harmonious development between human and nature, while meeting the growing needs for a better life and a beautiful environment. This echoes the advocacy of Paris Agreement for a climate-friendly low-carbon economy. The tremendous progress China has made in energy conservation and economic transformation resulted in part from integrating the management of climate change with China's sustainable development. The ultimate goal is a win–win scenario for the economy, people's livelihood, energy, environment, and CO_2 emission reduction. Xi Jinping's Thought on Ecological Civilization has become a key guiding ideology and a powerful driver for China's low-carbon transition.

The report of the 19th National Congress of the CPC called for intensified efforts to develop a system for building an ecological civilization, facilitate green development, establish a green, low-carbon, and circular economy. The policy measures for the string of targets and tasks set out in the report, "controlling environment pollution" and "building a beautiful China" are consistent with the goals of addressing climate change and mitigating carbon emissions, resulting in win-win solution. An energy revolution on both the production and consumption sides and the transition to a green and low-carbon economy are not only crucial measures for protecting the environment, improving the ecology and building a "beautiful China", but also key strategies for CO_2 emissions reduction and global climate change response. To realize the objectives which "the ecology and environment fundamentally improved, and the goal of a beautiful China basically achieved by 2035", having pollution prevention and control measures that solely rely on improving energy efficiency and "end-of-pipe" techniques are far from sufficient. The total consumption of fossil fuels such as coal and oil must be reduced so that pollution can be curbed at the source. This will also effectively cut CO_2 emissions. Capping and reducing the consumption of coal and oil has been an important step for the eastern coastal areas in the "blue skies" campaigns, and will aid in the reduction of CO_2 emissions. Addressing climate change and implementating emission reduction commitments are also powerful drivers and synergies for domestic environmental improvement. China is currently translating its new development philosophies into concrete actions by transforming the drivers of economic development through innovation, and reshaping the model of development through green initiatives. The key to steering the Chinese economy onto a track of high-quality development instead of merely focusing on high-speed growth is to transform the development model, to improve economic structure, unlock new drivers of growth, deepen supply-side structural reform, and to increase total factor productivity. These initiatives will help shift the model of development from extensive growth driven by additional inputs of labor and capital at the expense of resources and the environment toward an innovation-driven, green and low-carbon endogenous development path. They will also boost energy conservation, improve the energy utilization and output efficiency. They reflect a paradigm shift that fosters an enabling institutional environment for promoting low-carbon transition and reforms, setting the stage for accelerated low-carbon transition of the economic and energy systems.

China's accelerating low-carbon transitions in its economic and energy sectors not only aligns with the global agenda on low-carbon development, but also presents a great opportunity to enhance its global competitiveness and influence. The urgent call for global response to climate change will reshape the world's landscape in economic, trade, and technological competition. Advanced low-carbon and decarbonization technologies will become the frontiers of global technological innovation and competition. As the world undergoes profound changes, the body that masters advanced technology in the above-mentioned areas will gain initiative and enhanced competitive advantage. China's active pursuit of an energy revolution and low-carbon transition of the economy are integrated with the strategic blueprint for developing its own technological capabilities and low-carbon development drivers.China also seeks the opportunity to share its findings and experience in low-carbon transition in a world-wide context, enhancing global competitiveness and influence.

1.4   Research Objectives and Overall Structure

1.4.1   Context and Objectives

The report from the 19th National Congress of the Communist Party (CPC) of China put forward the goals, fundamental policies, and main tasks for realizing socialist modernization in the new era. The reports outlined the goals of developing China into a great modern socialist country that is prosperous, strong, democratic, culturally advanced, harmonious, and beautiful by the middle of the century, becoming a global leader in terms of composite national strength and international influence, and realizing the Chinese dream of national rejuvenation. The Paris Agreement set out the goal of holding the increase in global average temperature to well -below 2°C and pursuing efforts towards limiting it to 1.5°C. To achieve this target, all countries must increase their efforts to curb and reduce greenhouse gas emissions. The world should cut CO_2 emissions to near-zero by 2050, and strive to achieve carbon neutrality. Addressing the challenges of climate change is a common cause for all humankind, and China will play an important role in this process as a participant, contributor and torchbearer.

China's long-term low-carbon emission development strategies should support the attainment of domestic development and global response to climate change. By the middle of the century, while realizing the goal of building a great modern socialist country, China should also achieve deep decarbonization development fitted to the 2°C global target. To realize the harmonious coexistence of humans and nature as well as sustainable development, China should establish an industrial system of green, low-carbon, and circular development as well as a clean, safe, and efficient decarbonized energy system with new and renewable energy as pillars. When researching and formulating China's development strategies for 2050, we should keep in mind both our internal and international imperatives, promote both domestic and global ecological civilization, and achieve the coordinated governance and win–win collaboration of both sustained domestic development and global ecological safety. The 19th Party Congress presented the two-stage arrangement of socialist modernization. China's long-term low-carbon development strategy should be formulated and deployed in accordance with the two-stage arrangement. The period 2020–2035 marks the first stage of China's socialist modernization fulfilling the strategies and tasks to realize modernization, as well as the goals and plans for ecological civilization and building a beautiful China, China not only would fundamentally improve the ecological environment, but also would achieve and strengthen the NDCs and emission reduction commitments. As a result, it would meet standards for both environmental quality and CO_2 emission reduction. High-quality economic development would be facilitated in coordination with economic, energy, and environmental efforts to address climate change for a win–win scenario.

In the second stage of socialist modernization, from 2035 to 2050, after having basically realized modernization and entered a post-industrial society, China's economic and social development will be featured by quality, intensive endogenous growth, and China will have already peaked and lowered CO_2 emissions, which is more conducive to a low-carbon transition. During this stage, the need for global CO_2 emission reduction is more urgent, which should reach 6~7% on an annual basis. With substantive improvements in the environment and attenuated environmental stress, the goal and strategy of deep decarbonization will take precedence over the endogenous requirements of resource conservation, environmental protection and sustainable development in China, and more consideration should be given to assuming international responsibilities consistent with its ever-increasing national strength and international influence. China must take the goal of deep decarbonization through keeping global temperature rise to well -below 2°C and striving for 1.5°C as a key component of the overall strategy of building a modern socialist country, and assume its responsibilities and leadership in safeguarding the ecological safety of the Earth and the common interests of humankind, and contribute to the development of humanity. China should examine its due and potential responsibilities and obligations as determined by the goals and requirements of the Paris Agreement, gradually expand the scope of GHG reduction efforts from mainly CO_2 emissions from energy consumption to all six GHGs, and develop absolute reduction targets and measures aimed at 2050 that cover all GHGs across the entire economy. The Paris Agreement requests all parties to submit their 2050 low GHGs development strategies prior to 2020. This project will also provide support for the Chinese government in formulating and submitting the country's long-term low-emission development strategy.

1.4.2   Research Framework and Scenarios

Focusing on China's long-term low-carbon development strategy and pathways, this study encompassed a total of 18 topics (see Fig. 1.4), and pulled together the collective effort of domestic top research institutions in China. From a macroscopic point of view, areas under study include the medium- to -long- term socioeconomic goals and strategies, the coordinated development of the eastern, central and western regions and their respectivelow-carbon strategies and pathways in the process of urbanization, the impacts of international trade and industrial relocation, etc. From a sectoral perspective, the research covered low-carbon development strategies and pathways for the energy and power supply sector and the major end-use sectors such as industry, building, and transport. In addition, strategies, measures and pathways for non-energy related CO_2 and other GHG emission reductions are also presented under the section. Policy-related topics under study include energy-saving potential and pathways, the evaluation of advanced decarbonization technologies, infrastructure investment and financing, transformation of consumption patterns, and research on coordinated measures and effect analysis for the climate and environment. Furthermore, the project explores five comprehensive and integrated topics, including the formulation of vision statement for medium and long-term low-carbon emissions, policy and institutional support systems, and global climate governance and international cooperation, etc.

This project consists of both bottom–up and top–down approaches. Bottom–up approaches include scenario analysis and technical evaluation of energy consumption and CO_2 technologies of all sectors. Top–down approaches include a macro-model calculation and policy simulation. The study is oriented toward problem-solving and

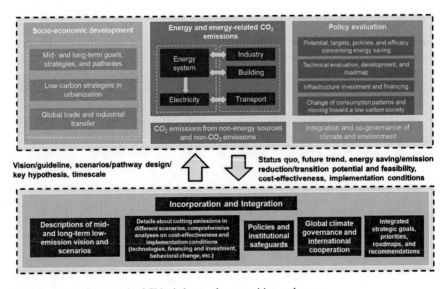

Fig. 1.4   Project framework of China's low-carbon transition pathways

in-depth analysis targeted at China's actual national conditions and characteristics of its development stage by focusing on the trends, policies, and pathways of long-term low-carbon development. The study has conducted policy simulation with the goal of building a great modern country and realizing deep decarbonization pathways, and has analyzed emission reduction pathways, technology support, costs, and prices driven by the long-term deep decarbonization goal. Four scenarios have been designed and all topics studied and analyzed based on these four scenarios.

1. Policy scenario: supported by China's NDCs submitted in 2015 under the Paris Agreement, action plans, and relevant policies based on the assumption that the current low-carbon transformation trends and policies remains.

2. Reinforced policy scenario: based on the policy scenario, this scenario further enhances the intensity and scope for reducing energy intensity (defined as energy consumption per unit of GDP) and CO_2 intensity (defined as CO_2 emissions per unit of GDP) and improves the proportion of non-fossil fuels in primary energy consumption. Under this scenario, exploring the potential for emission reduction, controlling total CO_2 emissions, strengthening policy support, and adapting to the enhanced and upgraded NDCs and actions under the Paris Agreement.

3. 2°C scenario: based on the goal of controlling global warming to within 2°C, this scenario considers the corresponding emission reduction pathways. In this scenario, analyzing the emission reduction measures and roadmaps driven by the deep decarbonization goals to be achieved by the middle of the century, and demonstrate and evaluate the associated technology and capital needs, costs and prices, and policy support.

4. 1.5°C scenario: based on the goals of controlling global warming to within 1.5°C and realizing carbon neutrality, this scenario demonstrates the possibilities and pathway options for the realization of net-zero emissions of CO_2 and deep reductions of other GHGs emissions by 2050, and evaluating the possible social and economic impacts.

Among the four scenarios, the reinforced policy scenario and the 2°C scenario are the primary points of focus of this study. Before 2030 and 2035, the impact on the fulfillment and updating of the NDCs against the background of the reinforced policy scenario is the main target for the analysis, with focus on the impact of emission reduction pathways driven by the 2°C goal on the targets and pathways for 2030 and 2035. After 2035, while realizing the goal of building a great socialist modern country, the emission reduction pathways and policy support driven by the 2°C goal become the main targets for study. This study also explores feasibility for realizing net-zero CO_2 emissions and deep emission reduction of other GHGs by 2050 to attain the 1.5°C goal. It is worth noting that the connections between the scenarios and the NDCs before 2035 are taken into consideration, as well as the coordination of low-emission goals and strategies in these two stages.

1.4.3   Overall Structure

This comprehensive report on the research on China's Long-term Low-carbon Development Strategies and Pathways, is based on the consolidation of the first 17 topics of the entire research project (the 18th topic is the comprehensive report), and focuses on the strategic objectives, key conclusions, roadmaps and policy recommendations. The report is divided into twelve chapters. The first chapter, the introduction, deals primarily with the context, theoretical framework, guidelines and overall design. The second chapter offers an overall analysis of long-term low-carbon emission development, and delves into the state and trends of low-carbon development both in China and internationally under the Paris Agreement.

Chapters 3–6 attempt to outline the scenarios of future emissions of GHGs, key emission reduction technologies and policy requirements across all sectors. Chapter 3 dedicated to end-use energy consumption and CO_2 emissions, is primarily concerned with the current state of development of the industry, building and transport sectors, decarbonization pathways, overall energy consumption and carbon emissions, and activity levels, as well as analysis of the potential costs, and crucial supporting policies. Chapters 4 and 5 focus on the power and energy system, and assess the future development and policy support needed for the power and energy system based on power demand from the end-use sectors. Chapter 6 mainly explores the current state of emissions of non-CO_2 GHGs, future emission scenarios and outcomes, and key emission reduction technologies and policy requirements. Chapters 7–9 consolidate and evaluate the three aspects of low-carbon technology, transition costs, and long-term low-carbon development pathways. Chapters 7 and 8 sort through critical zero-carbon and carbon negative technologies (including hydrogen energy technologies, energy storage technologies, CCS & BECCS, and zero-emission technologies in the production of steel, cement and chemical, etc.), and examine their current stages of development and future investment needs, etc. Chapter 9, devoted to an overall analysis and evaluation of pathways, integrates in a systematic manner the low-carbon strategies mentioned in Chapters 3–6, proposes an enhanced NDC pathway, 2°C pathway and 1.5°C pathway.

Chapter 10, titled strategic linchpins and policy safeguards, examines the existing low-carbon development policies and measures in China, and proposes key strategies to support the country's low-emission development in the future while taking into full account China's goals and strategies for building itself into a great modern socialist country by mid-twenty-first century. Chapter 11, dedicated to the topic of global climate governance and international cooperation, delves into the development, principles and new realities of the global climate governance system. Chapter 12 offers policy recommendations for China's mid—and long-term low-carbon development goals, strategies and implementation measures.

References

1. Intergovernmental Panel on Climate Change (IPCC) (2013) Climate change 2013: the physical science basis. Cambridge University Press, Cambridge
2. World Meteorological Organization (WMO) (2019) WMO statement on the state of global climate in 2018. WMO, Geneva
3. Intergovernmental Panel on Climate Change (IPCC) (2018) Special report on global warming of 1.5°C. an IPCC special report on the impacts of global warming of 1.5°C above pre-industrial levels and related global greenhouse gas emission pathways, in the context of strengthening the global response to the threat of climate change, sustainable development, and efforts to eradicate poverty. IPCC, Geneva
4. Intergovernmental Panel on Climate Change (IPCC) (2019) Climate and land: an IPCC special report on climate change, desertification, land degradation, sustainable land management, food security, and greenhouse gas fluxes in terrestrial ecosystems. IPCC, Geneva
5. Intergovernmental Panel on Climate Change (IPCC) (2019) Special report on the ocean and cryosphere in a changing climate. IPCC, Geneva
6. Climate Change Center, China Meteorological Administration (2019) Blue paper on climate change in China. China Meteorological Administration, Beijing
7. National Marine Information Center (2019) Ocean blue book of China's climate change in 2019. National Marine Information Center, Tianjin
8. Shijin W, Yongjian D, Cunde X (2018) Comprehensive influence of cryosphere change on economic and social systems and its adaptive management strategy. Journal of Glacial Permafrost 40(05):7–18
9. United Nations Climate Change (UNCC) (2017) Climate action plays central role in achieving the sustainable development goals (2017–07–18) [2020–04–30]. https://unfccc.int/news/cli mate-action-plays-central-role-in-achieving-the-sustainable-development-goals
10. Shindell DT, Faluvegi G, Seltzer K et al (2018) Quantified, localized health benefits of accelerated carbon dioxide emissions reductions. Nat Clim Chang 8:291–295
11. United Nations Environment Programme (UNEP) and Climate & Clean Air Coalition (CCAC) Air pollution in Asia and the Pacific: Science-based Solutions. UNEP and CCAC, Nairobi and Paris
12. United Nations Economic Commission for Europe (UNECE) Programmes (2016) The co-benefits of climate change mitigation (2016–01–22) [2019–08–28] http://www.unece.org/fil eadmin/DAM/Sustainable_Development_No._2__Final__Draft_OK_2.pdf
13. United Nations Environment Programme (UNEP) and the World Meteorological Organization (WMO) (2011) Integrated assessment of black carbon and tropospheric ozone. UNEP and WMO, Nairobi and Geneva
14. Urge-Vorsatz D, Herrero S T, Dubash N K and Lecocq F (2014) Measuring the co-benefits of climate change mitigation. Annual Review of Environ Resour 39: 549–582
15. Jiankun He (2019) New situation of global climate governance and China's countermeasures. Environmental Economics Research 3:1–9
16. Zhang X, Xie Q, Zeng N. Nature-based climate change solutions. Progress in climate change research (2020–02–17) http://kns.cnki.net/kcms/detail/11.5368.p.20200215.2003.002.html
17. Hochard JP, Hamilton S, Barbier EB (2019) Mangroves shelter coastal activity from cyclones. Proc Natl Acad Sci USA 116:12232–12237
18. Beck MW, Losada IJ, Menéndez P et al (2018) The global flood protection savings provided by coral reefs. Nat Commun 9:2186
19. Seddon N, Sengupta S, García M et al (2019) Nature-based solutions in nationally determined contributions: Synthesis and recommendations for enhancing climate ambition and action by 2020. IUCN and University of Oxford, Gland and Oxford
20. Le Quéré C, Andrew RM, Friedlingstein P, et al (2018) Global carbon budget 2018. Earth Syst. Sci. Data 1(4)
21. Lu F, Hu HF, Sun WJ et al (2018) Effects of national ecological restoration projects on carbon sequestration in China from 2001 to 2010. Proc Natl Acad Sci 115(16):4039–4044

22. National Bureau of Statistics (2019) 2018 residents income and consumption expenditure (2019–01–21) [2019–11–16] http://www.stats.gov.cn/tjsj/zxfb/201901/t20190121_1645791.html
23. Wu J (2014) Preface 1. // Ma Jun, Li Zhiguo et al. Economic policy for $PM_{2.5}$ emission reduction, 1–3. China Economic Press, Beijing
24. British Petroleum (BP) BP statistical review of world energy (2020–06–17) [2020–06–22] http://www.bp.com/statisticalreview

Chapter 2
The State and Trend of China's Low-Carbon Development

2.1 New Paradigm of Global Response to Climate Change Under the Paris Agreement

The Paris Agreement established the long-term temperature goal to hold global average temperature rise to well below 2°C above pre-industrial levels and to pursue efforts to limit the global temperature rise to 1.5°C. Based on a "bottom-up" approach, countries are asked to set their emission reduction targets and action plans independently, and a global stocktake is mandated every five years to assess and drive collective progress under the new global climate governance mechanism. The global community has now fully entered the implementation phase of the Paris Agreement, and arduous challenges and tasks await.

First, there is still a considerable gap between emission reduction commitments by countries and those necessary to limit temperature rise to well-below 2°C. By estimate, the annual emission gap by 2030 under the current NDCs would be 12–15 billion tons of CO_{2e} [1], and the world is on track to an average 2.7°C increase by 2100. There is a less than 5% probability of a temperature rise below 2°C, a greater than 25% probability of a temperature increase above 3°C, and a 10% probability of exceeding 3.5°C. Therefore, it is imperative that countries implement and enhance their emission reduction targets and climate actions [2].

Second, a growing number of studies and facts demonstrate that the negative impacts of climate change is more extensive, more severe, and accelerating at a greater speed than originally predicted. In the current high-emissions scenario without reinforced actions, there is the risk of temperature rise of 5°C by the end of this century with the chance of crossing the tipping points or risk thresholds in many areas, resulting in irreversible catastrophic consequences. For instance, heat waves, which currently feature an occurrence of less than 5% globally, will happen almost every year; both the probabilities of flooding and agricultural droughts may increase by ten folds; China's grain production will shrink by 20%, and glaciers by 70% [2]. In essence, if temperature rises too fast and too dramatic in the future, extremely high

© China Environment Publishing Group Co., Ltd. 2022
Institute of Climate Change and Sustainable Development of Tsinghua University et al.,
China's Long-Term Low-Carbon Development Strategies and Pathways,
https://doi.org/10.1007/978-981-16-2524-4_2

systemic risks to the economy, society, and ecosystem will pose significant threats to water and food security, urban infrastructure, and the public health system. To reverse the situation, active steps must be taken to improve the climate resilience of the economy, society and ecosystem, and prevent climate risks.

Third, the *IPCC Special Report on Global Warming of 1.5°C* highlighted the urgency of addressing climate change, and reinforced and advanced the goal of achieving "carbon neutrality" globally. The report, issued by IPCC in October 2018 as per the provisions of the Paris Agreement, not only provides scientific climate assessment, but also serves to inform decision-makers for negotiations and political consensus under the UNFCCC framework. The report shows the impacts and risks of climate change can be dramatically reduced at 1.5°C warming compared to 2°C. But that would entail much more urgent efforts in emission reduction: emissions should be down by 45% from 2010 levels by 2030; global net zero CO_2 emission and deep reduction of non-CO_2 GHGs is required in 2050. However, at the current pace of emissions, global average warming is likely to reach 1.5°C between 2030 and 2052, resulting in needing even more drastic reductions now. On the other hand, the cost of 1.5°C will be three to four times greater than the 2°C pathway. The *IPCC Special Report on Global Warming of 1.5°C* reinforced and advanced the global target of achieving carbon neutrality by the middle of this century and explored the pathways for long-term low-carbon emissions reductions, which is expected to influence the targets and choice of pathways as countries submit their 2050 long-term low-emission strategies by the end of 2020.

Fourth, developed nations and regions such as the EU have laid out specific timetables and action plans for the 1.5°C goal. The European Green Deal announced by the European Commission in December 2019 aims to make Europe the first climate-neutral continent, by pledging to cut GHGs by 50–55% by 2030 compared with 1990 levels, and to achieve net-zero emissions by 2050, and these objectives will be enshrined in legislation. The European Green Deal also proposes actions in seven strategic areas, including supplying clean, affordable and secure energy; mobilizing industry for a clean and circular economy; building and renovating in an energy and resource efficient way; accelerating the shift to sustainable and smart mobility; designing a fair, healthy and environmentally-friendly food system; preserving and restoring ecosystems and biodiversity; and a zero-pollution ambition for a toxic-free environment. In 2019, the British government passed legislation to cut GHG emissions to net zero prior to 2050, making the UK the first country to incorporate long-term climate targets into official laws and regulation. Although the Trump administration withdrew the US from the Paris Agreement, US climate actions have not come to a halt, and instead moved from the federal level to sub-state levels of states, cities, companies, and universities. More than 20 states in the US, cummulative accounting for more than 60% of the country's GDP and 55% of population, formed the United States Climate Alliance and pledged to honor their commitments under the Paris Agreement. States such as California, Nevada, New Mexico, and Washington have all passed legislations setting a goal of 100 percent carbon-free electricity by 2050 or earlier. In 2019, the New York State Legislature passed the Climate Leadership and Community Protection Act, which mandated net zero emissions in all sectors of the economy by 2050—reduction by 85% in GHG emissions from 1990 levels, with the remaining 15% being offset through reforestation, restoring wetlands, carbon capturing or other green projects. The more urgent the goal of carbon neutrality, the tighter the constraints on emissions for all countries to achieve sustainable development and the greater the pressure to bring about low-carbon transition.

Fifth, the world is giving higher priority to the coordinated governance of climate change, ecology and environment protection and sustainable development. The global community is becoming increasingly aware of the interlinkages between climate actions and the UN Sustainable Development Goals. Actions to mitigate and adapt to climate change will help promote local ecological protection and sustainable development, creating extensive synergies. Global climate efforts have triggered low-carbon transformation of the economic and energy systems, and ushered socioeconomic development onto a green, low-carbon cycle and sustainable development path. This is also the ultimate solution to improving regional ecology and environment and promoting the coordination of socioeconomic development with the environment capacity. Therefore, the increasing awareness of all countries on the urgency for energy revolution and low-carbon economic transition in the global response to climate change presents enormous opportunities for the pursuit of domestic sustainability goals. Efforts should be made to achieve the coordinated governance of energy, economy, environment, and climate to create a win–win scenario for all parties. In the meantime, greater attention should focus on the coordination of measures to mitigate and adapt to climate change.

Sixth, as the world is experiencing profound shifts unseen in a century, the sentiments of unilateralism, protectionism and self-interest of some developed countries have taken a stronger hold in climate negotiations, impeding the implementation of the Paris Agreement. At the 2019 UN climate change conference (COP25) held in Madrid, some developed countries sought to advance self-interests in a non-constructive manner during the negotiations while rejecting the concerns and demands voiced by the vast majority of other parties, especially the developing countries. They failed to advance various agendas in a comprehensive and balanced approach due to a lack of political will to reinforce pragmatic actions, and opposed to review and advance actions before 2020 while focusing solely on strengthening the 1.5°C temperature controlling goal. Consequently, the contracting parties were unable to settle on rules for market mechanism under Article 6 of the Paris Agreement, and to achieve balanced progress and estimated effects in negotiations on key areas including mitigation, adaptation, funding, technology, capacity building and transparency. COP25 ended with unfulfilled expectations, and the implementation of the Paris Agreement remains a serious challenge and daunting task.

Seventh, COVID-19 will significantly reshape the global climate process and competitive landscape between major powers. The pandemic has triggered a massive global economic downturn. Countries will give priority to restoring production, ensuring people's livelihood, and remediating deficiencies in the industrial chain in the post-pandemic era, which might result in weaker actions and policy measures for climate change and low-carbon transformation. Nevertheless, a consensus is also forming around "green economic recovery" in coping with the deeper global crisis of climate change. The response to climate change will remain one of the front-burner issues and critical battleground for major powers in their diplomatic efforts. Tackling climate change represents the shared interest of humankind and high ground of international moral, so there is also room for exchanges and cooperation among China, the US, the EU, and other regions in this regard. It is imperative for China to coordinate diplomatic policy and domestic green and low-carbon transition effectively.

2.2   The State and Outlook of China's Domestic Low-Carbon Transition

2.2.1   The New Situation of China's Low-Carbon Transition

The Chinese economy entered a "new normal" after 2013 when the country adopted new development philosophies, transformed the drivers of growth through innovation, transitioned to green development, and gave full play to the synergistic of environmental governance and CO_2 emission reduction. The results of China's energy saving and carbon reduction measures have exceeded expectations, setting the stage for achieving its 2030 NDC target.

The "new normal" prompted a shift of China's economic development from size and speed to quality and efficiency. GDP growth slowed down from an average 10.2% during 2005–2013 to 6.9% during 2013–2018. In the meantime, the acceleration of economic restructuring, industrial transformation and upgrading have saturated the market for energy-intensive raw-materials products and lowered energy elasticity. Between 2005 and 2013, energy elasticity stood at roughly 0.59 with an annual drop of 3.8% in energy consumption per unit of GDP, compared to 0.32 and 4.7% during 2013–2018. The dual reductions in energy elasticity and GDP growth led to a significant contraction in the growth of total energy demand. During the period from 2005 to 2013, China's energy consumption grew by 6% annually, and that growth slid to 2.2% during 2013–2018, which reflected the decreasing energy demand. When the GDP growth is held constant, the lower the elasticity of energy consumption, the greater the decrease in energy intensity. However, with constant energy elasticity, slower GDP growth would make for slower drop in energy consumption per unit of GDP. The above two factors could offset each other so that the drop in energy intensity will most likely remain at a level of 3.0–3.5% annually in the future.

While growth in China's energy demand has been slowing, new and renewable energies have kept a robust growth of roughly 10%. As the energy structure adjustment accelerates its pace, CO_2 emissions per unit of energy consumption have continued to fall. Prior to the "new normal", CO_2 emissions per unit of energy consumption fell by 0.57% on average annually between 2005 and 2013, as opposed to 1.38% between 2013 and 2018 under the "new normal". CO_2 emissions per unit

Table 2.1 Changes in energy consumption and CO_2 emissions under the "new normal"

	2005—2013	2013—2018
GDP growth (%)	10.2	6.9
Energy consumption elasticity	0.59	0.32
Average annual growth of energy consumption (%)	6.0	2.2
Annual decline of energy intensity of GDP (%)	3.8	4.7
Annual growth of CO_2 emissions (%)	5.4	0.8
Annual decline of CO_2 intensity of GDP (%)	4.4	5.7

of GDP also increased from 4.4% on average annually in 2005–2013 to 5.7% in 2013–2018, which signified speedier energy transition (see Table 2.1).

In the three years following 2017, the growth of energy consumption and CO_2 emissions has rebounded. In 2017 and 2018, total energy consumption rose by 3.0% and 3.3% respectively, and CO_2 emissions were up by 1.8% and 2.2% respectively. The Chinese economy grew by 6% and energy consumption by 3.3% in 2019. The total energy consumption was 4.98 billion tce in 2020, realizing the goal of limiting total energy consumption below 5 billion tce set in the 13th FYP period. Coal consumption has entered a downward trajectory since 2013. Despite the rebound in the past two years, total consumption has not exceeded the level of 2013. Generally speaking, China has moved into a peak platform period of coal consumption, which still might fluctuate in the future. But with a further slowdown in energy consumption growth and rapid expansion of renewable energy, coal consumption will no longer see steady growth. On the whole, as economic growth stabilizes and high-quality development becomes the norm, the growth of energy demand and CO_2 emissions will further ease. Chances are slim for a comeback of the strong growth experienced prior to 2013.

The COVID-19 put a severe damper on economic growth in 2020, dragging the average annual GDP growth during the 13th FYP period to below 6%. Energy consumptions per unit of GDP dropped by 13%, and CO_2 emission intensity by 18.8% during the 13th FYP, meaning that China has achieved the 13th FYP target of reducing CO_2 intensity by 18%.

2.2.2   Scenario Analysis of Energy Conservation and Carbon Reduction During the 14th FYP Period

The 14th Five-Year Plan period marks the first five years for China to embark on a new journey of building a modern socialist country in all respects. China would face a series of urgent tasks such as reviving the economy after the pandemic, ensuring people's livelihood and remedying the deficiencies of the industrial chain. Meanwhile, China is also in a critical period of transition to high-quality growth. During the 14th FYP period, China will adhere to new development concepts and speed up the creation of a clean, low-carbon, safe and efficient energy system and a green, low-carbon, circular, and sustainable economic system. New progress in energy conservation and carbon reduction is on the horizon.

The current capacity of high energy-consuming industries such as the modern coal chemical and petrochemical sectors might continue to expand into the early phase of the 14th FYP period, driving the growth in energy consumption. However, the latter part of the period will see no more capacity expansion in heavy and chemical industries, with a slowdown in the growth of energy consumption. As China presses ahead with industrial transformation and upgrading and high-quality economic development, it will witness consistent and steady improvements in energy conservation and carbon reduction during the 14th FYP period.

China will pursue the new development philosophies, intensify economic restructuring, industrial transformation and upgrading, and strive for high-quality development during the 14th FYP period, prompting increasingly faster energy conservation and carbon reduction. As for GDP growth during the period, the pandemic has wreaked on the international economy, industrial chain, and global supply and demand, which will hamper China's economic recovery and development. Most domestic research institutions project a roughly 5% of average annual economic growth under relatively optimistic circumstances during the 14th FYP period. Following the slowdown in GDP growth, the market for energy-intensive raw materials such as steel and cement will become saturated and begin to march downward, further pushing down the elasticity of energy consumption. The current stimulus package for reviving the economy, boosting new infrastructure and urbanization investment, spurring development of the digital economy and technological upgrading of traditional industries, and advancing the integrated development of intelligent urban infrastructure and low-carbon transitions will contribute to the decline of energy intensity.

Power generation costs for wind and solar will be comparable to that of coal during the 14th FYP period. Non-fossil fuels could maintain the 7% annual growth during the 14th FYP period; the non-fossil share of China's total primary energy consumption might reach roughly 20% in 2025; and the share of coal will fall further from 56.8% in 2020 to approximately 51% in 2025. Based on these projections, CO_2 emissions per unit of GDP will drop by roughly 19% during the 14th FYP period. By 2025, total energy consumption can be held under 55 tce, and total CO_2 emissions under 10.5 billion tons. See the table below for more details (Table 2.2).

The 14th FYP period is a critical stage for the green and low-carbon transition of China's economy and the attainment of high quality development. On one hand, China should keep up and drive the current momentum and achievements made in the energy conservation and carbon reduction by employing active energy conservation and carbon reduction targets as key instruments for the green and low-carbon transition and high-quality development of Chinese economy. On the other hand, China should boost the robust low-carbon transition since the implementation of the 11th FYP, maintain strategic determination amid economic slowdown, stick to energy conservation, carbon reduction and high-quality economic development, and holistically advance other related policies and measures aside from continued

Table 2.2   Analysis of the effects of energy conservation & carbon reduction by period

	12th FYP		13th FYP		14th FYP
	Planned	Actual	Planned	Estimated	Projected
Average annual GDP growth (%)		7.9		5.7	5.3
Decline in energy intensity of GDP (%)	16	18.5	15	13	14.7
Non-fossil share at end of period (%)	11.4	12.1	Approx. 15	15.8	20
Decline in CO_2 intensity of GDP (%)	17	21.7	18	18.8	19
Total energy consumption at end of period (billion tce)		4.30	< 5	4.98	5.5
Total CO_2 emissions at end of period (billion tCO_2)		9.29		9.9	10.4

efforts to establish quantitative energy-saving and carbon-reduction targets. Specific suggestions are as following:

1. Developing guiding targets for curbing total energy consumption and CO_2 emissions. In fact, the target for total energy consumption was formulated and implemented during the 13th FYP period. That, coupled with the target of a non-fossil energy share, virtually means that a CO_2 emissions target has been determined. It is imperative that target for curbing total CO_2 emissions be stated more explicitly during the 14th FYP period so that they are aligned with the policies and measures introduced to achieve CO_2 emissions peak before 2030, and gradually be integrated with and eventually replace the energy consumption cap. By capping total CO_2 emissions, a ceiling for energy consumption will not be set for local governments and enterprises. Instead, they are incentivized to develop and utilize more new and renewable energy, which will promote the green and low-carbon transition of the economy. A carbon emission allowance system for enterprises will gradually replace the allowance of energy consumption, and will be linked with the nationwide carbon trading market currently being developed. Total CO_2 emissions management is more aligned with international norms and practices than placing a cap on energy consumption.

   As uncertainties abound in the future growth of the economy and energy consumption, if GDP growth exceeds expectations, it will have less of an impact on energy intensity, but will significantly affect the total energy consumption target. Therefore, certain amount of leeway should be reserved for a looser target for total energy consumption than that for energy intensity reduction.

2. Encouraging optimal development zones to peak CO_2 emissions first. It was proposed in the 13th FYP that "China supports optimal development zones to reach the peak of carbon emissions first." China's CO_2 emissions could, in general, plateau after 2025. The more developed eastern coastal areas are well-positioned to take the lead nationwide in peaking emissions, with some

cities on track to peak carbon emissions by around 2020. Under this circumstance, the target for optimal development zones in the eastern coastal areas to reach emissions peak ahead of the nation should be included in the 14th FYP, and developed provinces and cities should be encouraged to formulate and implement the appropriate plans and policy measures.

3. Driving the peaking of CO_2 emissions in energy-intensive industries earlier. Industrial sector accounts for two-thirds of the country's total end-use energy consumption, and is the sector with the most energy-saving potential. With industrial restructuring, transformation and upgrading during the 14th FYP period, the industrial sector will be well-positioned to peak its carbon emissions earlier. The goal of peaking CO_2 emissions in the energy-intensive industries can be stated explicitly in the 14th FYP.

4. Controlling energy-related non-CO_2 greenhouse gas emissions. China's current emission mitigation efforts and commitments are mostly centered on energy-related CO_2 emissions. Starting from the 14th FYP period, the country should start to manage and rein in CO_2 and other greenhouse gas emissions from industrial production processes, agriculture, forestry, and waste management, and in particular the production and use of hydrofluorocarbons (HFCs). A monitoring, reporting, and verification system for all greenhouse gas emissions should be established with emission reduction measures and actions.

5. Deepening market-oriented reform of the energy management system. The boom of energy technologies and industrial innovation has accelerated the transformation of the energy industry, although institutional and policy obstacles remain to be overcome. China must plough ahead with reforms of the systems of energy management and pricing, and promote the development of distributed renewable energy. In the meantime, greater progress need to be made in the establishment of the emissions trading market. It is recommended that during the 14th FYP period, the coverage of China's national emissions trading market is expanded to include major energy-intensive industries as soon as possible and emissions allowance management is improved to promote the low-carbon transition of energy and economy through market-based approaches.

2.2.3 Challenges and Arduous Tasks Facing China for Rapid Low-Carbon Transition

China has scored remarkable and internationally acclaimed success in energy conservation and carbon reduction since the implementation of the 11th FYP. The country has pursued new development philosophies, sped up economic restructuring and industrial upgrading, and strengthened energy conservation and carbon reduction policies and measures under the new normal. Consequently, the rapid rises in energy consumption and CO_2 emissions have been reversed. The energy consumption and CO_2 emissions will, slowly but surely, continue their growth in the near future. Thus enormous efforts are still required to peak CO_2 emissions.

China's energy consumption per unit of GDP plunged by 42.4% from 2005 to 2019, down 3.86% year-on-year, much higher than that of developed countries over the same period. However, due to the unique industrial structure during this stage of industrialization and urbanization, the share of the industry, especially energy-intensive sectors, in the Chinese economy is higher than that of developed countries, and the manufacturing sector is also situated at the middle and low ends of the global industrial value chain, which are dominated by energy-intensive and low value-added products. This explains the fact that China's energy consumption per unit of GDP still stood at 1.6 times the global average and 2–3 times that of developed countries in 2017. Tremendous potential exists in the areas of energy conservation and CO_2 emission reduction as China presses ahead with economic restructuring and industrial upgrading, although the journey will take a considerable amount of time.

China is well endowed with coal resources, which have always predominated the energy mix. China has spared no effort to develop renewable energy, and the non-fossil share of China's energy mix more than doubled from 7.4% in 2005 to 15.3% in 2019 while the share of coal slid from 72.4 to 57.7%. China still faces tough tasks in terms of expediting the development of new and renewable energy sources, curbing and reducing coal consumption and speeding up low-carbon transition of the energy mix.

The peaking of CO_2 emissions means that sustained economic growth will be completely decoupled from carbon emissions. When fossil energy consumption has plateaued, increasing alternative energies for coal will not only slow down carbon emissions, but also curtail and reduce conventional pollutants from the source, promoting fundamental improvements in environmental quality. Peaking CO_2 emissions will be a landmark in the low-carbon transition of China's economic development as well as a historic milestone in the fundamental improvement of the country's ecological environment and the building of a beautiful China.

To ensure sustained economic growth while peaking carbon emissions, the decline in CO_2 emissions per unit of GDP must be greater than the growth of GDP, so that the increase in emissions brought by GDP growth is offset by a reduced emission intensity. Currently, China's economic growth remains at a level between 5–6% and is expected to continue tapering off after 2020. After the basic completion of the country's industrialization and urbanization drive around 2030, China will have joined the ranks of high-income countries in terms of national income per capita. Intensive growth will become the norm and GDP growth will slow further to roughly 4.5–5.0%, still higher than the average of developed countries as well as the global average. Therefore, efforts on energy conservation and substitution must be intensified so that the annual decline in emissions intensity of GDP will not be lower than 4.5–5.0%, and emissions peak will occur as soon as possible before 2030.

As China transitions to an increasingly low-carbon energy mix, it is expected to peak CO_2 emissions before it reaches the peak in total energy consumption. Another key precondition for emission peak is that the drop in CO_2 emissions per unit of GDP must be greater than the growth of energy consumption, so that the increase in emissions brought by growth in energy consumption is offset by a reduced CO_2 emissions per unit of energy consumption, and carbon emissions will plateau or trend downward. By around 2030 when GDP growth is projected to be 4.5–5.0%, the elasticity of energy consumption will also drop below 0.3, bringing the growth of total energy demand down below 1.5%. Yet the supply of non-fossil energy will need to sustain its growth of more than 6% on an annual basis in order for the increase in total energy demand to be met entirely by non-fossil sources. The annual reduction of CO_2 emissions per unit of energy consumed needs to exceed 1.5% and continue to rise, and CO_2 emissions will peak and then turn downward.

For China to peak CO_2 emissions around 2030, it needs to undertake greater efforts than its counterparts in the developed world. China will peak its CO_2 emissions earlier in the stage of development than developed countries, who only reached their emission peaks after industrialization when GDP growth was relatively modest (no higher than 3% in general) and the annual decline of CO_2 intensity of GDP was below 3%. To peak CO_2 emissions around 2030, China needs to secure a 4.5–5.0% annual decline in carbon intensity of GDP, which in turn entails painstaking efforts in energy conservation and low-carbon transition of the energy mix, and in particular the acceleration of new and renewable energy development and the reduction of CO_2 emissions per unit of energy consumed. The work on these fronts will play a critical role in driving the country to peak emissions.

In the long haul, the goals of holding the rise in global temperature below 2°C or even further to 1.5°C and achieving global net zero emission in the second half of this century or even by mid-century pose considerable challenges for China. The majority of EU members had peaked their CO_2 emissions in the 1980s while the United States and Japan also reached their peaks around 2005. There is a 50–70 years of transition from carbon peak to carbon neutrality for these countries. In comparison, China will only have a window of 20–30 years between its carbon peak in 2030 and the deadline for net zero emission. Therefore, more ambitious, rigorous, and faster emission reduction efforts are required for China, and greater challenges and costs lie ahead.

2.3 China's Long-Term Economic and Social Development Prospects

The 19th Party Congress put forward the goals, basic strategies and blueprints of China's socialist modernization drive. "By 2020, not only must we finish building a moderately prosperous society in all respects and achieve the first centenary goal," the report of the 19th Party Congress notes, "we must also build on this achievement to embark on a new journey toward the second centenary goal of fully building a modern socialist country."

In the first stage from 2020 to 2035, China will largely realize socialist modernization. The country's economic and technological progress will experience a giant leap, and it will become a global leader in innovation. There will be a fundamental

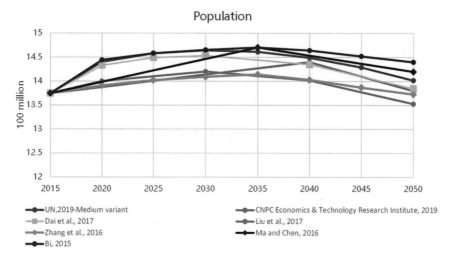

Fig. 2.1 Related studies on population projections [3–9]

improvement in the environment with the goal of building a beautiful China basically reached. In the second stage from 2035 to 2050, the country will develop into a great modern socialist country that is prosperous, strong, democratic, culturally advanced, harmonious, and beautiful. China's material, political, spiritual, social and ecological civilizations will be upgraded in all respects. China will rise to become a leading nation in terms of national power and global impact, and the goal of common prosperity for all will be basically realized, and a beautiful China will become a reality.

China's GDP per capita surpassed 10,000 USD in 2019 and is steadily closing the gap as an upper-middle-income country with a high-income country. This section presents the main trends of China's future economic and social development based on current national realities and the two centenary goals.

2.3.1 Population and Urbanization Rate

Existing studies have generally made similar predictions regarding the population trends of China (see Fig. 2.1). The country's population will hit a ceiling between 2030 and 2035 with 1.4–1.5 billion people, and then trend downward. According to a projection by the National Health and Family Planning Commission in 2017, China's population will likely reach a peak of 1.45 billion in 2030, after which it will drop to 1.4 billion by 2050.[1]

Since the beginning of this century, people over 60 years old in China has comprised more than 10% of the total population, which makes China officially

---

[1] http://news.cri.cn/gb/1321/2017/03/10/661s5236359.htm.

an aging society. According to the projection of *National Population Development Plan (2016–2030)*, the growth of the population aged 60 and older will accelerate sharply from 2021 to 2030, and its share in the total population will hit roughly 25% by 2030. The working-age population will shrink rapidly from 2021 to 2030, and by 2030, the share of workers aged 45–59 years is projected to reach around 36%. The dwindling labor force and population aging will hurt China's economic development. However, the improvement of population skills and the development of informationization and intellectualization will help offset the decline in the size of the labor force and ensure sustained economic and social development.

The level of urbanization is one of the key factors underlining China's economic development and energy consumption. Existing researches generally agree on the trends of China's urbanization in the coming three decades (see Fig. 2.2). Data shows that permanent urban residents exceeded 60% of the Chinese population in 2019, and by 2050, approximately 75% of the population will inhabit urban areas, an increase of nearly 20 percentage points compared to 2015. Looking at the trends by period, some studies suggest that the urbanization rate will surge before 2030 and see slower growth afterwards while others indicate that the growth will be more evenly distributed in different period. It can be assumed that China's urbanization rate will reach 65–70% in 2035 and 75–80% in 2050.

Projections of future population size and urbanization rate are presented below after incorporating results from various studies (Table 2.3).

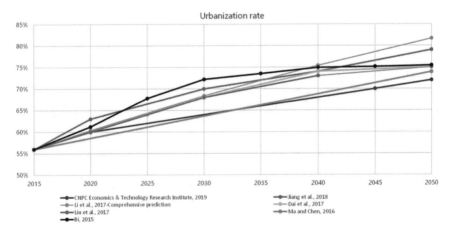

Fig. 2.2 Related studies on urbanization rate [4–6, 8–11]

Table 2.3 China's future population and urbanization rate

	2015	2020	2025	2030	2035	2050
Population (billion)	1.375	1.416	1.426	1.450	1.444	1.395
Urbanization rate (%)	56.1	60.4	64.6	66.9	68.5	75.0

2.3.2 Economic Growth and Changes in Industrial Structure

China's economic development is currently in the mid to late stage of industrialization and urbanization. The country's per capita GDP topped 10,000 USD in 2019 as it transitions from an upper-middle income economy into a high-income economy. The historical trajectory of development in developed countries also demonstrates that after the 10,000-dollar mark is crossed, economic growth will generally shift from the pursuit of quantity and scale to quality and efficiency; the share of the industrial sector will fall while that of the services will grow; the share of traditional manufacturing will shrink while that of high-tech industries is slated to rise; and economic growth will slow down but maintain steady growth as the economy transitions from extensive to intensive growth. China's economic development in the future will also by and large follow this pattern.

In the first stage of China's socialist modernization efforts from 2020 to 2035, the country's industrialization and urbanization will be largely completed. The economy will fall back to a medium growth and, in the meantime, overtake the US economy as the world's largest during this period. Investment growth will slow down but remain at a high level. Despite the shrinking of the labor force, human capital growth, labor transfer, coupled with reforms to enhance market vitality, will accelerate the growth of total factor productivity. With the completion of the massive infrastructure construction phase, the market for energy-intensive raw materials such as steel and cement will become saturated and trend downward, with a reduction in the share of the heavy and chemical industry in GDP. The manufacturing industry will move toward the mid-to-high end of the global value chain, with improvements in China's foreign trade structure and added value. The weight of modern services in the economy such as information and finance will experience steady growth, and the country will make notable strides in agricultural modernization. In the meantime, the consumption structure of Chinese society will undergo enormous changes as the caliber of demand rises, the middle class expands and the overall purchasing power of residents increases. Emerging consumption trends driven by the higher needs of the people such as culture and entertainment, health services, training, and tourism, will flourish. While capital accumulation will remain the main driver of economic growth, the role of consumption will take on greater prominence and the contribution of total factor productivity will continuously expand. Broadly speaking, economic growth at this stage will feature a slower speed but substantial improvements in efficiency and quality.

In the second stage of China's socialist modernization, from 2035 to 2050, the country will evolve into a post-industrial economy, and urbanization will also have stabilized and been largely achieved. This phase will be characterized by slower growth in investment, diminishing labor force and economic growth primarily driven by services and consumption. From the standpoint of total factor productivity, the world will witness a new wave of technological innovation with increasing penetration of new technologies in various sectors of the economy. The contribution of investment to economic growth will take a backseat to technological innovation, which will become the core driver of growth. The increase in total factor productivity will become the main ingredient of economic growth.

Projections on future GDP growth are given in Table 2.4 based on comprehensive consideration of factor inputs that affect China's potential growth and the laws of change as well as findings of other domestic researches (see Fig. 2.3). On the whole, China's future average annual GDP growth will feature a steady decline, dropping to roughly 3% by 2050, which is still slightly higher than that of the current developed countries. Growth is expected to experience fluctuations in the near term due to the impact of the pandemic.

Changes in future GDP growth will bear considerable impact on the reduction of CO_2 intensity, the schedule of peaking and the amount of emissions at the time of peaking. In the wake of the CO_2 emissions peak around 2030, economic growth will be decoupled from emissions, and changes in GDP growth will exert minor impact on long-term CO_2 emissions. This research opts for more optimistic estimates of pre-2030 GDP growth, or in other words, it chooses a scenario where the path to peaking emissions is more challenging. Industry, and in particular manufacturing, has always featured a large uptake in Chinese economy. The secondary industry made up 40% of GDP in 2018, higher than that of developed countries during their industrialization. China will maintain its status as a manufacturing powerhouse for a long time, and will not follow the footsteps of developed countries in the massive offshoring of manufacturing. The secondary industry will still represent considerable weight in the economy (see Figs. 2.4 and 2.5). The share of the tertiary industry will rise steadily to reach roughly 60% in 2030 and possibly more than 70% in 2050—on a par with that of developed countries in the current stage (Table 2.5).

2.3.3   Energy Consumption and CO_2 Emissions

Future energy demand and CO_2 emissions are not only related to factors such as economic and social development, industrial structure, and technological progress, but also dependent on strategic goals and policy decisions. The analysis of quantities and types of energy consumption and CO_2 emission trends is not built on projections of existing trends, but rather, on the scenario analysis of various strategic goals and policies. In particular, the pathway of emission reduction to reach the goal of the Paris Agreement to limit global warming to 2°C and 1.5°C will drive ambitious energy conservation and low-carbon transition of the energy mix. The strategy and goal of China's energy revolution in the future must not only provide a clean, safe, economical, and efficient energy supply for socialist modernization, but also be aligned with the CO_2 emission pathway of staying below 2°C. The long-term energy strategy ought to be both problem and goal-oriented by coordinating the two major interests of domestic sustainable development and global ecological safety, as well as the goals and tasks of the two stages of the country's development, namely the fulfillment of NDC ambitions in 2030 and the fundamental improvement of China's environment by 2035 during the first stage, and the building of a beautiful China

Table 2.4 Reference scenarios for future GDP growth

	2005–2010	2010–2015	2015–2020	2020–2025	2025–2030	2030–2035	2035–2040	2040–2045	2045–2050
Average annual GDP growth (%)	11.3	7.9	5.9	5.3	4.8	4.4	4.0	3.6	3.2

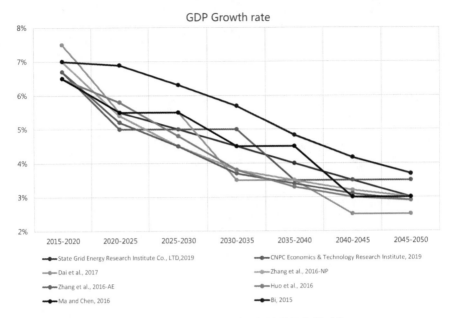

Fig. 2.3 Related studies on average annual GDP growth [4, 5, 7–9, 12, 13]

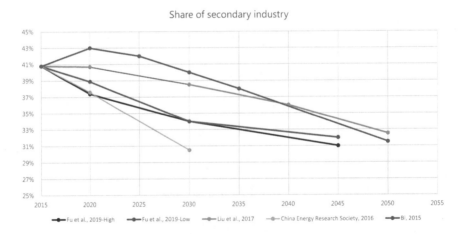

Fig. 2.4 Related studies on the share of the secondary industry [6, 9, 14, 15]

and the achievement of deep decarbonization by 2050 during the second stage. The Chinese government needs to coordinate its efforts and plan ahead, develop an energy revolution and low-carbon emissions strategy aligned with the two-stage plan of China's socialist modernization, strive toward a great modern socialist country and a beautiful China while pursuing a deep decarbonization pathway consistent with the target of near-zero emissions by mid-century under the global warming scenario

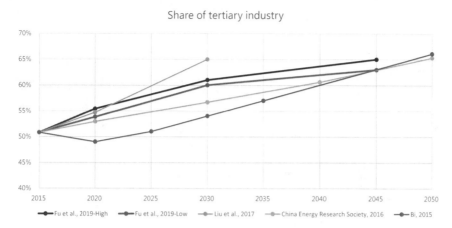

Fig. 2.5 Related studies on the share of the tertiary industry [6, 9, 14, 15]

Table 2.5 Reference scenarios for China's future industrial structure

	2015	2020	2025	2035	2050
Primary industry (%)	8.83	7.52	6.89	5.42	3.48
Secondary industry (%)	40.93	37.47	34.10	28.10	24.22
Tertiary industry (%)	50.24	55.01	59.01	66.48	72.30

of 2°C, make contributions commensurate with China's rising national strength and international profile, and spearhead the low-carbon transition of the global energy system and economy.

Most domestic research studies predict future energy demand and CO_2 emissions under the scenario of enhanced energy conservation and substitution. There is a dearth of studies on the 2°C or even 1.5°C target-driven scenario, where systematic research is needed. The findings of other domestic researches are illustrated in Figs. 2.6 and 2.7, which suggest that the pathway to achieve the 2°C goal is an extremely challenging one. Various studies have produced divergent estimates of CO_2 emissions in 2050 under different policy scenarios in the future. Most studies conclude that China's energy-related CO_2 emissions will reach between 4–10 billion tCO_2 by 2050, which is also a matter that will receive granular discussion in subsequent chapters of this report.

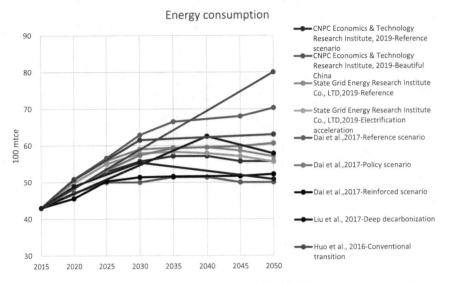

Fig. 2.6 Related studies on future energy consumption [4–6, 12, 13]

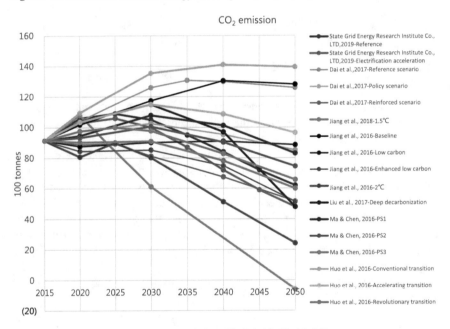

Fig. 2.7 Related studies on total CO_2 emissions [5, 6, 8, 10, 12, 13, 16]

References

1. United Nations Environment Programme. Emissions Gap Report 2019 (2019) UNEP, Nairobi. https://www.unenvironment.org/resources/emissions-gap-report-2019
2. China National Expert Panel on Climate Change (2019) UK-China co-operation on climate change risk assessment: developing indicators of climate risk. China Environmental Publishing Group, Beijing
3. United Nations. World Population Prospects 2019 (2019) [2020–04–24]. https://population.un.org/wpp/
4. CNPC Economics & Technology Research Institute. World and China energy outlook in 2050 (2019) Economics & technology research institute of China national petroleum corporation
5. Dai YD, Kang YB, Xiong XP et al (2017) Energy development and carbon emission scenarios towards 2050: Energy Transition and Low-carbon Development Roadmap for China. China Environment Press, Beijing
6. Liu Q, Chen Y, Teng F et al (2017) Pathway and policy analysis to China's deep decarbonization. China Popul Resour Environ 27(09):162–170
7. Zhang X et al (2016) Carbon emissions in China: How far can new efforts bend the curve? Energy Economics 54:388–395
8. Ma D, Chen WY (2016) Analysis of China's 2030 carbon emission peak level and peak path. China Popul Resour Environ 26(S1):1–4
9. Bi C 2015 Scheme and policies for peaking energy carbon emissions in China. Chinese population. Resour Environ 25(05): 23–24
10. Jiang K et al (2018) Emission scenario analysis for China under the global 1.5 degrees C target. Carbon Manag 9(5): 481–491
11. Li ST, Wu SM, Gao CL (2017) Prediction and analysis of China's urbanization speed. Dev Res 11:19–22
12. State Grid Energy Research Institute Co., LTD (2019) China energy and electricity outlook 2019. China Electric Power Press, Beijing
13. Huo J, Weng YY, Zhang XL (2016) Transition pathway toward a low carbon energy economy in 2050. Envrionmental Proj 44(16):38–42
14. Fu ZH, Sun YF, Weng MY, He HW (2019) Research on the strategy of a modern comprehensive transport system. People's Communications Press, Beijing
15. China Energy Research Society (2016) China energy outlook 2030. Economy & Management Publishing House, Beijing
16. Jiang KJ, He CM, Zhuang X et al (2016) Scenario and feasibility study for peaking CO_2 emission from energy activities in China. Clim Chang Res 03:167–171

Chapter 3
End-Use Energy Consumption & CO$_2$ Emissions

3.1 Industrial Sector

3.1.1 Current Development and Trend of the Industrial Sector

1. **State of Energy Consumption and CO$_2$Emissions of China's Industrial Sector**

As a key pillar of China's economic growth, the industrial sector constitutes the dominant source of energy consumption and CO$_2$ emissions. In 2018, the added value of China's industrial sector was 30.5 trillion RMB, accounting for 33.9% of GDP. Energy is consumed by the industrial sector for fuel combustion and raw materials, etc. [1]. In 2017, The end-use energy consumption of the industrial sector registered at 2.83 billion tce, comprising 64.8% of the total end-use energy consumption. Energy consumption of the industrial sector is mainly attributable to six energy intensive industries, including power, steel, cement, petrochemical, chemical and non-ferrous metal, which made up 75.1% of the total primary energy consumption of the industrial sector in 2017 [2] (see Fig. 3.1). Industrial CO$_2$ emissions include those from energy activities and industrial production processes. However, due to delayed survey and accounting, the statistics of CO$_2$ emissions of the industrial sector is relatively not up to date and inconsistent with the accounting standard of industrial output and energy consumption. The *First Biennial Update Report on Climate Change* reveals that the CO$_2$ emissions from fuel combustion in the industrial and construction sectors and from industrial production processes in China totaled 8.48 billion tons in 2012, or 91.0% of the total CO$_2$ emissions of the year (including land use changes and forestry), of which CO$_2$ emissions from fuel combustion of the industrial and construction sectors stood at 7.28 billion tons, which is 83.8% of the total CO$_2$ emissions from energy activities.

© China Environment Publishing Group Co., Ltd. 2022
Institute of Climate Change and Sustainable Development of Tsinghua University et al.,
China's Long-Term Low-Carbon Development Strategies and Pathways,
https://doi.org/10.1007/978-981-16-2524-4_3

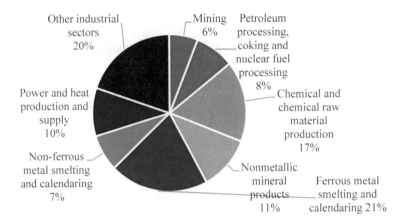

Fig. 3.1 Energy consumption mix of China's industrial sector (2017)

The expanding industrial output has resulted in rising energy consumption and carbon emissions of the industrial sector, but a notable slowdown has been observed. In terms of energy consumption, in 2017, the end-use energy consumption of the industrial sector was 2.83 billion tce, 1.9 times higher than in 2000 with an average growth of 6.5% year-on-year. [2]. From the perspective of carbon emissions, the published *National Greenhouse Gas Inventory* suggested total CO$_2$ emissions from fuel combustion in the industrial and construction sectors and from industrial production processes rose by 3.37 billion tons from 2005 to 2012, averaging 7.5% growth rate year-on-year, exceeding the 6.1% average year-on-year growth rate of industrial energy consumption in the same period [3]. Yet as China enters the advanced stage of industrialization and becomes a post-industrialized economy, a major slowdown in the growth of industrial energy consumption has been detected with a continuous decline in end-use energy consumption of China's industrial sector, which registered negative growth for two years in a row (2014–2016). Despite a bounceback in 2017, the figure was still 1.5 million tce lower than that in 2014. From 2011 to 2017, the industrial sector consumed an additional 186 million tce—a mere 30% of the added energy consumption in China [2] (see Fig. 3.2).

Improvements are continuously made in energy conservation and emission reduction policies with sustained enhancement in energy efficiency and optimization in energy consumption mix. Through continued supply-side structural reform in coal and steel industries, campaigns and implementation of key energy-saving projects such as comprehensive upgrading of coal-fired boilers for energy conservation and eco-friendliness, energy efficiency enhancement of motor systems, reduction and replacement of coal consumption, energy performance contracting, etc., China has shown a slowdown in the growth of major energy-intensive products and sustained improvement in industrial energy efficiency with some sectors ranking among the global leaders in this regard (see Fig. 3.3). From 2000 to 2016, energy efficiency grew by one-third for the steel industry and a half for the cement industry in China

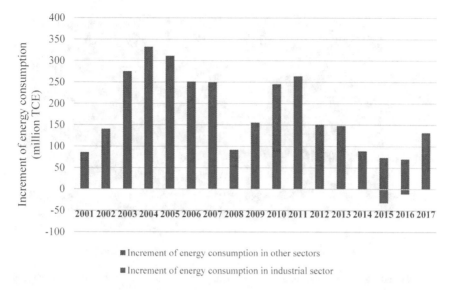

Fig. 3.2 Changes in added energy consumption in China and in the industrial sector from 2010 to 2017 (Data source: China Energy Statistical Yearbook)

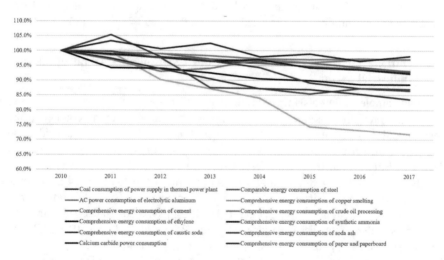

Fig. 3.3 Energy consumption in major energy-intensive products compared to 2010 level (Data source: Wang Qingyi, Energy in China 2018)

[4]. Meanwhile, through coal reduction and substitution, promotion of electrification and technical progress in production, notable changes have taken place in the end-use energy consumption mix of the industrial sector, with 2012 as a turning point, when a decline was seen in both the total use and the proportion of coal and coke in the industrial sector, while the consumption of natural gas and electricity had been on

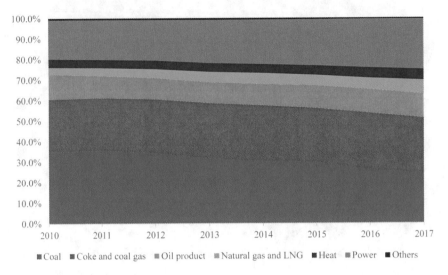

Fig. 3.4 End-use energy consumption mix of the industrial sector by category (Data source: China Energy Yearbook 2018)

a sharp rise. In 2017, the share of coal and coke consumption fell back to 44.8%, while the proportion bounced to 6.8% for natural gas and 24.6% for electricity [2] (see Fig. 3.4).

2. **Outlook on Energy Consumption and CO_2 Emission Trend of China's Industrial Sector**

With ample potentials in its development, China's industrial sector is facing mounting pressures on energy demand. As the biggest developing country in the world, China is still in the intermediate stage of industrialization with considerable space for further growth and growing pressure of energy consumption and carbon emissions of the industrial sector. Despite being a manufacturing powerhouse, China trails way behind the US, EU and Japan in terms of core technologies, quality, efficiency and indigenous innovation, etc., with its manufacturing sector on the lower end of the global arena. At the per capita level, China's industrial added value was 1.3 times of that of the world average yet merely 35.6% of the US in 2016 [5]. In the same year, China's per capita industrial end-use energy consumption was around 95% higher than the world average and 16% higher than OECD countries, but still 12% lower than US [4]. At the global level, industry remains the key pillar for economic growth with minimal sign of "de-industrialization". At the constant price of 2010, manufacturing as a percentage of global GDP climbed from 14.8 to 16.0% from 1991 to 2014 [5]. Despite a slowdown or even decline in total energy demand of the industrial sector in major developed countries, global industrial energy demand is set to grow continually. Studies have found that from 2017 to 2040, annual growth in energy demand of the world industrial sector will average around 1.3%, and major developed economies such as the US and EU might witness a drop in total demand, yet the share of industry in end-use energy consumption will slowly move along a rising curve [4]. Especially with the onslaught of COVID-19, increasing awareness of industrial security, independence and controllability is found in all countries, and the global industrial chain is approaching a new round of restructuring, which spells mounting strains for China to consolidate, shore up and empower its industrial chain.

The growth in energy demand will see a substantial slowdown as the industry shifts from extensive growth to quality development. As China transforms from a "world factory" to a manufacturing powerhouse and moves towards quality development, industrial energy demand is set to experience a tremendous slowdown despite its continued growth. In terms of industrial structure, as China's industrial economy advances towards medium and high-end, the mainstay of the industry will shift from traditional sectors such as steel, building materials, and textile towards emerging industries of strategic importance, high-tech and advanced manufacturing, with increasing decoupling of industrial development and energy consumption. Meanwhile, as China succeeds in building a moderately well-off society, its urbanization rush is likely to tide off and the demand for major energy-intensive products will be saturated, which will downsize a portion of the energy-intensive sectors. Besides, amid a sweeping new round of global industrial revolution, the massive application of new technologies such as industrial robotics, AI and industrial internet catalyzed by growing industrial digitalization, internet penetration and intelligence, the existing mode of industrial production and organizational structure will be disrupted, producing a far-reaching impact on the energy mix of the industrial sector. To illustrate, industrial robotics and 3D printing will tremendously boost the energy efficiency and the electrification of the industry while replacing the traditional way of production and manual labor arrangement.

As industrial electrification picks up its pace, the peak will occur sooner in carbon emissions than in energy demand of the industrial sector. CO_2 emissions of the sector hinge on energy demand, energy efficiency and energy mix, among other factors; with the slowdown in energy demand of the industry, improved electrification and constant changes in power mix of the society, industrial CO_2 emissions will peak before its energy demand. With emerging industries such as modern manufacturing on a fast track of development and fossil fuels such as coal being increasingly replaced in heavily polluted areas, energy demand of the industry will primarily come from electricity and natural gas with reduced demand for coal and more electric and lower-carbon energy mix. Studies have shown that electricity of non-energy-intensive sectors will comprise 47% of the end-use energy consumption in China by 2040 [4]. Among industrial sectors, building materials and textile would peak emissions prior to steel and non-ferrous metals, etc.; while the energy demand and CO_2 emissions of the petrochemical industry might peak later due to the rapidly growing output. In the meantime, the fast growth of petroleum and coal as raw materials of the petrochemical industry will also lead to an earlier peak in industrial CO_2 emissions than energy demand.

The transition towards low-carbon industry remains unpredictable, with deep emission reduction as a daunting task. Ranking the top in industrial output, China is at the forefront of a new round of industrial revolution with tremendous emission reduction potential in employing innovative technologies and promoting integrated development of industrialization and low-carbon energy, electrification and information technologies. However, despite the sustained enhancement in energy efficiency of the industry, the development model featuring high input, consumption, and emissions has not been essentially reversed. Recently, many key petrochemical and coal chemical projects, including coal-to-aromatics, coal-to-ethanol, and coal-to-hydrogen projects, are under planning due to multiple factors, which brings additional pressure on controlling industrial energy consumption and carbon emissions. Meanwhile, the prevalent overcapacity and inter-regional imbalance in industrial development generate substantial challenges for energy-intensive sectors to widely reduce intensity. With a sizable existing industrial capacity and the limited operation time of many carbon-intensive plants, a premature retirement would incur massive sunk asset losses and potential unemployment risk. In addition, the lock-in effect stemming from the existing market stronghold of carbon-intensive technologies and industries is another stumbling block for the rapid expansion of low-carbon or zero-carbon technologies. Industrial transformation is highly associated with high-quality economic and social development, calling for an overhaul in the pattern of production and consumption across the society. Studies have suggested that postponing the goal of industrial transformation to 2035 would mean an extra 100 million tce consumption and 260 million tons of more CO_2 emissions from the industrial sector in 2040 [6]. Besides, as globalization grows, changes in international division of labor and industrial capacity footprint, uncertainties in the policies and market mechanisms of the carbon market and carbon tax and emerging trends such as CCS technology would produce major impact on future carbon emission trends from the industry.

3.1.2 Key Findings of Scenario Analysis

1. **Total End-use Demand**

Under the policy, reinforced policy, 2 °C and 1.5 °C scenarios, the end-use energy demand of the industrial sector will still rise from the 2015 level, yet at a slower pace until peaking in around 2030 before a continual decline.[1] Under the constraints of the 2 °C and 1.5 °C scenarios, the end-use demand of the industrial sector is to peak prior to 2025 and drop by approximately 24–32% by 2050 from that of 2015.

As seen in Fig. 3.5 and Table 3.1, a more elaborated analysis reveals that in the policy scenario, the end-use energy demand of the industrial sector peaks at around 2030 at approximately 2.67 billion tce before sliding to roughly 2.44 billion tce by

---

[1] Note: in this study, processing and converting petrochemical and coal chemical are included in the industrial sector. The statistics of end-use energy consumption of the industrial sector include energy consumed as fuel and raw materials and for hydrogen generation in the future, etc., but exclude energy consumed by means of transportation such as vehicles owned by plants, or coal and natural gas consumption by captive power plants—slightly different from the statistics released by National Bureau of Statistics.

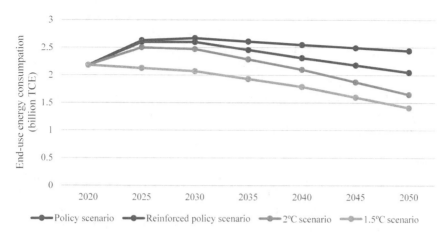

Fig. 3.5 End-use energy demand analysis of the industrial sector under different scenarios

Table 3.1 Analysis of end-use energy demand of the industrial sector (billion tce)

	2020	2030	2040	2050
Policy scenario	2.18	2.67	2.55	2.44
Reinforced policy scenario	2.18	2.60	2.31	2.05
2 °C scenario	2.18	2.47	2.10	1.65
1.5 °C scenario	2.18	2.07	1.79	1.41

2050; in the reinforced policy scenario, the demand hits the peak at around 2025 at about 2.6 billion tce before a drop to 2.05 billion tce in 2050; in the 2 °C scenario, the peak occurs in 2020–2025 at 2.5 billion tce and declines to around 1.65 billion tce by 2050; and in the 1.5 °C scenario, the peak arrives between 2020 and 2025 at roughly 2.2 billion tce before a reduction to 1.41 billion tce by 2050.

2. Industrial CO_2 Emissions

In the four scenarios, total CO_2 emissions from industrial energy activities and production processes peak prior to 2030 before a continuous drop. CO_2 emissions from industrial energy activities peak in 2030 in the policy scenario, compared to around 2025 in the reinforced policy scenario or the 2 °C scenario and 2020–2025 in the 1.5 °C scenario.

As shown in Fig. 3.6 and Table 3.2, in the policy scenario, the peak of total industrial CO_2 emissions stands at roughly 5.71 billion tons and the figure falls to around 4.61 billion tons in 2050, when CO_2 emissions from industrial energy activities reaches 3.69 billion tons and emissions from industrial processes reach around 920 million tons. Compared to 2020, total CO_2 emissions from the industrial sector diminish by approximately 9.4%, of which emissions from energy activities contributed a 2.1% reduction in 2050.

In the reinforced policy scenario, total industrial CO$_2$ emissions are reduced to roughly 3.42 billion tons by 2050, of which 2.62 billion tons is from energy activities and 800 million tons from industrial processes; total industrial CO$_2$ emissions are cut by 32.8% in 2050 relative to 2020, and a 30.5% emission reduction is observed in energy activities.

In the 2 °C scenario, total industrial CO$_2$ emissions fall to around 1.67 billion tons by 2050, of which roughly 1.2 billion tons are from energy activities and around 470 million tons from industrial processes; compared to 2020, total industrial CO$_2$ emissions are cut by around 67.2% in 2050, and emissions from energy activities are down by 68.2%.

In the 1.5 °C scenario, total industrial CO$_2$ emissions reduces to approximately 710 million tons by 2050, of which 460 million tons are from energy activities and 250 million tons from industrial processes; compared to 2020, total industrial CO$_2$ emissions are reduced by 86.1% in 2050, of which 87.8% emission reduction is from energy activities.

3.   CO$_2$ Emissions from Industrial Process

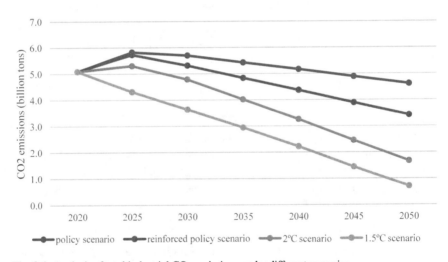

Fig. 3.6 Analysis of total industrial CO$_2$ emissions under different scenarios

Table 3.2 Analysis on total industrial CO$_2$ emissions (billion tons)

	2020	2030	2040	2050
Policy scenario	5.09	5.71	5.17	4.61
Reinforced policy scenario	5.09	5.31	4.37	3.42
2 °C scenario	5.09	4.77	3.26	1.67
1.5 °C scenario	5.09	3.64	2.22	0.71

Table 3.3 Analysis of CO_2 emissions from industrial process (billion tons)

	2020	2030	2040	2050
Policy scenario	1.32	1.17	1.04	0.92
Reinforced policy scenario	1.32	1.10	0.94	0.80
2 °C scenario	1.32	0.95	0.70	0.47
1.5 °C scenario	1.32	0.88	0.56	0.25

In industries such as steel, cement, building material, and chemical, CO_2 emissions from both energy activities and industrial processes can be cut by substitution of raw materials or fuels, optimization of technologies and processes, improvement in system energy efficiency, and development of innovative low-carbon products. By 2050, CO_2 emissions from industrial processes is down by approximately 30% from 2020 levels under the policy scenario, 39% under the reinforced policy scenario, 64% under the 2 °C scenario and 81% under the 1.5 °C scenario (Table 3.3).

Emission reduction of industrial processes is, on one hand, attributed to industrial restructuring, technical renovation and the development of alternative raw materials and fuels. For example, lime may be replaced by raw materials of lower carbon intensity such as carbide slag, blast furnace cinder, fly ash and steel slag in cement production to cut emissions. In flat glass production, CO_2 emissions from the mixing process can be halved by replacing dolomite and limestone with MgO and CaO; while in the coal chemical industry, new gasification techniques such as pressurized coal-water slurry gasification and pressurized coal powder gasification help to reduce CO_2 emissions by a large margin. On the other hand, a constant drop is observed in demand for energy-intensive products owing to industrial restructuring and product quality upgrading. For instance, under the policy and reinforced policy scenarios, cement demand falls by around 30% from 2020 to 2050, while even sharper declines of 62% and 71% are expected under the 2 °C and 1.5 °C scenarios respectively (Fig. 3.7).

4. **Industrial Energy Mix and Electrification**

All four scenarios point to a clear trajectory of decarbonization and electrification in the industrial energy mix, with smaller share of fossil fuels such as coal and petroleum and larger uptake of electricity in end-use energy, and distributive renewable electricity, renewable heating, hydrogen, and biomass are deployed on a wider scale (Fig. 3.8). The share of non-fossil energy and electricity in end-use industrial energy demand in 2050 will reach 34.3% in the policy scenario, 45.2% in the reinforced policy scenario, 66.7% in the 2 °C scenario, and 85.1% in the 1.5 °C scenario.

Fossil energy sees a continuous decline in the end-use industrial energy demand with its usage moving from fuel to feeding stock. For example, the industrial demand for coal, which is mainly used for raw materials and for heating in certain processes, is continuously shrinking. Compared to 2020, the industrial demand for coal drops by 14.3, 36.2, 77.1, and 92.4% respectively by 2050 under the policy scenario, reinforced

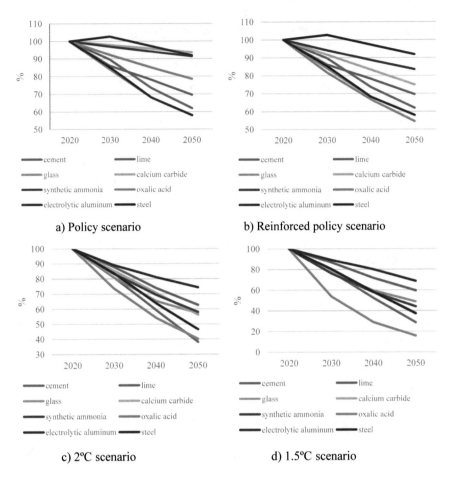

a) Policy scenario

b) Reinforced policy scenario

c) 2°C scenario

d) 1.5°C scenario

Fig. 3.7 Forecast and analysis of demand for major energy-intensive products under different scenarios (2020 = 100)

policy scenario, 2 °C scenario and 1.5 °C scenario. Oil demand experiences a minor reduction of 4.9% in 2050 from 2020 levels under the policy scenario, and even more dramatic decline under the reinforced policy scenario, 2 °C scenario, and 1.5 °C scenario of 39.0, 65.9, and 82.9% respectively. Much uncertainty lies ahead for industrial gas demand with a surge of 220, 110, and 70% respectively under the policy, reinforced policy, and 2 °C scenarios yet a drop of 40% under 1.5 °C scenario.

It's essential to reinforce the electrification transformation during the shift toward a low-carbon industry so that it will become the predominant source of energy. Specifically, the industrial electrification rate of industrial sector reaches 31.0%, 39.9%, 58.2%, and 69.4% respectively in 2050 under the policy scenario, reinforced policy scenario, 2 °C scenario, and 1.5 °C scenario. Such progress in China has visibly outpaced the historical development in general as the period between 2000

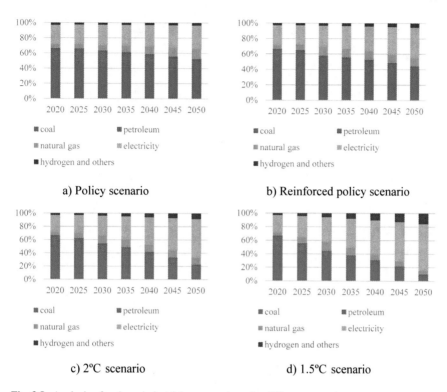

Fig. 3.8 Analysis of end-use industrial energy mix under different scenarios

and 2015 saw an annual increase of industrial electrification of 0.5%, and under the 2 °C and 1.5 °C scenarios, such increase would jump to 0.9 and 1.4% respectively (Table 3.4).

The expanding demand for alternative industrial fuels and raw materials such as hydrogen and renewable energy has become a key lever for further carbon reduction in the industrial sector. Under the policy scenario, hydrogen and renewable energy comprise 3.3% of end-use industrial energy consumption by 2050, and the percentage is 5.3%, 8.5%, and 15.6% respectively for the reinforced policy scenario, 2 °C scenario, and 1.5 °C scenario (Table 3.5). Fossil fuels would be the main

Table 3.4 Analysis of end-use electrification of the industrial sector (%)

	2020	2030	2040	2050
Policy scenario	25.7	26.1	28.1	31.0
Reinforced policy scenario	25.7	27.8	32.3	39.9
2°C scenario	25.7	30.0	41.2	58.2
1.5°C scenario	25.7	37.0	49.8	69.4

Table 3.5 Analysis of end-use energy mix of the industrial sector (%)

Policy scenario	2020	2030	2040	2050	Reinforced policy scenario	2020	2030	2040	2050
Coal	48.2	44.0	40.9	36.7	Coal	48.2	41.3	37.3	32.5
Oil	18.2	19.8	17.9	15.9	Oil	18.8	17.0	15.8	12.1
Natural gas	4.6	7.5	10.2	13.1	Natural gas	4.6	10.8	10.6	10.2
Electricity	25.7	26.1	28.1	31.0	Electricity	25.7	27.8	32.3	39.8
Hydrogen and others	2.8	2.6	2.9	3.3	Hydrogen and others	2.8	3.1	4.0	5.3
2°C scenario	2020	2030	2040	2050	1.5 °C scenario	2020	2030	2040	2050
Coal	48.2	39.7	29.6	14.5	Coal	48.2	35.1	24.0	5.7
Oil	18.8	15.0	12.4	8.5	Oil	18.8	9.6	7.2	5.0
Natural gas	4.6	11.7	11.0	10.3	Natural gas	4.6	12.5	8.7	4.3
Electricity	25.7	30.0	41.2	58.2	Electricity	25.7	37.0	49.8	69.5
Hydrogen and others	2.8	3.6	5.8	8.5	Hydrogen and others	2.8	5.8	10.3	15.6

hydrogen source in the near and mid-term future before replaced by renewables or nuclear in the mid- to long term.

5. **Energy Conservation and Efficiency Enhancement**

To proactively tackle climate change, energy conservation and efficiency enhancement should become the "primary energy" for meeting the rising energy demand of the industrial sector. By 2050, the industrial output is expected to grow by around 2.6 times despite a reduced weight of industrial production in GDP. Through continuous optimization of its internal structure, improvement in added value in the products, and boost of energy efficiency of products and systems, the energy consumption of industrial output will continue to shrink. Under all four scenarios in terms of end-use energy consumption, a fall of 66%, 72%, 78%, and 80% in energy consumption of industrial output is expected between 2015 and 2050, which is an annual drop of 3.0%, 3.5%, 4.2% and 4.5%, respectively. In comparison, the period of 2000 to 2015 saw an annual decrease of 4.4%, meaning that long-term ambitious actions are needed for industrial structural improvement and energy efficiency enhancement in order to achieve the low emission goal of the industry.

In terms of technical energy efficiency improvement, significant potential is expected in energy-intensive industries such as steel, cement, and electrolytic aluminum; yet under the deep emission reduction scenario, the potential mainly arises from technical transformation. In the policy scenario, with a full achievement of the potential of existing feasible technologies in energy-intensive sectors, the unit energy consumption is reduced by 10–20% by 2050; in the reinforced policy scenario, major energy-intensive sectors see a further reduction of 10–20% in unit energy consumption by 2050 with the global leadership in technology and continual optimization of technical route; in the 2 °C scenario, world-leading technologies and

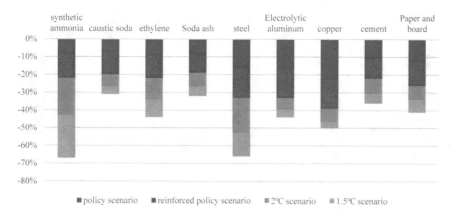

Fig. 3.9 Decrease in unit energy consumption of energy intensive products under different scenarios

complete technical innovations prompt a further drop of 10–15% by 2050. While in the 1.5 °C scenario, energy is saved mostly from a paradigm shift in the choice of technologies and raw materials, including the wide use of hydrogen (see Fig. 3.9).

3.1.3 Cost–Benefit Analysis of Investment in Different Scenarios

1. **Assessment of Energy Conservation and Carbon Reduction Potential in Industrial Sector**

As the world's largest manufacturer and a major developing country, China boasts an independent industrial system covering a full spectrum of sectors. Standing at the forefront of a new round of technological revolution and industrial transformation, it's also challenged by the coexistence of advanced and obsolete production capacity and the imbalance of industrialization. Much awaits to be done in the promotion of mature low-carbon technologies and cutting-edge technological innovation for energy conservation and emission reduction of the industrial sector, where huge potential also exists in demand reduction, energy mix improvement, and upgrading of smart technologies. A comparison between the policy scenario and the 2 °C scenario is conducted as an example (see Fig. 3.10).

Under the reinforced policy scenario, the industrial energy demand is set to be 2.04 billion tce in 2050, which can be reduced to 1.62 billion tons under the 2 °C scenario through a bunch of technological and policy measures. In terms of activities, 50 million tce can be cut in energy demand of energy-intensive sectors through reducing the output of energy-intensive products while 45 million tons can be saved from that of non-energy-intensive sectors—a total of 95 million tons or 22.6% of the total energy conservation. In terms of technical route, optimized production processes, especially

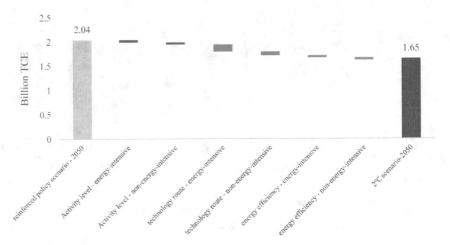

Fig. 3.10 Sources of potential of achieving the 2 °C scenario by 2050 (as compared to Reinforced Policy Scenario)

the massive deployment of deep decarbonization technologies in energy-intensive sectors helps reduce energy demand by 150 million tce, plus a 80-million-tons cut in non-energy-intensive sectors—a combined 230 million tons are saved, or 54.8% of the total energy conservation. In terms of energy efficiency, energy-intensive sectors are on track to reduce its energy use by 45 million tce by boosting energy efficiency while 50 million tons can be saved in non-energy-intensive sectors, adding up to 95 million tons or 22.6% of the total energy conservation.

2.   Cost–Benefit Analysis of Investment Under Different Scenarios

Now, the industrial sector boasts the greatest potential for emission reduction among all end-users, but the potential is likely to narrow down in the future with the rapid proliferation of energy-saving and carbon-reduction technologies and equipment. In the current stage, much potential still exists in emission reduction at low or negative cost in the industry, yet considering the fact that China shall remain the world's biggest manufacturer over a long period of time, plus the inertia of industrial development and barriers of energy-saving and low-carbon technology progress, the rising cost of industrial emission reduction and increasing difficulties in deep emission reduction are inevitable. A look at the sources of emission reduction potentials shows that technical transformation in industrial production, development in energy efficiency technologies, and application of renewables mostly depend on an overhaul of the industrial investment structure and direction; however, a high level of electrification and hydrogen scale-up would rest on the supply of low-cost green electricity and hydrogen, of which the investment and benefit analysis goes beyond the industrial sector itself. Besides, cutting redundant demand for industrial products depends on the transition of China's development model, the progress in China's industrial productivity as compared to other countries, the implementation of industrial and restructuring policies and shifts in consumption pattern, etc. It would be difficult to carry out an accurate and quantitative assessment of the direct investment cost and benefit.

With these factors in mind, this research focuses on analyzing the main investment increment under the four scenarios, in other words, extra investment associated with shrinking demand due to restructuring will not be considered while investment in changes of technical routes, energy efficiency improvement, and energy or material substitution will be the focus of the study. A bottom-up method is employed and a production capacity resetting method is used for key sectors such as steel, cement, and petrochemical to compare the accumulative changes in investment in production capacity associated with production scale, technology and process improvements, by 2050 under different scenarios; for other industrial sectors, the accumulative investment demand for higher energy efficiency and electrification under different scenarios are compared through case studies of investment in energy conservation and electrification. For the industrial application of hydrogen, the statistics concerning investment cost is scant in existing studies and demo projects, while it remains unclear whether investment in hydrogen generation and transmission is included. In this study, a hypothesis is made that industrial hydrogen consumption is mainly generated from low-carbon energy without taking into account the investment demand for hydrogen generation.

Given the complexity of sectors, products and technical processes involved in the industrial sector, priority-setting and categorization shall be adopted in calculating energy-saving and decarbonization investment i.e. focusing on energy-intensive sectors such as steel and cement with non-energy-intensive sectors as secondary consideration, and categorizing mainstream energy-saving and decarbonization technologies into "average energy efficiency" (currently average performance), "advanced energy efficiency" (currently advanced performance), "leading energy efficiency" (globally leading performance) and "near theoretical efficiency" (close to theoretical energy efficiency) based on technical features, investment benefit, and application prospects, in order to explore the investment required for varied levels of penetration and structure of production capacity. Under the policy scenario, corporate energy efficiency witnesses continued improvement driven by technical progress, and the sector-wide energy efficiency migrates from "average" to "advanced" by 2050; under the reinforced policy scenario, cost-effective and energy-saving technologies are both adopted for the sector-wide energy efficiency to reach "leading" level by 2050; under the 2 °C scenario, cutting-edge and ground-breaking technologies and processes are deployed, helping most businesses to hit "near theoretical" energy efficiency with zero-carbon energy and carbon-free processes utilized to some extent; under the 1.5 °C scenario, high cost processes that drastically lower carbon emissions will be adopted on a massive scale.

Taking cement as an example, according to the *Global Roadmap of Cement Sustainable Development in 2050* published by IEA, low carbon development is to be achieved in cement industry through enhancing energy efficiency, promoting alternative fuels, and applying CCS technologies. Building on the aforementioned research approach and scenario setting, and the investment needed for plants and processes of varied types of energy efficiency, the incremental investment for the cement industry

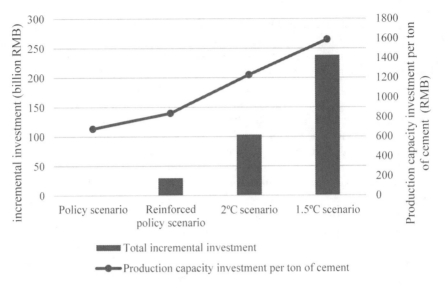

Fig. 3.11 Incremental investment needed for different scenarios for cement industry

and the investment required per ton of cement under different scenarios are calculated (see Fig. 3.11).

Under the reinforced policy scenario, the energy efficiency reaches "leading" level with alternative fuels making up 30% of the fuel mix, thus increasing the investment needed per ton of cement capacity to 840 RMB from 680 RMB under policy scenario, while the incremental investment under the reinforced policy scenario and policy scenario is 30 billion RMB. Under the 2 °C scenario, energy efficiency rises to "near theoretical level" with the penetration of 50% for alternative fuels and 30% for CCS, and the investment per ton of cement capacity rises to 1,200 RMB, an extra investment of 100 billion RMB compared to reinforced policy scenario. Under the 1.5 °C scenario, CCS penetration grows to 80% while alternative fuels rise up to 70% of the fuel mix, driving up investment per ton of cement capacity to 1,590 RMB with approximately 240 billion RMB more investment than the 2 °C scenario. Specifically, the application of CCS entails an extra investment of around 250 billion RMB while retrofitting for alternative fuels requires an extra 40 billion RMB, accompanied by a drop in energy efficiency investment of 56 billion RMB due to reduced capacity.

3.1.4 Key Pathways and Policy Enablers

1. **Structural Transformation and Shrinking Demand**

The development of service-oriented economy and inherent restructuring of the industry will promise more economic output with less energy consumption. Due to

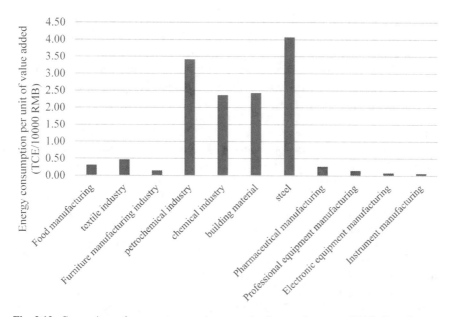

Fig. 3.12 Comparison of energy consumption per unit of output by sector (2015) (Data Source: National Statistics Bureau and calculations by the research team)

the huge discrepancy of energy productivity among industries and sectors, generally speaking, the service sector requires less energy than the industrial sector, light industry less than heavy industry, and emerging industries less than traditional ones when generating the same economic output (Fig. 3.12). Changes in industrial structure, organizational structure and positioning of product value chain of a country all exert tremendous impact on the energy efficiency of a country in a broad sense. The replacement of traditional industries by emerging industries will spur economic restructuring and a fresh round of economic growth, during which the internal restructuring and industrial modernization will boost the efficiency of energy allocation among different sectors as the energy consumption per unit of output in high value-added industries is much lower than that of energy-intensive and traditional industries, providing a strong buttress for the overall energy productivity of the industrial sector. Studies have found that during the 11th Five-Year Plan period, restructuring only contributed 4% of energy conservation of the industry, but the figure surged to 34% during the 12th Five-Year Plan period and is on track to approach 50% during the 13th Five-Year Plan period [7].

Elimination of wastage due to massive demolition and construction and improvement in material strength and quality will cut the demand for energy-intensive products from the source. Statistics show that the demolished housing area in China is around $37 \sim 41\%^2$ of newly built commercial residential houses, and the demolished

---

[2] In 2002, a total of 120 million m^2 of urban houses were demolished in China, or 37.5% of newly completed commercial residential buildings of 320 million m^2 of the same year; in 2003, a total

buildings of short life cycle account for around 23% of newly built ones in urban areas [8]. The demolished buildings in China are 25–30 years of age on average despite their designed service life of 50–100 years. Such massive demolition and construction have been the cause of severe waste of energy and resources and environmental pollution. Based on 2.5–3 billion m^2 of newly completed buildings annually and an annual demolition of short-lived buildings comprising 25% of such, an extra 40 to 50 million tons of steel and 220 to 260 million tons of cement are consumed annually. Taking into account the energy consumed throughout the construction process, an extra energy consumption of approximately 120 million tce is recorded each year— around 5% of the current energy consumption countrywide. What's also generated from this process are 400 million tons of waste, accounting for roughly 40% of total waste throughout the year. Apart from waste reduction efforts, wider acceptance of high-strength products will significantly minimize the use of materials, which represents a major step to build a resource-efficient and environmentally-friendly society. It not only contributes to energy saving and emission reduction, but also produces apparent economic benefits as such move reduces construction spending and the overall cost, and is pivotal for facilitating the restructuring and upgrading of related industries. Taking high-tensile steel bar as an example, such twisted steel bar of a tensile strength of 400 MPa or above features desirable strength and overall performance. 12%–14% of steel consumption can be reduced by replacing steel bar of 335 MPa with that of 400 MPa; while a further 5–7% reduction will be possible if steel bar of 400 MPa is replaced by that of 500 MPa. The use of high-tensile steel bars in high-rise or wide-span buildings is even more effective as the use of steel bars can be cut down by 30%.

The export product mix shall be optimized to reduce the export of "embodied energy". China being the world's biggest trading nation has seen massive export of "embodied energy" in its huge export to the overseas market. Many researchers in China and elsewhere have analyzed and calculated the embodied energy export in China's foreign trade by the input–output method, concluding that such embodied energy export has expanded by 3.3 times from 338 million tce in 1995 to 1.438 billion tce in 2010, making up 25.8 and 44.3% of the total energy consumption of the year. With embodied energy import associated with general commodities and energy products deducted, the net export of embodied energy soared from 35.67 million tce in 1995 to 415 million tons in 2010, a 10.6-fold increase, and the share in total energy consumption of the year spiked from 2.7 to 12.8% [9]. The massive growth in net export of embodied energy arising from the rapid expansion of export constituted a key driver of the upsurge in China's energy consumption in recent years.

2.   Technical innovation and Circular Economy

Switching from traditional carbon-intensive production mode towards low-carbon or zero-carbon techniques and products while improving product quality and life span would deliver high potential for energy conservation and emission reduction. Taking

---

of 161 m^2 of urban houses were demolished in China, or 41.3% of newly completed commercial residential buildings of 390 million m^2 of the same year.

cement as an example, replacing cement ingredients or fuel with low-grade materials such as urban waste and industrial waste and by-products can dramatically cut down energy consumption and emissions. In the chemical fertilizer industry, using blue-green algae to produce nitrogen or nitrate fertilizers can effectively reduce CO_2 emissions from ammonium synthesis. In steel industry, CO_2 emissions per ton of steel using DRI is only 1/3 of that of BF-BOF steelmaking process, and the integrated electrolysis technology and renewable power process—still under development—could go one step further to enable near-zero carbon emission [10]. Meanwhile, among ten innovative energy-saving and decarbonization technologies yet to be commercialized in electrolytic aluminum industry, such as inert electrode, wet cathode, and multipole slot technology, techniques including carbothermic reduction of alumina and carbothermic reduction of kaolin are able to save 20–40% energy relative to traditional electrolysis techniques [11]. Besides, polymer (such as ethylene) production from CO_2 electrolysis process is also under development [12].

Revolutionizing traditional production and consumption patterns, actively developing circular economy, and reducing industrial output from primary resources constitute an integral part in driving down energy demand and carbon emissions of the industrial sector. According to a study by the International Resource Panel, by 2050, the development of circular economy would reduce global resources exploitation by 28%, which, coupled with climate policies and actions, can save approximately 63% of global CO_2 emissions while ensuring a 1.5% growth in global economic output [13]. Studies have shown that in energy-intensive industries such as steel, cement, electrolytic aluminum, and plastics, 3.6 billion tons of CO_2 emissions can be prevented each year by developing circular economy. Taking plastics as an example, with every million ton of plastics recycled, CO_2 emissions from 1 million cars will be offset [14]. With the aggregate stock of various products reaching a high level in China, vast potentials are promised in expanding recycling. Research findings have indicated that by 2050, scrap steel can take up 31–40% in steelmaking; alternative materials such as mixed materials can account for 35% in cement output; 40% of primary aluminum can be produced from recycled aluminum while 81% of paper and carton output can be based on recycled paper [15]. By expanding plastics recycling, demand for ethylene production is set to decline by 1/3 by 2050 [10]. Yet expansion of recycling is challenged by technological and market barriers as well as policy inadequacies, etc. Waste landfill is a case in point: landfill cost in China is only 7% of that in UK, incentivizing a preference for landfill for most plants in China over the possibilities of reduction and recycling [16].

3. **Technological Advancement, Energy Conservation, and Efficiency Boost**

The spread of advanced energy-saving technologies and equipment helps to continuously tap into the potential of energy conservation and emission reduction in the industrial sector. Despite notable improvement in overall industrial technologies in China, great potential exists in enhancing overall energy efficiency via leading and mature technologies. Meanwhile, with technical progress in medium and low-heat residual heat and pressure utilization and industrial intelligence, new possibilities for energy conservation and decarbonization keep cropping up. In the steel industry, analysis finds that from 2010 to 2020, the potential for energy conservation and decarbonization arising from technological advancement has exceeded that from restructuring [17]. Meanwhile, through broader application of 36 typical energy-saving technologies, the steel industry saw its energy consumption reduced by 7.8% and CO_2 emissions by 10% in 2012, where secondary heating, power generation with residual heat from sintering and TRT are among the most promising technologies of energy conservation and decarbonization. Meanwhile, synergistic benefits such as pollutant reduction and water conservation also bear a major impact on the technical economic viability. 10 technologies are cost effective measured by energy-saving benefit alone, while over two thirds are cost effective if co-benefits of energy conservation, emission reduction and water-saving are taken into account [18]. A "bottom-up" study on the potential of 28 energy-saving technologies in China's steel industry has demonstrated that from 2010 to 2050, the comprehensive energy consumption per ton of steel can be cut by 30–31% for blast furnace steelmaking and by 65–76% for electric furnace steelmaking [19]. Further analysis of the emission reduction potential and cost of the steel industry finds that the extension of mature energy-saving technologies promises the best potential and cost-effectiveness, which can deliver a total emission reduction of 818 million tons between 2015 and 2030 and energy savings of 216.9 billion RMB; besides, numerous negative-cost emission reduction technologies are available in the steel industry by 2030 [20]. In the cement industry, a clear prospect of emission reduction exists in phasing out obsolete production capacity and promoting existing mature and efficient technologies. The alternative fuel technology for cement kiln will demonstrate great potential between 2020 and 2030 while new alternative cement technology will be very attractive despite its immaturity at present [21].

A higher level of precise management empowers smart energy use and energy saving via big data. The growing popularity of "Internet plus" and IT application make real-time monitoring, precise analysis, and dynamic adjustment of energy consumption possible for businesses. In the future, China should place greater priority to improve the intelligence digitalization of its industrial development, with whole-process management of the entire energy consumption in production and better utilization of big data for energy conservation. For instance, the eTelligence at Cuxhaven in Germany's energy internet pilot project "E-energy" allows real-time power supply and demand to be released on the internet platform based on CCHP. To illustrate, in case of wind power surplus, power companies would send a low price alert to major industrial users, who would turn on the cooling and heating tanks to balance energy supply and demand and reduce power bill. For another example, many of the comprehensive energy service projects widely launched in China enable online monitoring of energy consumption by installing monitors and sensors on key equipment in a bid to materialize on-line monitoring of energy use, compile and analyze data, identify abnormalities and reduce waste. In a word, digitalization and intelligence application in the energy sector make for systematic, precise and digital energy conservation and stand to be the "amplifier" of energy efficiency in the future.

4. Electrification and Changes in Energy Mix

Promoting electrification of the industry while replacing fossil energy with low-carbon alternatives offer considerable opportunities for emissions reduction. For sectors with greater potentials such as non-ferrous metal and chemicals, the co-benefit of low-carbon electricity may bring about deeper decarbonization. But some researchers argue that as China's current power mix is predominated by coal with a high CO_2 emission intensity, improved industrial electrification in the short term makes little difference to emission reduction. In terms of thermal demand alternatives, tremendous potentials exist with low-carbon electricity-based heat pump and electric boiler technologies as alternatives for low-temperature industrial heat demand, as heating demands at lower temperature (below 100 °C) can be met by heat pump, whose energy efficiency is 2 to 3 times higher than traditional boilers; while for mid-temperature heat demand (100 °C–500°C), alternative fuels such as biomass, low-carbon electricity and hydrogen are viable options. Studies have revealed by 2040, heat pump technology would represent a cost-effective solution to cater 6% of industrial thermal demand worldwide [4]. Besides, EAF, ultrared heating, and induction heating also provide opportunities, albeit at a limited scale. But for the industry's demand for higher temperature, technological revolution is essential if low-carbon electricity is to replace traditional thermal boilers.

Hydrogen and biomass are pushing their borders in petrochemical, chemical, and steel industries and are expected to underpin deep decarbonization across the industrial sectors. Expanding the use of industrial hydrogen, especially in combination with hydrogen generation from renewable electricity, could enable substantial reduction in carbon emissions from industrial production processes and energy consumption. In the steel industry, DRI technology allows pellets to be directly reduced under lower temperature with green hydrogen as the reducing agent to produce sponge iron (directly reduced iron), with vapor and extra hydrogen released from the top of boiler, where the vapor can be reused after condensation and washing, thus achieving deep decarbonization. According to the statistics obtained by SSAB, CO_2 emissions per ton of steel produced with long process technique is 1600 kg (the average level ranges between 2000 and 2100 kg in other European countries) and electricity consumption is 5385 kWh; while CO_2 emissions per ton of steel produced with HYBRIT technology is only 25 kg with 4051 kWh of electricity consumed. In petrochemical and chemical sectors, combining technologies such as renewable electricity and hydrogen generation from water electrolysis, or developing cutting-edge technologies such as hydrogen generation from methane decomposition can reduce carbon emitted from the traditional SMR process. For biomass, studies have shown that increased use of biomass can replace 15–20% of energy consumption from industrial raw materials and fuels by 2050; for example, biodiesel can be a source of bionaphtha for refineries while ethylene can be produced from biopetroleum through dehydration, the latter already industrialized in Brazil [5]. Yet as things stand, cost remains a main barrier that impedes wider application of alternative energy in the industry, which might require synergistic transformation of technological process and infrastructure system.

5. **Emission Reduction in Industrial Process and Development of Negative Emission Technologies**

CO$_2$ emission reduction from industrial process can be achieved mostly through alternative materials or fuels and system efficiency upgrade, where limitations abound in technological options, and deep emission reduction calls for improved technical process or innovation in materials, etc. Taking cement as an example, replacing limestone with newly developed adhesive materials can reduce CO$_2$ emissions from industrial process. Some new adhesives can even react with CO$_2$ to bring down carbon intensity in cement production by 30–90%. Meanwhile, some industrial companies are currently exploring the use of high temperature solar heat to enable lower emissions from alumina calcination in the aluminum industry and limestone calcination in the cement industry by developing solar rotary kiln [22]. Hydrogen also stands to be a high achiever in reducing emissions from industrial processes. Electrolyzed water based on renewables can be used to produce hydrogen-rich chemicals such as ammonia or methanol for various industries such as precursors (as in nitrogen fertilizers), reducing agents (as in low-emission steelmaking) and fuels, while hydrogen and CO$_2$ can be used for producing olefins and aromatic hydrocarbons.

Embracing negative emission technologies such as CCS will make a significant difference to the emission reduction target. For instance, in the cement industry, up to over 90% of CO$_2$ emissions can be captured by post-combustion capture (chemical absorption or membrane separation), oxy-fuel combustion (full or partial), e, or calcium cycle, etc. In the steel industry, the ratio exceeds 85% via chemical or physical absorption, membrane separation, and oxy-fuel combustion. In coal chemical industries such as ammonia synthesis, methanol production and coal to olefins, CO$_2$ captured can be used to produce chemical products such as carbonates, borax, dicyandymide, hydroxy, etc., which is expected to prevent over 300 million tons of CO$_2$ emissions.

3.2 Building Sector

3.2.1 Status and Trend

1. **Overview of Building Energy Consumption in China**

Building sector represents a major energy consumer, consisting of residential buildings as well as public and commercial buildings. It refers to the energy consumed to provide such services as heating, air-conditioning, ventilation, lighting, cooking, domestic hot water, electric appliances and elevator/escalator, etc. Energy consumption of residential buildings mainly refers to energy consumed during the operation phase of residential buildings, including energy consumed from the use of various devices in the building to cater for daily life, study and rest, etc., mostly to meet the

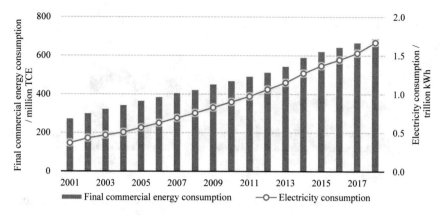

Fig. 3.13 Final commercial energy and electricity consumption of the building sector in China (2001–2018)

needs of residents and enhance services. Energy consumption of public and commercial buildings, on the other hand, refers to the energy consumed in the operation phase of public and commercial buildings to provide their designed functions. Such buildings refer to all non-residential buildings except for industrial production facilities, such as office buildings, schools, hospitals, shopping malls, hotels, stadiums and gymnasiums, theaters and transportation hubs, etc. Types of energy consumed in building operation mainly include electricity, coal, natural gas and heat from centralized heating system. Currently, building sector accounts for around one third of the world's total energy consumption [23].

China's building sector has witnessed soaring energy consumption in recent years, as illustrated in Fig. 3.13 [24]. In 2018, the final commercial energy consumption of building sector stood at 690 million tce (primary energy consumption converted as 1 billion tce),[3] making up approximately 20% of total final energy consumption across the country; commercial energy consumption and non-commercial biomass totaled 780 million tce (energy consumption of biomass at around 90 million tce). In 2018, electricity consumption of building sector was 1.7 trillion kWh, an electrification rate of 30%. The energy consumption and energy use intensity per capita has more than doubled compared to 2001.

---

[3] Building energy consumption includes direct fuel consumption, heating consumption and electricity consumption. Generally speaking, the final energy consumption is the sum of these three categories calculated with electrothermal equivalent method. In this section, considering the scale of centralized heating in North China where the heating source upgrading has long been the main task of energy conservation efforts in China, which is not discussed in the chapter of energy supply. In this chapter, direct heat consumption is converted into primary energy consumption (i.e. coal, natural gas, electricity etc. needed to generate heat) to sum up, i.e. building energy consumption in this chapter refers to the sum of direct energy consumption, electricity consumption and heat consumption converted into primary energy consumption, calculated with electrothermal equivalent method. In carbon emission calculation, direct carbon emission only includes the part attributable to direct fuel consumption in buildings; while carbon emissions from indirect heating and electricity are correlated with heating and electricity consumption respectively.

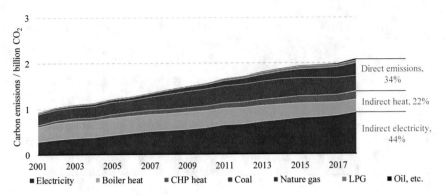

Fig. 3.14 Carbon emissions associated with fossil energy in the building sector in China (2001–2018)

In 2018, carbon emissions associated with fossil fuel consumption in building sector in China totaled 2.09 billion tCO_2[4] [24] as illustrated in Fig. 3.14, of which indirect carbon emissions from electricity consumption was 920 million tCO_2, or 44% of carbon emissions associated with the building sector; carbon emissions attributable to heat consumption in centralized heating in North China comprised 22%. From the countrywide perspective, carbon emissions per capita from building sector was 1.5 tCO_2/cap.

Continuous adjustment and growth of building scale is one of the key drivers for the spike in building energy consumption in China: on one hand, the ever increasing building area produces massive future demand for energy consumption in building operations, as more energy is needed to accommodate more buildings to function and provide services; on the other hand, a tremendous amount of building materials are used for massive construction works, and large quantities of energy consumption and carbon emissions result from the production of building materials.

China has seen a boom in new building construction since 2001 with over 1.5 billion m^2 of completed floor area annually and a whopping 2.89 billion m^2 in 2014 [25] (Fig. 3.15). However, with changes in the macroeconomic landscape, the annually completed floor space has diminished since 2015. Yet the massive newly completed floor area has resulted in continued growth in the building stock in China. In 2018, total building area in China was approximately 60.1 billion m^2, including 24.4 billion m^2 of urban residential buildings, 23 billion m^2 of rural residential buildings and 12.8 billion m^2 of commercial and public buildings (C&P buildings) [24].

Despite the surge in energy consumption in the building sector of China, its intensity is still rather low compared to developed countries, as illustrated in Fig. 3.16.

From a historical point of view, energy consumption per capita in most of the developing world has experienced a rapid climb before leveling off or a steady growth, but the eventual figure of the plateau varies greatly among countries. Currently the

---

[4] Carbon emission coefficient of electricity is defined as 553 gCO_2/kWh.

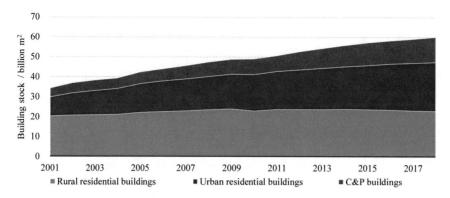

Fig. 3.15 Building area in China (2001–2018)

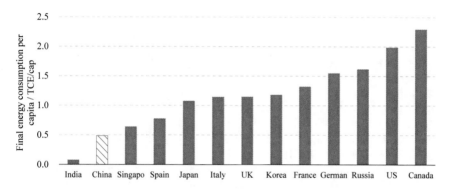

Fig. 3.16 Comparison of energy use in buildings per capita between China and other countries (2017) [26]

energy consumption per capita in China is roughly 0.5 tce/cap, close to that of South Korea in the 1980s and of Japan or Italy in the 1960s. As China is still undergoing relatively fast economic growth, there is possibility for another spike in building energy consumption in the future. In other words, China is now at a crucial juncture of deciding the pathway toward an energy-efficient building sector, and the choice would mean a great deal to the future trend, which directly bears on the trajectory of the total energy consumption and carbon emissions of China.

2. **Building Energy Consumption by Sub-sectors**

Given the differences in winter heating pattern, building types in urban and rural areas, lifestyle as well as people's activities in residential and public buildings and energy-powered equipment between southern and northern regions in China, building energy consumption can be grouped into four sub-sectors, namely north urban heating (NUH), urban residential buildings energy consumption (UR buildings, excluding NUH), commercial and public building energy consumption (C&P

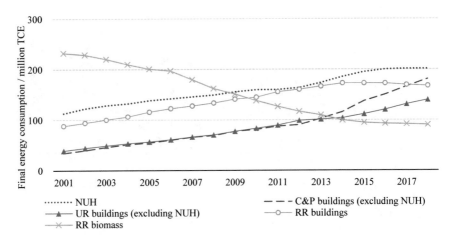

Fig. 3.17 Year-on-year changes in total energy consumption by category (2001–2018)

buildings, excluding NUH), and rural residential buildings (RR buildings) energy consumption [24].

Each accounting for around one fourth of total building energy consumption in China, these four categories each features its own characteristics. A look at their changes between 2001 and 2018 shows a substantial growth in total energy consumption of all categories except for biomass in rural regions that saw continuous decline, as illustrated in Fig. 3.17 [24].

Changes in total energy consumption and intensity between 2001 and 2018 by category and the main causes are provided below:

Energy consumption for NUH is rising but with a continuous drop in consumption intensity, which signals the progress made in building energy conservation and clean heating in recent years as the average energy consumption per unit of floor area has fallen from 23 kgce/m^2 in 2001 to 14 kgce/m^2 in 2018, to which major contributors include better building insulation, higher proportion of efficient heat sources and improvement in heating system efficiency.

Total energy consumption of C&P buildings (excluding NUH) registers a four-fold increase, accounting for the biggest share among the four categories with the consumption intensity per unit area on constant rise from 8 kgce/m^2 to 14 kgce/m^2 and electricity use per square meter climbing from 35 kWh$_e$/m^2 to 63 kWh$_e$/m^2. The growth in total area of public buildings and proportion of large-scale public buildings and the expanding demand for energy are the main contributors.

Total energy consumption of UR buildings (excluding NUH) has more than doubled, with average energy consumption per household rising from 253 kgce/hh to 469 kgce/hh and electricity consumption per household from 794 kWh$_e$/hh to 1809 kWh$_e$/hh. The increase is mainly attributable to the growing demand driven by domestic hot water, air-conditioning and home appliances, etc.; with energy-saving lamps widely used, no significant increase in lighting energy consumption is observed in households while the energy consumption intensity of cooking remains largely unchanged. Yet the growing energy consumption of winter heating and home air-conditioning in regions with hot summer and cold winter has sparked extensive debates on the path of energy conservation for heating and air-conditioning.

Commodity energy consumption has doubled in RR buildings, accompanied by a continuous and rapid drop in the use of biomass. Greater access to electricity, better income and increased number and use of home appliances in rural households are the main drivers for the surging electricity consumption per rural households. Meanwhile, a growing portion of biomass is replaced by other commercial energy, resulting in the sharp drop in the share of biomass in domestic energy consumption in rural regions. An analysis on the development of energy consumption in rural residential households indicates no major change in total consumption per household but a steep reduction in the share of biomass; the commercial energy consumption and electricity consumption per household rose from 459 kgce/hh to 1131 kgce/hh and from 435 kWh$_e$/hh to 1770 kWh$_e$/hh respectively while biomass consumption per household nosedived from 1212 kgce/hh to 612 kgce/hh.

3. Forecast in Related Studies

Much has been studied in the energy consumption of the building sector in China, of which some results are illustrated in Fig. 3.18.

As is illustrated above, under reference scenario, it is generally believed that building energy consumption in China will experience sustained growth by over 60%; under the scenario where ambitious energy-saving and emission-cutting measures are taken, energy consumption of the sector is generally believed to reach a plateau or start to fall around 2030 until it reaches 600–800 million tce by 2050, close to the current energy use.

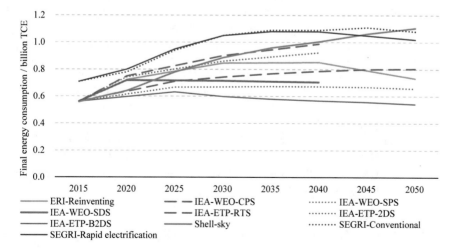

Fig. 3.18 Forecast of energy consumption of China's building sector by some organizations [22, 27–30]

3.2.2   Scenario Analysis

1.   **Scenario Design**

The following four scenarios are set for this research:

a.   Policy scenario
The policy scenario is based on the assumptions that the occupant behavior modes in China will move towards an intermediate level and that the building stock and energy intensity will stay on the current track of growth, the energy consumption per capita by 2050 would draw close to energy-efficient countries in the developed world, with energy-saving policies in the sector introduced and implemented on schedule. However, stricter policies arising from the needs to tackle the climate change, revolutionize the energy system and curb environmental pollution are excluded, and no guidance is provided for people to change their lifestyle.

b.   Reinforced policy scenario
The reinforced policy scenario assumes that the building stock and energy intensity would grow to some extent yet under further control, with energy consumption per capita still lower than the overwhelming majority of developed countries by 2050, that energy efficiency policies for buildings would be stricter than the current ones under China's climate commitment and some form of encouragement are provided to change people's lifestyle.

c.   2 °C scenario
The 2 °C scenario is based on the assumptions that existing constraints in carbon emissions, resources and environment in China are fully considered for energy consumption in the building sector and that frugality is preserved to the maximum provided that people's fundamental demand for a satisfactory livelihood is met, the policies will be reinforced across the board with policy measures proposed to cap energy consumption and carbon emissions and their intensity; energy-saving technologies will sustain steady development with significantly enhanced efficiency and slight increase of total energy consumption. The growth of building stock and energy intensity will gradually decline and total carbon emissions from the building sector will peak before 2030 for China to meet the 2 °C temperature target.

d.   1.5 °C scenario
Based on the 2 °C scenario, 1.5 °C represents a more ambitious target to be achieved through a more proactive electrification journey with less use of fossil fuels and continual improvement in energy efficiency, thus further cutting direct emissions by 2050 and enabling China to meet the 1.5 °C target.

For clarification, variation in key parameters in different scenario settings is shown in Table 3.6.

Table 3.6 Parameters set for different scenarios

	Policy scenario	Reinforced policy scenario	2 °C scenario	1.5 °C scenario
Policy development of building energy conservation	Maintain the building energy conservation policies during the 13th Five-Year plan period	Strengthen development and implementation of policies for total building energy consumption control on top of those during the 13th Five-Year plan period	Take ambitious actions in implementing energy conservation policies to make energy-saving buildings a reality	Go further with certain measures than 2 °C scenario
Building energy use and carbon emissions planning	Energy consumption per capita drawing close to more frugal developed countries	Further control on energy consumption growth and energy consumption per capita lower than most developed countries; Total carbon emission peaks before 2030	No significant growth in energy consumption per capita, provided that quality of life and service are not compromised; Total carbon emissions peak as early as possible before 2030	No significant growth in energy consumption per capita, provided that quality of life and service are not compromised; Carbon emissions continue to decline from the level of 2 °C scenario
Building stock control	No control of building stock	Conduct feasibility study of the scale of new buildings, with total built area kept under 80 billion m^2 by 2050	Stricter review and feasibility study of the scale of new buildings, with total built area kept under 74 billion m^2 by 2050	Stronger control of building scale on top of 2 °C scenario, with total building stock kept under 72 billion m^2 by 2050

(continued)

2. **Results of Scenario Analysis**

A scenario analysis of China's building stock, energy consumption and carbon emissions by 2050 is conducted based on the above assumptions. 2050 Energy consumption by category is shown in Tables 3.7, 3.8 and Fig. 3.19.

Under the policy scenario, building energy consumption would continue its rise until reaching roughly 990 million tce in 2050, with electrification rate accounting for 50% of energy consumption and 10% coming from coal consumption; under the reinforced policy scenario, energy consumption of the sector reaches a plateau by 2030 and starts to fall after 2045, reaching approximately 850 million tce in 2050, with electrification making up 56% of energy consumption and coal less than 5%;

Table 3.6 (continued)

	Policy scenario	Reinforced policy scenario	2 °C scenario	1.5 °C scenario
Heating source planning for North China	Gradually phase out coal-fired boilers as heating source and replace it with CHP as dominating heating source, while keeping a number of gas boilers	Decline in the proportion of gas-fired boilers while prioritizing industrial residual heat and various types of CHP	Further decline in the proportion of gas-fired boilers while fully utilizing industrial residual heat and various types of CHP; for buildings lacking in excess heat resources, high-efficient heat pumps should be prioritized	Phase out gas-fired boilers of various types on top of 2 °C scenario
Guidance in lifestyle changes	Discourage luxurious lifestyle	Encourage frugal and green lifestyle and behaviors; discourage forms of buildings and equipment that fail to fit into a frugal lifestyle	Vigorously provide guidance in frugal and green lifestyle and behaviors; plans that fit into green lifestyle should be prioritized in building design and equipment selection	Same as under 2 °C scenario
Electrification	No encouragement of electrification	Encourage electrification	Vigorously promote electrification for it to exceed 60% in 2050	Further promote electrification for it to exceed 75% in 2050
Biomass utilization in rural residences	No encouragement of efficient biomass utilization in rural regions	Encourage using efficient biomass technologies for cooking and heating in rural regions	Press ahead with efficient utilization of biomass in rural regions, prioritizing the use of efficient biomass technology for cooking and heating in rural residences	Same as under 2 °C scenario

Table 3.7 Energy consumption by type under reinforced policy scenario in selected years

Unit	Fuel type	2015	2020	2025	2030	2035	2040	2045	2050
Million tce	Coal	291	223	171	129	101	77	56	37
Million tce	Oil	1	0.3	0.1	0	0	0	0	0
Million tce	Natural gas	163	255	310	336	339	330	312	285
Million tce	Biomass	95	67	50	40	40	43	47	54
Trillion kWh	Electricity	1.34	1.87	2.25	2.56	2.9	3.28	3.58	3.87
Million tce	End-use energy consumption	714	775	807	820	837	853	855	851

Table 3.8 Energy consumption by category under 2°C scenario in selected years

Unit	Fuel type	2015	2020	2025	2030	2035	2040	2045	2050
Million tce	Coal	291	223	141	79	51	33	19	6
Million tce	Oil	1	0.3	0	0	0	0	0	0
Million tce	Natural gas	163	255	277	271	256	235	207	178
Million tce	Biomass	95	67	53	58	65	68	71	77
Trillion kWh	Electricity	1.34	1.87	2.23	2.51	2.8	3.08	3.39	3.68
Million tce	End-use energy consumption	714	775	745	716	716	715	714	713

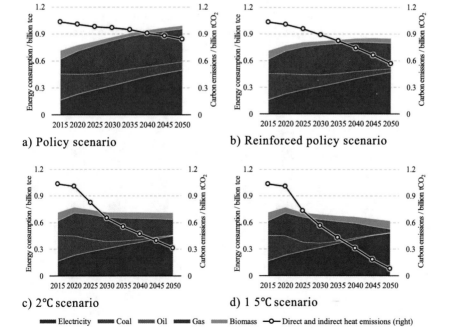

Fig. 3.19 Energy use (by fuel type) and carbon emissions under different scenarios

Table 3.9 Energy consumption and carbon emissions in selected years

	Energy consumption (million tce)			Direct + Indirect Thermal Emissions (million tCO$_2$)		
	2020	2030	2050	2020	2030	2050
Policy scenario	775	860	995	1000	969	834
Reinforced policy scenario	775	820	851	1000	888	562
2 °C scenario	775	716	713	1000	650	306
1.5 °C scenario	775	692	621	1000	565	81

under 2 °C scenario, energy consumption of the sector begins its downward trajectory from 2020 to around 710 million tce in 2050, with electrification comprising 63%, the use of coal almost phased out and biomass increasing to 77 million tce thanks to the expanded penetration of efficient use of new biomass resources; under the 1.5 °C scenario, energy consumption of the sector drops to 620 million tce, with 78% of electrification and biomass energy of around 90 million tce.

A drop is detected in the aggregate of direct carbon emissions and indirect emissions from heating under all scenarios (Table 3.9). By 2050, direct carbon emissions stand at roughly 831 million tCO$_2$ under the policy scenario, 562 million tCO$_2$ under the reinforced policy scenario, 306 million tCO$_2$ under the 2 °C scenario and 81 million tCO$_2$ under the 1.5 °C scenario.

Energy consumption by 2030 and 2050 under different scenarios are illustrated in Fig. 3.20. As is shown in the Figure, building energy consumption is set to continue its growth for a period of time compared to 2015. With the implementation of relevant

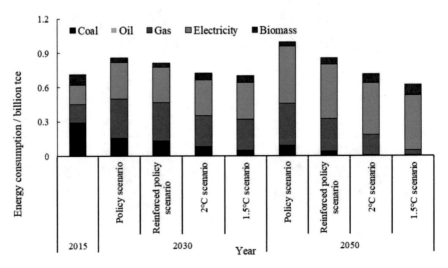

Fig. 3.20 Comparison of energy consumption by 2030 and 2050 under different scenarios

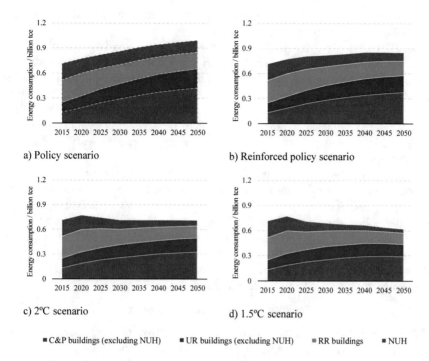

a) Policy scenario

b) Reinforced policy scenario

c) 2°C scenario

d) 1.5°C scenario

■ C&P buildings (excluding NUH)   ■ UR buildings (excluding NUH)   ■ RR buildings   ■ NUH

Fig. 3.21 Energy consumption under different scenarios (by category)

policies, total energy consumption of buildings would go down, with less use of coal and natural gas and bigger share of electricity and increased biomass consumption.

3. Development Trend of Energy Consumption by Category

The energy consumption by category under different scenarios is illustrated in Fig. 3.21.

With continuous improvement in building performance and optimization of heating source mix in northern China, energy consumption of NUH is expected to fall continually, with a drop of approximately 30% by 2050 under the policy scenario and 70% under the 2 °C scenario. In the future, it is advised to prioritize the use of excess heat from industrial production and various forms of CHP for heating in northern urban China, and heat pump is recommended for regions where such excess heat is out of reach. Meanwhile, further progress should be seen in overheating due to excessive or unevenly distribution of heating.

As the demand for improvement living standards and services still prevails in China, energy consumption of both UR and C&P buildings (excluding NUH) continues to rise. Under the policy scenario, energy consumption of urban residential and public buildings experiences a two-fold and one-fold increase by 2050 respectively, compared to 1.5 times and 50% under the 2 °C scenario. For these two types of buildings, proper guidance of service demand and behavioral patterns could obviously reduce energy consumption.

For C&P buildings, new building construction in the next stage mainly caters to the demand for better public services and the development of tertiary industry. Proper control on the number of large-scale and energy-intensive buildings and active adoption of energy consumption quota are expected to curb the growth of energy usage. For residential buildings, energy consumption for cooking and lighting has already entered a plateau and the construction of new buildings going forward are mostly energy consumed by air-conditioning, heating in regions with hot summer and cold winter and home appliances.

Growing urbanization is accompanied by a decline in rural population. Despite the constant rise in energy consumption per household, total energy consumption of rural residences is on the decrease, dropping by 25% and 45% by 2050 respectively under the policy scenario and the 2 °C scenario. Energy-efficiency retrofitting of rural houses will go a long way toward reduced energy consumption, especially in the northern region. Another key contributor is biomass. Under the 2 °C scenario, efficient biomass recycled from the fields is preferred by rural residents to meet their heating and cooking needs in an efficient manner.

4. Carbon Emission Peaking and Near-Zero Emission

The foregoing analysis shows that direct emissions from buildings have plateaued in recent years and kept falling after 2020. With indirect emissions taken into account, however, carbon emissions from buildings are on the rise in recent years. In the context of changes in emissions from electricity under different scenarios, total emissions from the building sector are illustrated in Fig. 3.22.

The figure shows that under the policy scenario, total carbon emissions from the building sector peaks around 2030 at close to 2.3 billion tCO₂ before starting its decline in 2035 to approximately 1.56 billion tCO₂ in 2050; under the reinforced policy scenario, total carbon emissions hit the peak around 2025 at approximately

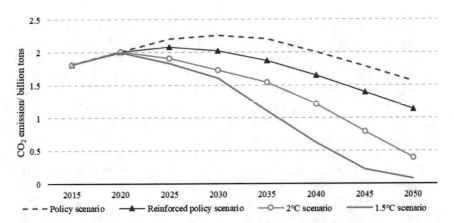

Fig. 3.22 Direct and indirect total carbon emissions of the building sector under different scenarios

2.1 billion tCO_2 and fall back to around 1.13 billion tCO_2 in 2050; under both the 2 °C and the 1.5 °C scenarios, the peak occurs around 2020 at respectively 2.01 billion and 2 billion tCO_2, followed by steady drop in carbon emissions from the building sector under the 2 °C, gaining speed from 2035 until coming close to 400 million tCO_2 in 2050; under the 1.5 °C, carbon emissions begin a steep fall from 2030 and slow down after 2045 until reaching roughly 80 billion tCO_2 in 2050.

To make the 1.5 °C scenario a reality and bring about near zero emission, the following efforts are needed: ambitious move towards higher energy efficiency and changes in energy consumption mix for buildings; encouragement of green lifestyle and employment of efficient and green technologies to minimize energy demand; a much less use of fossil fuels in heating, cooking and domestic hot water; higher penetration of electrification and rational and efficient use of biomass to reduce carbon intensity of energy consumption. At present, compared to other scenarios, no apparent technological barriers exist for the building sector to achieve near-zero emission, the challenge are mostly associated with the proliferation of technologies, like pushing for changes and upgrades of heating sources in the north, changes in residents' cooking habits and utilization of biomass, etc. Meanwhile, under this scenario, the proportion of centralized heating in northern urban regions will see a significant decline because of shrinking coal-fired power and full usage of excess heat. Higher level of electrification means that natural gas pipelines are no longer necessary for buildings, and these changes will trigger a shift in the demand for related infrastructure.

3.2.3 Evaluation of Emission Reduction Technologies and Pathways

This section will provide an introduction of key emission reduction measures and their potentials in each of the four categories of energy consumption.

1. **NUH (Northern Urban Heating)**

The main solution lies in reducing the actual heat needed for buildings, and upgrading the heat source mix and system efficiency [31].

The former includes higher standards of building envelops to minimize the heat demand for newly constructed buildings and retrofitting the existing building envelops. Notable progress has been made through these two steps during the past three decades, yet there is still room for improvement. Meanwhile, with the urbanization drive in China, the existing buildings are gradually overtaking new construction, which means a growing importance of retrofitting existing buildings for enhanced energy efficiency.

To improve the heat source mix, it is advised to prioritize low-grade excess heat from CHP and industrial production. And for buildings where low-grade excess heat is unavailable, heat pump is recommended. In recent years, with the steady expansion of clean heating in northern China, CHP is rapidly gaining prominence, so are gas heating, heat pumps and industrial excess heat, but such excess heat is yet to be fully harnessed. Meanwhile, with maturing technology of long-distance heat transfer, location is less of a constraint for excess heat utilization, which also makes much more economic sense. Persistent energy-saving renovation and clean heating penetration, technology development, comprehensive enhancement in system and equipment efficiency and remediation for overheating heating will also contribute to energy conservation and emission reduction on this front.

Energy-saving potentials of key measures under different scenarios are illustrated in Fig. 3.23.

2. Public and commercial Buildings (Excluding NUH)

Energy consumption in this category is set to rise further. But the growth can be slowed down through enforcing energy saving and emission reduction measures. In light of the overall principle of capping energy consumption and intensity, the whole-process building management with curbing actual energy consumption as the main target will be the central and most effective step to curtail energy consumption in this category and achieve energy saving and emission reduction [32].

Energy-saving feasibility study should be conducted on building design and renovation plan, where the demand for indoor environment, building stock control, etc. should be discussed, and passive technologies and renewables such as solar PV should be fully utilized to optimize the design, save energy and reduce emissions.

For the building equipment and systems, there are already cases of high energy consumption accompanied by high energy efficiency as a result of the lopsided pursuit of higher equipment efficiency, i.e. behavioral changes triggered by equipment replacement, eventually leading to much more energy consumption. Therefore, the development of equipment system should address both efficiency and its compatibility with use cases. Meanwhile, to boost electrification of public and commercial buildings, domestic hot water, disinfection and vapor generation equipment should gradually switch to electrothermal pump and direct electrothermal appliances, etc.

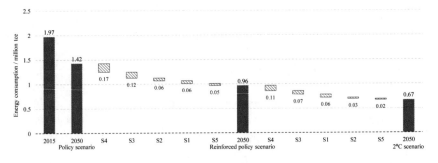

S1: control building stock; S2: higher standards for new buildings; S3: retrofitting of existing buildings; S4: improvement of heat source efficiency; S5: higher pipe network performance.

Fig. 3.23 Energy-saving potentials of various measures in urban heating in north china (2015–2050)

Examples have proven that for healthcare facilities with special need for vapor, production of hot water or vapor with distributed electrothermal approach should be encouraged as it consumes less energy than centralized gas-fired boilers due to less heat loss during transmission.

In addition, optimizing operation management represents another major energy-saving pathway in this category. This includes encouraging users and operators to adopt more energy-efficient ways of use, and promoting EPC, etc.

Energy-saving potentials of each measure are illustrated in Fig. 3.24.

3. Urban Residential Buildings (Excluding NUH)

Like public buildings (excluding NUH), growing energy consumption is also expected of this sector, and interventions are needed to mitigate the surge [33].

To upgrade the design of residential buildings, possible measures include promoting passive energy-saving technologies (such as the use of natural ventilation and lighting, improving building envelope and providing shade on the external façade of the building, etc.), improving residential community planning and using more renewable energy such as solar PV.

Studies have shown that lifestyle mostly explains why energy consumption in China is significantly lower than the developed world. To enable energy-saving and low-carbon development, its essential to sustain the current green lifestyle. Meanwhile, technologies adaptable to green living shall be developed to avoid luxurious behavioral changes due to technological option. For example, the mode of using heating and cooling only when needed should be encouraged, and the use of energy-intensive home appliances shall be discouraged.

Currently, residential buildings are still heavily reliant on gas for cooking and domestic hot water, prompting greater efforts to increase electrification in the energy mix. Various types of electric cooking utensils and electric heating for domestic hot water are mature enough to meet residential needs, and guidance in lifestyle,

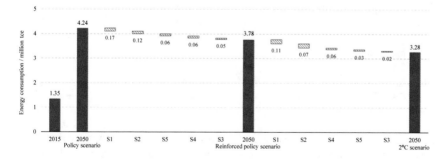

S1: control building stock; S2: optimize operation management; S3: energy-saving design of new buildings; S4: retrofitting of existing buildings; S5: optimize and upgrade equipment and system

Fig. 3.24 Energy-saving potentials of various measures for public buildings (Excluding NUH) (2015–2050)

especially in the transformation of traditional ways of cooking, is vital to promote these technologies.

Potential of each measure is illustrated in Fig. 3.25.

4. **Rural Residential Buildings**

In view of the type of buildings, people's lifestyle and resources endowment in rural regions, the notion of "coal-free villages" might be an option in the north and "eco-villages" in the south to enable energy-efficient and low-carbon development in rural regions [24].

Compared to urban buildings, rural residences have much room for improvement in terms of building performance, giving rise to the need of green retrofitting to enhance the energy efficiency. In northern regions, better building envelop is needed to provide more effective insulation, and passive solar power technology can be deployed to cut the heating demand in winter; while in southern regions, more effective insulation is also desired for building performance through passive cooling technologies.

Now, many rural households are still using traditional biomass or bulk coal for heating and cooking, which is inefficient and causes indoor air pollution. It is recommended to switch to more efficient appliances for heating and cooking, such as efficient and clean biomass combustion.

Moreover, optimized energy mix and better use of renewables are also important steps. Northern regions are advised to reshape the system of rural energy supply with full use of biomass and solar power, among other renewables to phase out coal in domestic use. Southern regions, on the other hand, are encouraged to be coal-free and maximize the use of all renewables.

Potentials of each measure are illustrated in Fig. 3.26.

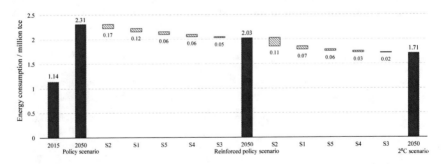

S1: control building stock; S2: promote green lifestyle; S3: assess energy-saving design of new buildings; S4: retrofitting of existing buildings; S5: optimize and upgrade equipment and system.

Fig. 3.25 Energy-saving potentials of various measures for urban residential buildings (excluding NUH) (2015–2050)

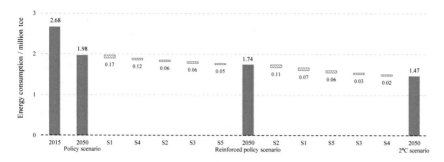

S1: control building stock; S2: promote green lifestyle; S3: upgrade building performance; S4: optimize and upgrade equipment and system performance; S5: improvement of energy system and fully use of renewable energy.

Fig. 3.26 Energy-saving potential of various measures in rural residential building (2015–2050)

3.2.4 Implementation Pathway & Policy Recommendations

Joint efforts are indispensable for multiple stakeholders to adopt policy measures from different angles to secure green and low-carbon development of the building sector. Given a sound planning of the building size, a robust development roadmap for low-carbon buildings should be devised with energy demand reduction and energy structure improvement as the shared targets and the development of energy technologies and guidance for lifestyle as major approaches.

1. **Setting a Cap on Total Building Stock**

Total building stock is a significant factor for energy consumption and building emissions, and appropriate planning to cap total building stock in the future are necessary for low carbon development of the building sector [34].

One the one hand, the overall building stock must be checked and the total floor area of civil buildings should be kept under 74 billion m². Currently the building area per capita in China is close to and even exceeded the level of some developed countries in Europe and Asia. Even with the continuous growth of urbanization in the future, the demand for new buildings is limited based on the current building floor area per capita. So new construction projects and new buildings must be put under strict control.

On the other, massive demolition and construction projects must be curtailed with advancement in maintenance technologies to improve the ratio of maintenance and upgrading of buildings. During the last two decades, an enormous number of houses and infrastructure projects have been built in China, and going forward, the focus should shift from massive construction to massive maintenance, renovation and function upgrades, i.e. from demolition and construction to life cycle extension and quality upgrade.

2. **Building an Energy Conservation System with "Dual Controls"**

A "dual control" system—policies for capping both total amount and intensity for building energy conservation—should be created with actual energy consumption as the core criteria for assessment. It is advised to encourage the introduction of policies and regulations concerning building energy conservation at local levels on a case-by-case basis, thus putting into practice the energy consumption and intensity control measures on the ground. Meanwhile, to facilitate the execution of control measures, *Standard for Building Energy Consumption* should be revised and updated and used as the master standard for creating a supporting standard system [35].

Clear and consistent building energy consumption data is the basis of data-driven building energy saving efforts. It is recommended to implement *Classification and Presentation of Building Energy Use Data* in a timely manner and improve the calculation of various metrics. At the same time, enhancements on the accuracy, availability and transparency of the rules governing building energy use data would provide better support for the energy-efficient and low-carbon development of buildings [36].

3.   Realizing Low Carbon Structural Transformation

It is crucial to move toward a new pattern of building energy use and form of system, and incorporate restructuring efforts in the building sector for low-carbon transformation in the building development goals [37].

Major steps should be taken to promote all-round electrification, PV buildings and absorption of renewable power to accommodate the uniqueness of future energy supply. Mature technologies are available to meet end-use building demand through electricity, but guidance needs to be provided for people's lifestyle and behavior modes, especially for the change of traditional cooking method. Roofs and vertical façades of buildings exposed to sunlight should be fully harnessed to roll out PV buildings and increase the renewable power generation of the buildings themselves. It is suggested to vigorously develop flexible power consuming buildings integrating DC distribution, distributive power storage and smart charging piles in cities and towns to alleviate power shortage. In rural regions, in contrast, PV roof and wind and solar power penetration in DC microgrid should be developed in full scale with changes in power consumption modes.

Large quantities of fossil fuels are currently used for NUH. The heating potential of low-grade residual heat from CHP and industrial production should be fully tapped as the primary heat source for NUH. Clean heating renovation based on natural gas boilers should not be furtherly encouraged. Instead of determining power generation by heating supply in thermal power stations, large-capacity power storage installations and electrothermal conversion equipment such as heat pump should be built in heat source facilities to materialize the coordination of heat and power.

Ambitious actions should be encouraged in rural regions to develop and efficiently use biomass energy and put in place a system of biomass acquisition, processing and sales. Through compression molding of solid fuels and biogas scale-up, local biomass resources can be fully tapped to meet local demand for cooking, domestic hot water and part of heating. This could curb GHG emissions from compost and straw returning to field while creating opportunities for products and industries in biomass collection, processing and application.

4. **Providing Guidance for Green Lifestyle**

Variance in usage patterns is the key reason for a much-lower building energy intensity in China than in the developed countries. Apart from encouraging green lifestyle, it's also a priority for building energy-saving efforts to design and construct buildings and systems that are compatible with the traditional way of energy-saving for residents [38].

The fact that building energy intensity is lower in China than in the developed world is best explained by the green behavior pattern that should be encouraged instead of jumping on the bandwagon of a luxurious lifestyle. In line with *Master Plan of Building a Green Life Campaign*, guidance for green lifestyle should be reinforced during campaigns to build green communities as well as green buildings, and building energy-saving awareness should be raised in other green initiatives in collaboration with other ministries and commissions.

The need for energy-saving technologies stems from behavioral pattern, which is in turn subject to the guidance of technologies. While arousing people's awareness for a green lifestyle, buildings and systems compatible with user habits should be designed and constructed. Building designers should prefer nature based concepts and advocate the preference for distributive AC systems. Technology assessment should prioritize behavioral pattern consistent with a green lifestyle.

5. **Developing Feasible Low-Carbon Technologies**

Suitable energy-saving and low-carbon technologies hold the key to energy efficient development of the building sector. Scenario and pathway analyses point to the need for further breakthroughs in key energy-saving technologies, including clean heating for northern China, DC building technology enabling flexible power use in buildings, building and equipment system design and operation methods compatible with behavioral pattern and renewable energy utilization (such as efficient biomass utilization, integrated PV building technology), etc.

As evidenced by many cases, the one-sided pursuit for new technologies and high efficiency may result in high energy efficiency and high energy consumption, eventually causing buildings to consume more energy [38]. Therefore, technological decisions should be made with outdoor weather conditions and corresponding behavioral pattern taken into account and actual energy consumption as a key metric for evaluation.

The following roadmap for future low-carbon building development is shaped based on the above analysis, as illustrated in Fig. 3.27.

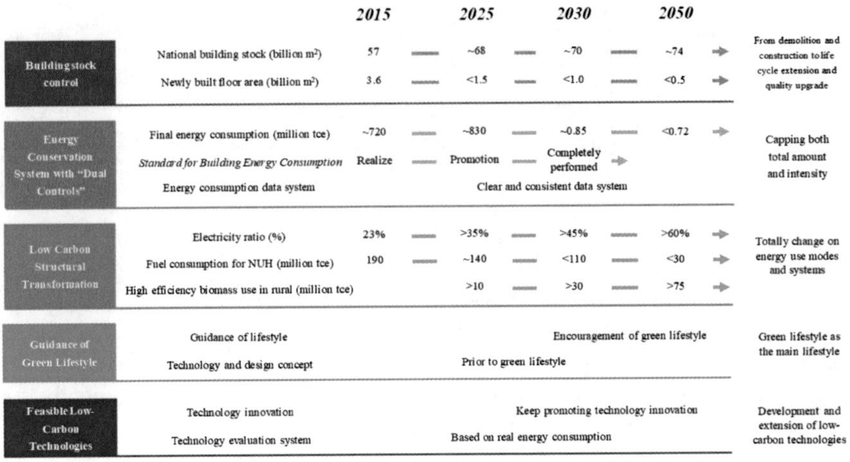

		2015		2025		2030		2050		
Building stock control	National building stock (billion m²)	57	—	~68	—	~70	—	~74	⇒	From demolition and construction to life cycle extension and quality upgrade
	Newly built floor area (billion m²)	3.6	—	<1.5	—	<1.0	—	<0.5	⇒	
Energy Conservation System with "Dual Controls"	Final energy consumption (million tce)	~720	—	~830	—	~0.85		<0.72	⇒	Capping both total amount and intensity
	Standard for Building Energy Consumption	Realize	—	Promotion	—	Completely performed	⇒			
	Energy consumption data system				Clear and consistent data system					
Low Carbon Structural Transformation	Electricity ratio (%)	23%	—	>35%	—	>45%	—	>60%	⇒	Totally change on energy use modes and systems
	Fuel consumption for NUH (million tce)	190	—	~140	—	<110	—	<30	⇒	
	High efficiency biomass use in rural (million tce)			>10	—	>30	—	>75	⇒	
Guidance of Green Lifestyle	Guidance of lifestyle					Encouragement of green lifestyle				Green lifestyle as the main lifestyle
	Technology and design concept			Prior to green lifestyle						
Feasible Low-Carbon Technologies	Technology innovation				Keep promoting technology innovation					Development and extension of low-carbon technologies
	Technology evaluation system			Based on real energy consumption						

Fig. 3.27 Development roadmap of buildings' low-carbon transition

3.3 Transportation Sector

3.3.1 Development & Trend of the Transportation Sector

3.3.1.1 Traffic & Carbon Emissions

China has scored enormous achievements in its transportation sector as a leading country in the scale and capacity of transportation infrastructure. As illustrated in Figs. 3.28 and 3.29, a sharp rise is seen in cargo and passenger turnover with the rapid economic growth and social development. Despite an array of proactive measures in low-carbon development with the adoption of more efficient means of transportation including vehicles, trains, vessels and airplanes, the sector still see a surge in the quantity of energy consumption and total CO$_2$ emissions and an increasing proportion in total energy use.

Cargo turnover in China has undergone a steep increase in China, reaching 19.4 trillion tons-km in 2019, a 198% increase compared to 2005.

Passenger turnover and urban passenger traffic are also on the fast track of increase. In 2019, intercity passenger turnover registered at 3.54 trillion people-km, 2.61 times of that in 2005. Meanwhile, urban passenger transport is on the rise—126.6 billion trips were made in 2005, which grew to 231 billion in 2019, up 82%.

Total energy consumption of transportation rose from 231 million tce[5] in 2005 to 505 million in 2019, an increase of 119% and an annual growth of 5.8%, with its weight in China's total energy consumption growing from 9.1% in 2005 to 10.7% in 2018.

---

[5] Electricity consumption in the transportation sector is converted into coal equivalent with electrothermal method: 1 kWh = 122.9 gce.

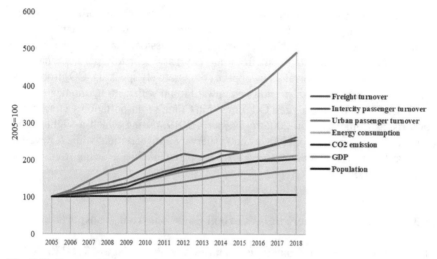

Fig. 3.28 Exponential increase in activities, energy consumption and CO_2 emissions in the transportation sector in China (2005 = 100)

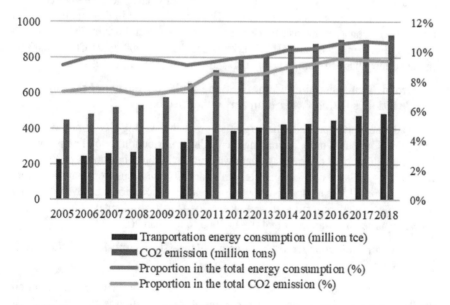

Fig. 3.29 Energy consumption and CO_2 emissions of transportation in China and their proportion

CO_2 emissions from the sector soared from 457 million tons in 2005 to 954 million tons in 2019, up 108% with an annual growth of 5.3%. The proportion of direct CO_2 emissions from the transportation sector in total CO_2 emissions in China climbed from 7.3% in 2005 to 9.5% in 2018.

Since 2005, CO_2 emissions from passenger and cargo transportation has been rapidly expanding, of which cargo transportation—whose total CO_2 emissions are still on the rise—is responsible for more carbon emissions than any other category of transportation. CO_2 emissions from intercity passenger transportation has been growing since 2005, albeit at a slower pace comparatively speaking. CO_2 from urban passenger transportation also shows an upward trend with a strong momentum.

Road transport is the biggest CO_2 emitter in China's transportation sector, proportion rising from 72.0% in 2005 to 84.7% in 2019, during which a slight drop is observed in emissions from railway and waterway transportation; while for aviation transportation, the increase in CO_2 emissions is remarkable, growing from 4.0% in 2005 to 6.9% in 2019.

3.3.1.2   Main Characteristics of Carbon Emissions from Transportation

1.   **Carbon Emissions from Transportation as a Percentage of Total Emissions Are Low, but are Rapidly Rising**

Soaring energy consumption and carbon emissions have increasingly become a headache amid a boom in transportation. The strong growth in demand has pushed up its energy consumption as a percentage of total energy end-use in China from 5% in 1980 to 9.1% in 2005 and 10.7% in 2019. Its share in total direct CO_2 emissions in the country has also jumped from 6.1% in 2005 to 9.3% in 2019.

2.   **Road Transportation Comprises the Bulk of Carbon Emissions, but Aviation is Rising Faster than Any Other Sector**

Road transport represents the largest share in the surging CO_2 emissions from transportation, but aviation is making increasing contribution to total emissions. In 2019, the share of road, railway, waterway and aviation transportation accounted for 84.7, 1.4, 6.9 and 6.9% respectively in total CO_2 emissions from transportation. During the past 15 years, the contribution of road transportation has hovered around 85% while that of railway and waterway experienced a minor decline. A strong growth is noted in aviation from 4.0% in 2005 to 6.9% in 2019.

3.   **Carbon Emissions from Gasoline and Diesel Consumption are the Mainstream**

In 2019, the transportation sector took up 10.7% of total energy consumption in China, and 88.3% of energy consumption of transportation is attributable to oil products. CO_2 emissions from gasoline, diesel and kerosene accounted for 35.18, 52.53 and 6.88% respectively of energy-based CO_2 emissions from the transportation sector in 2019, of which gasoline and diesel added up to 87.72%, an overwhelming majority of the total emissions.

4. **Indirect Carbon Emissions from Electricity Consumption Can't Be Ignored**

Despite a modest share of transportation in electricity consumption in the current stage, a steep rise is expected with the future development of railway transport, urban rail transit, tram, trolleybus, EVs and electric vessels. A full life cycle analysis indicates that CO_2 emissions from power generation that corresponds with electricity consumption should not be overlooked.

3.3.1.3 Future Trend of Carbon Emissions of the Transportation Sector

As is illustrated in Fig. 3.30, a range of studies have demonstrated that CO_2 emissions from China's transportation sector will witness rapid growth in the short- and mid-term and gradually slow down in longer term. By 2050, annual CO_2 emissions under the baseline scenario will surge to 1.8 to 2.4 billion tons, a strong growth momentum projected by the majority of forecasts and scenario studies, which signify that without proactive and continuous mitigation policies in place, CO_2 emissions are likely to double or triple from the current level [39–43].

These studies show that with the completion of industrialization and upgrading of industrial structure, the mix of transportation will be continually optimized with increased proportion of low-carbon transportation, innovation and wider application of low-carbon transportation equipment and technologies, which stand to dramatically reduce GHG emissions from the transportation sector.

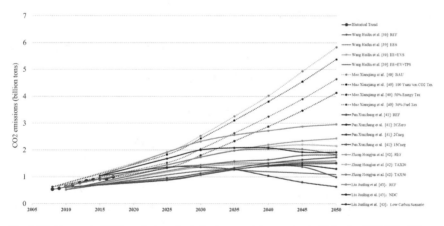

Fig. 3.30 Forecast of CO_2 emissions from the transportation sector in China by 2050 [39–43]

3.3.2   Future Development and Scenario Setting

3.3.2.1   Transportation Demand, Structural Changes and Technological Development in the Future

1.   **Learnings from International Experience**

Currently, passenger and cargo transportation and their associated energy consumption and carbon emissions in China are still on the increase, standing in stark contrast to the decline after a phase of growth in many developed countries, some of whom have put in place a full-fledged transportation infrastructure system when China is still in the phase of rapid development. In parts of the developed world, urbanization and motorization have been brought to fruition with minor growth in private car ownership. By comparison, urbanization and motorization are pressing ahead in full speed in China with a rapidly increasing number of private cars on the road and a peak is not yet in sight. Despite the differences in the stage and characteristics, the transportation sector in China does share a similar path of development with the developed nations, and their course of development as well as the experiences and lessons for low-carbon transportation are valuable references for China [44].

International experience suggests that the decline in carbon emissions from passenger transportation faces tough challenges and proper guidance should be offered to encourage rational consumption. The relative trend of changes in carbon emission intensity from passenger transportation worldwide demonstrates the unit energy consumption may grow with rising demand for comfortable travels, resulting in the continuous increase in carbon emission intensity [45]. Between 2007 and 2018, the private car ownership in China skyrocketed from 57 million in 2007 to 204 million in 2019, with a average growth rate of 8.8% year-on-year. The exponential growth has led to a sharp rise in carbon emissions. In 2013, private cars overtook taxis as the biggest contributor to carbon emissions in urban passenger transport. With relatively less room and more challenges of reducing the carbon emission intensity of passenger transportation in China, it's essential to come up with a holistic plan to strike a perfect balance between energy conservation, emission reduction and service improvement. To this end, public awareness for a sensible consumption pattern holds the key, which should neither compromise convenience, safety and comfort of travels for the sake of emission reduction nor pursue maximum comfort at the cost of excessive carbon emissions.

Vast potentials, however, are observed in cargo transportation where efficiency improvement is paramount. Throughout the past decades, China's economy has flourished with a remarkable rise in demand for bulk goods such as steel and cement, among other materials for infrastructure, which are primarily transported by road. For example, cargo turnover of road transportation experienced an annual growth of 8.6% from 2005 and 2019, and 90% of road transportation depends on fossil fuels for energy, prompting continual increase in carbon emissions. On the other hand, demand for urban home delivery has apparently picked up with the advent of e-commerce, from which carbon emissions as a percentage of freight transport has nearly doubled from 2010 to 2019. In the future, modern logistics should be given much greater priority with enhanced organized transportation and efficiency to lower the energy intensity and carbon emissions from freight transport.

Optimized transportation structure, intermodal transportation, adoption of advanced means of transport and low-carbon energy technologies and smart transport solutions are the major avenues for low-carbon development of the transportation sector. By employing a mixture of measures such as planning, pricing, investment, publicity, education and organization, passenger and freight transport should be encouraged to switch from carbon-intensive options such as cars and trucks to railway, waterway and public transportation with less carbon emissions. High-speed rail and intercity rail should be built to link cities and light rail, maglev and suburban railway are recommended in urban environment for more intensive city or city cluster to form with the aid of energy-efficient, large-capacity and high-speed transportation modes, thereby shaping an intensive and low-carbon transportation supply system. The development and application of new transportation technologies, alternative fuels, IT and smart transport should be valued, especially when it comes to alternative fuels and new transportation technologies, together with a clear roadmap for technology innovation.

2. **Analysis of Transportation Demand and Structure**

In freight transport, demand will change gears from high speed to a slower but steady growth, of which the transport of bulk cargo will peak prior to 2030 while the demand for high-value, decentralized, small-sized and time-sensitive cargo deliveries will pick up rapidly; low-carbon policies will also steer towards an enhanced freight transport structure.

Chinese economy has shifted from a high-speed growth to high-quality development with more progress in supply-side structural reform. The growth in demand for cargo transport will slow down, but as industrialization and urbanization forge ahead, such demand will maintain a medium growth before 2030. China will embark on a new journey of new industrialization underpinned by technological innovation from 2030 to 2040, during which the growth in industrial output and cargo transport will slacken off. Cargo turnover will move toward the peak after 2040 and slowly decline afterwards.

In 2018, China took the world's top spot in the output and transport of bulk goods. With greater strides in terms of supply-side structural reform, shift in development mode, economic structure and transition of growth drivers, the heavy chemical industry is set to report a slower growth, cutting the demand for bulk goods such as coal, iron ore and steel. It is expected that the demand for transporting bulk commodities will reach a high plateau from 2020 to 2030 before a decline after 2030 as urbanization and massive infrastructure works come to fruition. Given the sheer size of China's territory and population and uneven resources endowment, no drastic reduction is expected even after the peaking of bulk commodity transport, which would be kept relatively stable in the short term.

Logistics is becoming more "personalized" thanks to internet-based customization and the shift from centralized to decentralized reinforced control of industrial production, among other new trends, which has rapidly driven up the demand for decentralized and small-sized cargo transportation. Meanwhile, the demand for small-sized, multiple-lot and high-value freight transport is rising as a result of improved living and consumption standards, which also requires faster, more convenient and more timely delivery and spurs the growth of cargo transportation by air and road. A remarkable increase is noted in the share of high added-value and lightweight products, whose value per unit of transport far exceeds bulk commodities. These changes in the category of goods will drive sustained growth of containers and express delivery transportation.

With continuously upgraded road network and the advantage of "door-to-door" delivery, road will remain the primary means of cargo transportation. The weight of railway and waterway transportation is subject to policy measures: forceful low-carbon transportation policies will shift part of the cargo transport from road to railway and waterway, resulting in an increased share of railway and waterway turnover; incentives for railway-waterway intermodal transportation will also foster new opportunities in railway and waterway transportation.

With economic restructuring taking place in China, bulk goods transportation by road such as coal, grain and mineral ores will gradually move towards railway and waterway, while road will undertake more small-quantity, multiple-lot, lightweight, high added-value and time-sensitive cargo transport. On the other hand, with the increase in the length of highway and improvement in logistic delivery system, the average distance for road cargo transport will be longer. Passenger transportation will see the total demand rapidly pick up, but the growth would be dragged down as urbanization slows down. The booming city clusters will spur the growth in inter-city passenger transport while urban passenger transportation is set to grow further, driving the development of private cars and public transport.

More often than not, rapid urbanization is accompanied by high total demand and growth of passenger transport, while in later period of urbanization, the demand is featured by a high volume but minor growth. Currently in the medium phase of urbanization, China is expected to enter a transition period by 2030 when industrialization and urbanization move from medium to medium-to-late phase with a mid- to high-speed growth in passenger transport. Between 2030 and 2050, China's urbanization will gradually mature, which means a slower growth in passenger transport as urbanization loses its momentum.

An important embodiment of the urbanization drive in China, city clusters will see strengthened economic and social ties, and the share of intercity passenger travels between major city clusters grows in total passenger transport across the country.

Demand for urban travels will continue to rise with ongoing urbanization and economic growth. The emergence of new technologies and new models such as autonomous driving and carpooling will encourage more people to travel by car for speed and comfort; when better infrastructure becomes available and awareness of low-carbon and green mobility gets stronger, public transportation might also be a preferred option.

3. **Development Trend of Transportation of Typical Freight Categories in China**

The composition of freight categories for railway, inland waterway, coastal water-borne and expressway artery transportation is shown in Table 3.10. The transport of different categories of cargo varies in proportion among these means of transportation, with bulk goods accounting for over 90% of railway and 52.6% of expressway [46].

Table 3.10 Artery transport cargo mix by category and proportion of bulk freight in China

No.	Category	Proportion of Bulk Freight (%)		
		Railway	Inland river and coastal waterborne transport	Expressway
1	Coal and coal products	61.39	25.38	8.19
2	Oil, natural gas and their products	3.5	5.4	4.8
3	Metal ores	10.79	8.83	0.96
4	Steel and non-ferrous metals	5.66	5.23	12.73
5	Mineral building materials	3.58	21.83	13.76
6	Cement	0.92	3.72	3.94
7	Non-metal ores	2.26	2.34	3.47
8	Wood	0.63	0.18	2.07
9	Grains	2.77	1.97	2.68
10	Chemical fertilizers and pesticides	2.2	0.24	0.94
11	Salt	0.39	0.11	0.16
12	Machinery equipment and electric appliances	0.17	1.05	7.09
13	Chemical raw materials and products	1.31	1.74	6.08
14	Light industry and pharmaceutical products	0.6	0.87	17.39
13	Agricultural, forestry, livestock and fishery products	0.1	0.28	8.52
16	Others	3.63	20.83	7.22
17	Total	100	100	100
Sub-total of Categories 1–9		**91.5**	**74.88**	**52.6**

Data source: data collected by the research team

The following forecast is conducted on the trends of transportation of coal, steel and smelting materials, building materials, grains, containers and express deliveries (Table 3.11):

1. Coal transportation will remain stable in short term and report substantial decline in mid and long term.
2. Transportation of smelting materials, especially steel, will experience a steady fall [47].
3. Transportation of building materials such as cement will show a downward trend, but a slight increase will occur with wood.
4. Total grain transportation sees no major change with continuously improving transportation mode.
5. A sustained growth is expected for container transportation.
6. Transportation of express deliveries will sustain rapid increase.

4. **Development in Transportation Technologies**

Future innovations in transportation technologies will speed up green and low-carbon development of the transportation sector. With higher penetration of electrification in railway, lightweight vehicles and fuel and power saving technologies, wider acceptance of autonomous driving, eco-driving and fleet, upgrading of engine and vehicle manufacturing technologies in road transportation, application of precise flight management technology and biofuels in aviation, remarkable progress will be made in transportation technologies, which will in turn contribute to higher energy efficiency and emission reduction. Faster market penetration of clean energy and new energy vehicles and vessels and accelerated use of aviation biofuels will catalyze cleaner and lower-carbon energy supply in transportation.

Progress in railway transportation technology is mainly manifested in the level of electrification and lightweight vehicles. In light of the progress in railway electrification rate, lightweight trains, power and fuel efficiency and smart management, etc., energy consumption from railway is on track to drop by 15–30% in 2030 and 35–50% in 2050.

Advancement in road transportation technologies is mainly demonstrated by better fuel economy. For over a decade, fuel consumption of passenger vehicles in China has continued to fall. According to the statistics from MIIT, corporate average fuel consumption (CAFC) was 7.02 L/100 km in 2015, followed by an average growth of 1.7% annually after the implementation of Phase I Standard of Fuel Economy in 2006. *Made in China 2025* and *Energy-Saving and New Energy Vehicle Technology Roadmap* have specified targets for reduction in fuel consumption for varied passenger car technologies, which have been supported by multiple measures adopted by the national government. Phase IV Standard of Passenger Vehicles has entered into effect in 2016 with stricter fuel consumption standards in the future. Given the technological upgrades in engine and vehicle manufacturing, eco-driving, truck fleeting and road network optimization, energy consumption by road transportation is set to fall by 35–50% in 2030 and 55–70% in 2050.

Table 3.11 Forecast of typical cargo categories

Category		Current production (million tons)	Production forecast (million tons)			Railway transport (million tons)			Share of railway transport as major means (%)		Main origin of production	Main source of demand in future	Travel distance (km)
		2018	2030	2050		2015	2050		Now	2050	2050	2050	2050
Coal and coal products		3,650	3,000–3,500	1,000–1,500		2,176	300–500		70	85	Shanxi, Inner Mongolia, etc	Central/West China	500 +
Smelting materials-Steel		1,100	650–750	320–450		3,500	1,500–2,000		5	50	North China/East China/Northeast China	Countrywide	500 +
Building materials	Cement	2,200	1,800	800							Countrywide	Countrywide	312
	Glass	860 million weight cases	700 million weight cases	600 million weight cases							Countrywide	Countrywide	
	Wood	90 million m²	100 million m²	120 million m²		24	50			10	Import/Northeast China/Southwest China	Countrywide	500 +
Grains		657	650	700		100	150		70	90	Countrywide	Countrywide	200 +
Container transport		2,700	3,900	5,400		213	2,160		15	40	Countrywide	Countrywide	300 +
Express delivery		50 billion pcs	80 billion pcs	130 billion pcs					2	40	Countrywide	Countrywide	500 +

Data source: analysis and forecast by research team

Technological progress in waterway transportation is mainly reflected in vessel building standards. With vessel scale-up technology and vessel type standardization, its energy efficiency will increase by 20–50%.

Such progress in civil aviation, on the other hand, is most remarkable in terms of flight management technology and biofuels. With the development of delicacy flight management technology, biofuel application and development and use of new engines/aircrafts, energy efficiency of aviation is projected to improve by 20–70%.

China has seen an upsurge in new energy vehicles (NEV) in terms of production and sales, which boast tremendous potential for future growth. From 2011 to 2018, annual sales of NEV shot up from less than 10,000 to 1.256 million units, making China the world's biggest NEV market. The production and sales of fuel cell vehicles (FCV) hit 1527 units. The industry has grown to over a hundred billion yuan from a few billion at the outset. The industrial chain, consisting of upstream mineral resources and key raw materials, mid-stream core components such as battery, motor and controller and downstream vehicle and charging piles, has evolved and matured, with the emergence of many world-leading businesses. As the market takes off, the end sales and penetration will continue to climb. From the second half of 2017, many countries have announced plans and timetables to ban the sale of oil-fueled vehicles. Studies have shown that in light of the development stage of Chinese carmakers, industrial structure and transportation conditions, China is expected to impose the ban (except for trucks for special purposes and heavy-duty trucks) around 2030 to pave the way for NEV.

NEV will be quickly universalized in the future and become the overwhelming market mainstream by 2050. *Energy-Saving and New Energy Vehicle Technology Roadmap* issued by the national government in 2016 stated that NEV, including EV, will partially replace vehicles with internal combustion engines by 2025, and its ownership will account for 2–3% of the total, and such replacement will be scaled up by 2030 to 5–10%. According to CATARC, the manufacturing cost of BEV will be on the par with the average HEV by 2025 and further reduced to the level of traditional vehicles by 2033, after which BEV will enjoy the best cost-competitiveness among its peers, and the market will be the lever for wider acceptance of NEV. Based on consumer choice, it is expected that annual sales of BEV and NEV will respectively make up around 60 and 70% by 2050, when NEV will be the overwhelming market mainstream, of which FCV will comprise a bigger share in the category of coaches and mid- to heavy-duty trucks.

Biomass energy represents the future of clean energy for aircrafts. Studies have indicated that compared to conventional fuels, the use of biofuels during cruising will cut CO_2 emissions by 60–98%, with which biofuels can be directly mixed for use. IATA has unveiled a forecast report and introduced plans and roadmap in this connection. Given that biofuels are fully adaptable to the existing aviation system and will not cause erosion of rubber sealing elements of fuel pipes, no major changes will be needed in aircrafts or ground systems. It is expected that 30% of aviation fuels around the world will be biofuels by 2030, and 50% by 2040.

3.3.2.2 Introduction of Scenario Setting

Building on the above analysis, one should see that with new climate targets such as the "carbon peak" and the strategic move to build a strong nation through transportation development, transport is moving ever closer to the goal of low-carbon growth.

Meanwhile, with the application of mobile internet, IoT, cloud computing, big data and other new technologies, disruptive innovations have taken place in NEV, energy storage and autonomous driving, and the notion of "internet + " has permeated into all areas of transportation, which has given birth to new business models and revolutionized the transportation landscape.

CO_2 emissions from transportation will be subject to multiple factors in the future, such as the model and pattern of transportation, energy efficiency of conveyance, relations between public transportation and private cars, tendency in transportation demand, etc.

Four scenarios, namely the policy scenario, reinforced policy scenario, 2 °C scenario and 1.5 °C scenario, are set in this study. Energy consumption and carbon emissions of the transportation sector are calculated with the metrics in the four scenarios for comparison and analysis. Parameters used by the research include emergence of new business models such as shared mobility, change in consumer mindset, penetration of new energy means of transportation, application of autonomous driving, optimization of transportation structure, technological advancement in transportation and spatial layout of infrastructure, etc. (see Table 3.12).

3.3.3 Analysis of Different Scenarios

3.3.3.1 Energy Consumption and Carbon Emissions

As shown in Fig. 3.31 and Tables 3.13 and 3.14, given the current development, total energy consumption of the transportation sector undergoes rapid growth until around 2040, and peaking around 2035 will not happen unless strong policy interventions are conducted.

As is shown in Fig. 3.32, Tables 3.15 and 3.16, given the current development, total carbon emissions of the transportation sector grow rapidly until around 2040, and peaking around 2030 will not happen unless strong policy interventions are conducted.

Under the policy scenario, total carbon emissions are on continuous rise. Despite the decline in carbon emissions per unit of turnover/passenger traffic, total CO_2 emissions grow continuously as the growth in passenger and cargo turnover and urban passenger traffic outpaces the decrease in unit energy consumption until it peaks at 1.16 billion tons in 2040.

Table 3.12 Main parameters and characteristics of scenario setting

	Policy scenario	Reinforced policy scenario	2 °C scenario	1.5 °C scenario
Transportation technology	Energy efficiency of transportation increases by 30% in 2050 from 2015 levels	Energy efficiency of transportation increases by 40% in 2050 from 2015 levels	Energy efficiency of transportation increases by 50% in 2050 from 2015 levels	Energy efficiency of transportation increases by 65% in 2050 from 2015 levels
New energy vehicle	Share of NEV ownership reaches 1.2, 10 and 40% in 2020, 2030 and 2050 respectively, and NEV takes up respectively 0.2, 3 and 10% of cargo transportation vehicles	Share of NEV ownership reaches 1.3, 15 and 50% in 2020, 2030 and 2050 respectively, and NEV takes up respectively 0.2, 4 and 15% of cargo transportation vehicles	Share of NEV ownership reaches 1.5, 20 and 65% in 2020, 2030 and 2050 respectively, and NEV takes up respectively 0.2, 5 and 20% of cargo transportation vehicles	Share of NEV ownership reaches 1.5, 20 and 85% in 2020, 2030 and 2050 respectively, and NEV takes up respectively 0.2, 5 and 60% of cargo transportation vehicles
Green urban mobility	In 2050, public transport ratio in all travel modes reaches 45%, the mileage of rail transport is 50,000 km, private car ownership registers at 340/1,000 people with a carpooling ratio of 10%	In 2050, public transport ratio in all travel modes reaches 50%, the mileage of rail transport is 60,000 km, private car ownership registers at 320/1,000 people with a carpooling ratio of 15%	In 2050, public transport ratio in all travel modes reaches 55%, the mileage of rail transport is 65,000 km, private car ownership registers at 280/1,000 people with a carpooling ratio of 20%	In 2050, public transport ratio in all travel modes reaches 60%, the mileage of rail transport is 70,000 km, private car ownership registers at 270/1,000 people with a carpooling ratio of 25%

(continued)

Table 3.12 (continued)

	Policy scenario	Reinforced policy scenario	2 °C scenario	1.5 °C scenario
Optimized transportation structure	After restructuring of the cargo transportation mix, in 2050, share of cargo transportation by railway, road, waterway and pipeline registers at 16%:65%:16%:1%:2% respectively. After restructuring of the passenger transportation mix, in 2050, share of passenger transportation by railway, road, aviation and waterway is 38%:34%:27%:1% respectively	After restructuring of the cargo transportation mix, in 2050, share of cargo transportation by railway, road, waterway and pipeline registers at 19%:59%:18%:1%:3% respectively. After restructuring of the passenger transportation mix, in 2050, share of passenger transportation by railway, road, aviation and waterway is 40%:31%:28%:1% respectively	After restructuring of the cargo transportation mix, in 2050, share of cargo transportation by railway, road, waterway and pipeline registers at 23%:53%:21%:1%:2% respectively. After restructuring of the passenger transportation mix, in 2050, share of passenger transportation by railway, road, aviation and waterway is 42%:27%:30%:1% respectively	After restructuring of the cargo transportation mix, in 2050, share of cargo transportation by railway, road, waterway and pipeline registers at 24%:51%:22%:1%:2% respectively. After restructuring of the passenger transportation mix, in 2050, share of passenger transportation by railway, road, aviation and waterway is 43.5%:26.5%29%:1% respectively

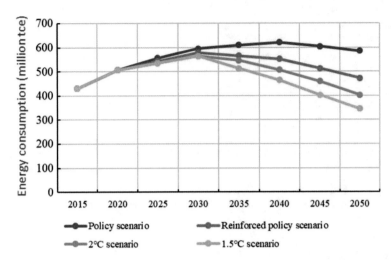

Fig. 3.31 Energy consumption of transportation sector by scenario in China between 2015 and 2050 (million tce)

Table 3.13 Energy consumption of transportation sector by scenario in China between 2015 and 2050 (million tce)

Total	Policy scenario	Reinforced policy scenario	2 °C scenario	1.5°C scenario
2015	430	430	430	430
2020	514	514	514	514
2030	599	590	583	583
2050	585	471	402	346

Under the reinforced policy scenario, with the optimization of transportation vehicle structure, technological innovations and proper allocation of resources, total CO$_2$ emissions from the transportation sector continue to grow and peak around 2035 before declining.

Under the 2 °C scenario near zero emission, regional and urban–rural transportation can be better coordinated with higher efficiency through strengthened connection of comprehensive transportation hubs; growing application of "internet + " in transportation enhances the level of intelligence in transportation; an optimized transport structure rationalizes passenger and cargo transportation of railway, road, waterway and civil aviation; more modern transportation equipment and better fuel mix are made possible through technological progress; more ambitious penetration of NEV, including EV, among other measures, enables a peak of total CO$_2$ emissions from the sector around 2030 at 1.075 billion tons.

Under the 1.5 °C scenario net zero emission, through further development of intermodal transportation, transportation improvement, accelerated use of electrification and hydrogen and application of low-carbon technologies in aviation and

Table 3.14 Energy consumption of transportation sector by category in China between 2015 and 2050 (million tce)

	Year	2015	2020	2030	2050
Policy scenario	Total	430	514	599	585
	Gasoline	132	158	179	193
	Diesel	242	277	306	267
	LPG	0	1	2	6
	Natural gas	15	17	18	14
	Electricity	16	23	33	40
	Hydrogen	0	0	0	1
	Aviation fuel	25	38	58	58
	Biofuel	0	0	2	7
Reinforced policy scenario	Total	430	514	590	472
	Gasoline	132	158	173	111
	Diesel	242	277	299	210
	LPG	0	1	2	8
	Natural gas	15	17	19	11
	Electricity	16	23	37	68
	Hydrogen	0	0	0	1
	Aviation fuel	25	38	57	52
	Biofuel	0	0	3	11
2 °C scenario	Total	430	514	583	402
	Gasoline	132	158	165	69
	Diesel	242	277	295	160
	LPG	0	1	2	10
	Natural gas	15	17	20	8
	Electricity	16	23	41	97
	Hydrogen	0	0	0	2
	Aviation fuel	25	38	57	21
	Biofuel	0	0	3	35
1.5 °C scenario	Total	430	514	583	340
	Gasoline	132	158	165	21
	Diesel	242	277	295	47
	LPG	0	1	2	10
	Natural gas	15	17	20	4
	Electricity	16	23	41	191
	Hydrogen	0	0	0	18
	Aviation fuel	25	38	58	4
	Biofuel	0	0	3	46

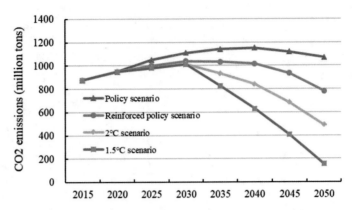

Fig. 3.32 CO$_2$ Emissions from the transportation sector in China from 2015 to 2050 (*Note* international aviation is not included in aviation, and ocean shipping is not included in water transport)

Table 3.15 Total carbon emissions from the transportation sector in China from 2015 to 2050 (million tons)

Year	2015	2020	2030	2050
Policy scenario	877	991	1155	1109
Reinforced policy scenario	877	99,100	1112	804
2°C scenario	877	991	1075	550
1.5 °C scenario	877	991	1037	172

ocean navigation, total CO$_2$ emissions from transportation would peak around 2030 at 1.04 billion tons before a drastic decline to around 172 million in 2050.

A comparative analysis of the four scenarios suggests that compared to the policy scenario, energy carbon emissions from the transportation sector are down by 10.2% in 2040 under the reinforced policy scenario; the emissions are down by 7.8% in 2030 under the 2 °C scenario and by 84.7% in 2050 under the 1.5 °C scenario.

3.3.3.2   Analysis on the Trend Towards Electrification

Under the policy and reinforced policy scenarios, oil products—primarily diesel and gasoline—are the biggest energy consumers in the transportation sector through 2050. As indicated in Table 3.17, under the 2 °C scenario, and 1.5 °C scenario in particular, more ambitious measures would be adopted to expand electricity consumption in transportation as a replacement of oil products, including promoting the penetration of NEV and railway electrification. Major steps would be taken to boost EV development for road transportation to embrace electrification.

"Pure electric drive" should be deemed as the mainstream technical pathway. Owners of PHEVs are to be encouraged to use the motor-driven (CD) mode for their

Table 3.16 CO_2 Emissions from transportation in China from 2015 to 2050 (million tons)

Year		2015	2020	2030	2050
Policy scenario	Total	877	991	1155	1109
	Road	748	814	914	945
	Railway	12	17	23	16
	Aviation	48	73	107	89
	Waterway	68	86	110	58
Reinforced policy scenario	Total	877	991	1112	804
	Road	748	814	881	649
	Railway	12	17	21	17
	Aviation	48	73	109	79
	Waterway	68	86	101	59
2 °C scenario	Total	877	991	1075	550
	Road	748	814	846	426
	Railway	12	17	21	13
	Aviation	48	73	108	62
	Waterway	68	86	100	48
1.5 °C scenario	Total	877	991	1037	172
	Road	748	814	816	119
	Railway	12	17	20	2
	Aviation	48	73	104	2
	Waterway	68	86	96	48

Table 3.17 Electricity consumption of transportation by scenario (billion kWh)

	Policy scenario	Reinforced policy scenario	2 °C scenario	1.5 °C scenario
2020	220.6	220.6	220.6	220.6
2030	286.4	368.1	418.7	561.9
2050	324.3	553.1	790	1592.2

rides. Due considerations should be given to the penetration of energy-saving vehicles and hydrogen-based FCV as well as alternative fuels. Moreover, the construction of charging infrastructure shall be accelerated to prepare for the electrification of road transportation.

3.3.3.3 Key Measures for Net Zero Emission

Analysis has suggested that under the 2 °C scenario, CO_2 emissions from transportation by 2050 would stand around 550 million tons, of which aviation and water transport encounter more challenges in emission reduction because of their dependence on traditional petroleum-based fuels. A host of interventions must be done in order to achieve net zero emission.

CO$_2$ emissions from the transportation sector can be further reduced to around 100 million tons via energy efficiency enhancement and massive use of biofuels and hydrogen in aviation and water transportation. Key measures include replacing 125 million tons of refined oil with 31 million tons of biofuels and 19 million tons of hydrogen in the transportation sector to avoid approximately 370 million tons of CO$_2$ emissions.

3.3.3.4 Summary of Scenario Analysis

If the current trend—rapid economic and social development and accelerated industrialization and urbanization—is to continue, total carbon emissions from the transportation sector is bound to rise sharply until around 2040. Only with powerful policy tools can the peak occur around 2030. A swift reduction in emissions can be made possible through further development of intermodal transport, enhanced transport structure and faster penetration of electrification and hydrogen.

Under the 2 °C or 1.5 °C scenario, total carbon emissions from transportation hit the peak around 2030 at roughly 1 billion tons. Under the policy or reinforced policy scenario, refined oil represents the biggest component of energy consumption in transportation until 2050. The 2 °C or 1.5 °C scenario, however, the soaring electricity consumption in transportation, replaces a large chunk of refined oil, which would quickly decrease the total emissions onwards.

The massive use of biofuel and hydrogen in aviation and ocean freight transportation helps drive down CO$_2$ emissions from transportation to approximately 170 million tons in 2050, close to net zero emission.

3.3.4 Goals, Pathways and Policy Recommendations for Low Carbon Development

3.3.4.1 Goals and Pathways of Low Carbon Development

Table 3.18 provides the roadmap for the migration toward low-carbon development, with metrics including total carbon emissions, spatial layout, vehicles, transportation structure, transport services and low-carbon governance, etc. 2025, 2035 and 2050 are selected as the milestones.

Carbon emissions from the transportation sector is expected to peak in 2035 with a clean, low-carbon, multi-model and efficient modern comprehensive transportation system taking shape. The infrastructure capacity of various means of transportation is adequate but not excessive. The overall coverage of infrastructure network further

Table 3.18 Low carbon development targets of the transportation sector

No.	Metric category	Metrics	2025 target	2035 target	2050 target
1	Total carbon emissions	Carbon emissions from transportation sector	/	Peak	Near zero emission
2	Transportation equipment	Proportion of additional new energy and clean energy equipment	50%	70%	100%
		Electrification rate of railway	75%	85%	100%
3	Transportation mix	Share of railway freight transport	Increase	/	19%
		Share of waterway freight transport	Increase	/	18%
4	Transportation services	Share of public transport travels	45%	55%	50%
		Share of slow traffic	25%	30%	30%
5	Low-carbon governance	Low-carbon transport governance	Further improvement in low-carbon transport strategic planning and regulatory and standard system	Establishment of a market-driven approach for energy-saving and emission-reduction	Establishment of a modern, green and low-carbon governance system

expands with weakness largely disappearing and the accessibility and smoothness of domestic travel substantially improved. A modern, user-friendly and efficient comprehensive hub system are in place with efficient connection between different means of transportation, more convenient passenger transfers and much greater efficiency in cargo transshipment. Passenger and cargo transportation structure are rationalized with growing freight transport via railway and water; eco-friendliness of infrastructure, resources utilization efficiency and supporting services for clean energy transportation are substantially enhanced, and the proportion of clean and efficient transportation vehicles ranks top among global leaders.

By 2050, near zero emission should become a reality in the transportation sector, and the green and low-carbon development of the sector is aligned with the call to build a strong and modern socialist nation and a transportation power-house. The synergy of technology, management and innovation should shape low-carbon transport to make clean and low-carbon vehicles and industrial chain, and improve management efficiency across the board. A low-carbon, integrated and three-dimensional transportation network is created through the greening of the whole process and full life cycle of transportation. An all-day green development system of transport infrastructure emerges, compatible with the environmental capacity of resources and in harmony with work, life and the ecosystem. Cleaner and lower-carbon means of transport gain further prominence with new energy or clean energy applied in almost all new means of transport.

3.3.4.2 Strategic and Policy Recommendations

To build a clean, low-carbon, intermodal and efficient integrated modern transportation system, the following recommendations are made for the scientific development of low-carbon strategies and policies for the transportation sector:

1. Ensure an earlier peak of energy demand of the transportation sector. Fully harness the role of electricity, hydrogen and biofuels as alternative fuels; further improve the transportation mode and demand management, peak energy consumption of transportation somewhere between 2030 and 2035 and enable a remarkable drop in consumption in the wake of the peak.
2. Set the targets for carbon emission peak of the transportation sector. Given that road transports remains the principal source of energy consumption and direct CO_2 emissions and boasts tremendous potential in energy replacement, sometime around 2030 might be defined as the target for CO_2 emission peak for the entire transportation sector.
3. Explore the pathway for near zero carbon emission in transportation. Actively tap into the potential for technology application in subsectors facing greater challenges in carbon reduction; devise a net zero emission plan in order for aviation and water transport system to massively adopt biofuels and hydrogen around 2050.
4. Vigorously promote NEV development. Support groundbreaking EV technologies and plan ahead for the charging infrastructure; speed the penetration of EVs in the passenger car segment; facilitate a ban on gasoline passenger vehicles by 2040; speed the electrification of coaches and trucks; meanwhile, secure a spike in EV and FCV ownership.
5. Stricter fuel economy standards for cars and other means of transportation to gradually enhance transport energy efficiency. Improved fuel economy represents one of the very effective steps to curb energy consumption and GHG emissions in the near, mid and long term.
6. Optimize transportation structure and enable a shift of long-distance passenger and freight transport towards railway and waterway, which shall remain the backbone of mid- to long-distance transportation of bulk commodities; further

utilize assembly/evacuation ports and reduce the use of heavy-duty diesel trucks for long-distance transportation of bulk commodities; transport not suitable for road, such as long-haul passenger duty, should switch to aviation or railway to reduce inappropriate need for road.

7. Enforce transport demand management across the board and encourage green and low carbon mobility. Urban public transportation should be prioritized in the planning, land approval, finance and right of way; accelerate the construction of BRT, bus lanes and rail transit, and slow traffic such as bicycle and pedestrian lanes; large-capacity public transport system should be developed; exercise differentiated transport management and hold down the rise of private car ownership; enhance management and control measures through driving up the user cost of private cars and discourage the use of private cars on the road.

8. Reinforce financial incentives such as green transport finance and taxes. Employ policy packages comprising market-based policy measures, i.e. carbon tax, license tax, congestion charge and parking fee as well as mandates and control measures, i.e. biofuel standard, fuel economy/carbon standard in order to ensure the coordinated and effective management of demand, energy and CO_2 emissions of the transport sector.

References

1. National Bureau of Statistics (2019) China statistical yearbook 2019. China Statistical Press
2. National Bureau of Statistics (2018) China energy statistics yearbook 2018. China Statistics Press
3. People's Republic of China (2016) First Biennial update report on climate change
4. IEA (2018) World energy outlook 2018. Paris
5. UNIDO (2018) Industrial development report 2018. Vienna
6. IEA (2017) World energy outlook 2017. Paris
7. Energy Research Institute, National Development and Reform Commission (2017) Research on the energy conservation of the 13th five-year plan
8. Research Center of Building Energy Conservation, Tsinghua University (2016) China building energy conservation annual development research report 2016. China Building Industry Press
9. Xie Jianguo, Jiang Peishan (2014) Measurement and decomposition of embodied energy consumption in China's foreign trade. Economics (Quarterly) 7:1365–1392
10. McKinsey & Company (2018) Decarbonization of industrial sectors: the next frontier
11. Springer C, Hasanbeigi A (2016) Emerging energy efficiency and carbon dioxide emissions reduction technologies for industrial production of aluminum. LBNL-1005789
12. Donald S, Thomas R, Matthew W (2018) Carbon Monoxide gas diffusion electrolysis that produces concentrated C_2 products with high single-pass conversion. Joule
13. UNEP (2017) Resource efficiency: Potentials and economic implications
14. Material Economics (2018) The circular economy
15. Kermeli K, Graus W, Worrel E (2014) Energy efficiency improvement potentials and a low energy demand scenario for the global industrial sector. Energ Effi 7:987–1011
16. Tao Y (2017) How policies work to foster industrial symbiosis: a comparison between UK and China. University of Cambridge, Cambridge
17. Wen Z, Meng F, Chen M (2015) Estimates of the potential for energy conservation and CO_2 mitigation based on Asian-Pacific Integrated Model (AIM): the case of the iron and steel industry in China. J Clean Prod 65:120–130

18. Ma D, Chen W, Xu T (2015) Quantify the energy and environmental benefits of implementing energy-efficiency measures in China's iron and steel production. Future Cities Environ 2015(1):1–7
19. Zhang Q, Hasanbeigi A, Price L, Lu H, Arens M (2016) A bottom-up energy efficiency improvement roadmap for China's iron and steel industry up to 2050. LBNL 1006356
20. An R, Yu B, Li R, Wei Y (2018) Potential of energy savings and CO_2 emission reduction in China's iron and steel industry. Appl Energy 226:862–880
21. Wen Zongguo, Li Huifang (2018) Potential and roadmap of China's industrial energy conservation and carbon reduction. Financial Think Tank 11:93–106
22. IEA (2017) Energy technology perspective 2017. Paris
23. IEA & UNEP (2019) 2019 Global status report for buildings and construction. IEA, Paris
24. Research Center of Building Energy Conservation, Tsinghua University (2020) China building energy conservation annual development research report 2020. China Building Industry Press
25. Statistics Department of Fixed Asset Investment (2018) National Bureau of Statistics. China construction industry statistical yearbook. China Statistics Press, Beijing
26. IEA (2020) Data and statistics. Paris. https://www.iea.org/data-and-statistics/data-tables?country=WORLD
27. Energy Research Institute, National Development and Reform Commission (2016) Reshaping energy: China, Construction volume. China Science and Technology Press, Beijing
28. International Energy Agency (2019) World energy outlook 2019. International Energy Agency, Paris
29. Shell (2018) Shell scenarios sky: Meeting the goals of Paris agreement. Hague, Shell
30. State Grid Energy Research Institute Co., Ltd. (2019) China energy and electric power development outlook 2019. China Electric Power Press, Beijing
31. Research Center of Building Energy Conservation, Tsinghua University (2019) Annual development research report of China building energy conservation 2019. China Building Industry Press
32. Research Center of Building Energy Conservation, Tsinghua University. (2018) Annual development research report of China's building energy conservation 2018. China Building Industry Press
33. Research Center of Building Energy Conservation, Tsinghua University (2017) Annual development research report of China building energy conservation 2017. China Building Industry Press
34. Peng Chen, Jiang Yi, Qin Youguo, et al (2018) Low-carbon buildings and low-carbon cities. China Environmental Publishing Group, Beijing
35. Institute of Standard Quota of Ministry of Housing and Urban-Rural Development, etc. (2018) Guidelines for the implementation of national standard "civil building energy consumption standard". China Building Industry Press, Beijing
36. Research Center of Building Energy Conservation, Tsinghua University (2016) Research report on supporting technology system of building energy conservation. Beijing
37. China Society of Urban Science (2020) China Green building 2020. China Building Industry Press, Beijing
38. Research Center of Building Energy Conservation, Tsinghua University (2016) Speculative ideas of Chinese building energy conservation. China Building Industry Press
39. Wang HL, Ou XM, Zhang XL (2017) Mode, technology, energy consumption and resulting CO_2 emissions in China's transport sector up to 2050. Energy Policy 109:719–733
40. Mao XQ, Yang SQ, Liu Q et al (2012) Achieving CO_2 emission reduction and the co-benefits of local air pollution abatement in the transportation sector of China. Environ Sci Policy 21:1–13
41. Pan X, Wang H, Wang L et al (2018) Decarbonization of China's transportation sector: in light of national mitigation toward the Paris Agreement goals. Energy 2018(155):853–864
42. Zhang H, Chen W, Huang W (2016) TIMES modelling of transport sector in China and USA: Comparisons from a decarbonization perspective. Appl Energy 162:1505–1514
43. Liu Junling, Sun Yihe, Wang Ke, et al (2018) Research on medium—and long-term low-carbon development path of China's transport sector. Progress in Climate Change Research 14(5):513–521

44. China Automotive Energy Research Center, Tsinghua University (2012) China automotive energy outlook 2012. Science Press, Beijing
45. Feng Zhenhua, Wang Xuecheng, Zhang Haiying (2019) Zhang Haiying. Research in transportation 5(4):37–45
46. Chang'an University (2018) 2017 China expressway traffic statistical survey and analysis report. People's Communications Publishing House, Beijing
47. Zhang Yanfei (2014) Research on regional layout adjustment of China's steel industry. China Academy of Geological Sciences, Beijing

Chapter 4
Power Sector

The global shift towards low carbon energy faces the dual challenges of continually growing energy consumption and carbon emission reduction [1]. The power system is transforming from fossil fuels to low-carbon non-fossil fuels as the main energy source of the power system, which will be instrumental to the goal of energy transition. In terms of future demand, the proportion of electric power is expected to grow in global energy consumption, with a 90% increase by 2040 from now, and one fifth of the growth will come from China [2]. On the supply side, the trend of clean electricity will continue with renewables and nuclear energy representing the mainstream of power generation, replacing fossil energy. The future of thermal power plants will, in large part, hinge on the cost and scale of application of carbon capture and storage (CCS); while for the power grid, the large share of intermittent solar and wind power will bring unprecedented challenges in stability and flexibility for the power system. From the standpoint of end users, on the other hand, the electrification rate will continue to rise in final energy consumption, replacing fossil fuels [1][2][3].

The power sector is the key of the decarbonization of China's energy system. With the world's largest power generation sector, China produced nearly one fourth of the world's power in 2018 with its power sector consuming approximately 50% of the country's coal, and carbon emissions from power generation and heating supply accounts for around 40% of energy-related carbon emissions in China. Various studies have indicated that the power sector needs to take the lead in carbon neutrality and even negative emission by the middle of the century. The core and biggest challenge in China's power sector transition lies in its massive coal-fired power units, which made up 48% of the world's total installed coal-fired capacity in 2018. These power units in service are technically advanced with a short service life of 12 years on average, employing nearly 4 million people, including those in the upstream coal industry. Coal power and coal-related industries are major contributors to local economy. For example, coal-related industries accounted for 17% of GDP of Shanxi province in 2018. It is foreseeable that the decarbonization of China's power industry will have a far-reaching impact on the society, economy and technology and so

© China Environment Publishing Group Co., Ltd. 2022   109
Institute of Climate Change and Sustainable Development of Tsinghua University et al.,
China's Long-Term Low-Carbon Development Strategies and Pathways,
https://doi.org/10.1007/978-981-16-2524-4_4

on. Therefore, it is urgent to find the pathway that is technically feasible, safe and affordable under the temperature rise control targets of 2°C and 1.5°C set by the Paris Agreement.

4.1   Power Demand Forecasting and Research Methods

Electric power consumption in China totaled 6.85 trillion kWh in 2018. From the supply side, 69.1% of this came from fossil fuels and 30.9% from non-fossil fuels, of which 26.7% was from renewables and 4.2% from nuclear power. On the consumption side, the share of electric power in end user energy consumption is on a constant rise in recent years, reaching 25.5% in 2018 with an average annual increase of 0.65% since 2015 [4]. By end-use consumption mix, the industrial sector remains the biggest electric power consumer, which, despite the fluctuations in recent years, comprised 70.4% in the total end user electric power consumption in 2017; a growing momentum is observed in the proportion of the building sector in end user power consumption in recent years, reaching 27.8% whereas the agricultural sector took up a mere 1.8%. In future, the level of electrification of end-use sectors is expected to significantly increase in China, with electric power gradually becoming the most important component in terminal energy mix.

4.1.1   End-Use Sectors' Demand Forecast

Under China's long-term deep decarbonization strategy, CO_2 emissions will peak around 2030, and energy consumption will gradually enter and remain at the plateau from 2035 to 2050 under different scenarios. During the plateau period, total energy consumption tends to stabilize, and economic and social development is gradually decoupled from energy and resource consumption. Meanwhile, end-use sectors, such as industrial, transport, and building, need to achieve deep decarbonization. While energy conservation and energy efficiency keep intensifying, electrification also accelerates to replace coal, petroleum, and other fossil energy sources of direct combustion and utilization. Hydrogen consumption or production using renewable and nuclear power will also help to increase the share of electricity in end-use energy consumption and the share of energy used for power generation in primary energy consumption, resulting in a faster growth rate of electrification than that of energy consumption.

According to the forecast on future power demand by various research institutes, the power consumption in 2050 will range between 11.7–14.4 trillion kWh [1][2][5–12] under the 2°C scenario, and between 14.4–15.2 trillion kWh under the 1.5°C scenario. In this study, a bottom-up method is applied in the power demand forecast, i.e. future power demand is calculated mainly based on the future predictions of end-user sectors such as industrial, building, transport, etc. Under all scenarios, end users will see a remarkable improvement in electrification with a growing trend of power consumption, of which the industrial sector will be the cornerstone in driving electrification. As is shown in Table 4.1, under the policy scenario, reinforced policy scenario, 2°C scenario and 1.5°C scenario, the power demand in 2050 will respectively reach 11.384 trillion kWh, 11.91 trillion kWh, 13.13 trillion kWh and 14.27 trillion kWh, an increase of 0.66 times, 0.74 times, 0.92 times and 1.08 times from the power supply in 2018.

Under the 2°C scenario, power demand will rise notably in all sectors, where the industrial sector is to become the main driver for power demand, with an increase of 3.21 trillion kWh from 2017 to 2050, which is followed by the building sector with a growth of 1.81 trillion kWh and the transport sector where electric power will replace part of fossil fuels as a major part of the energy mix. The transport power demand is expected to climb from close to zero to 0.79 trillion kWh; the power loss in the agriculture sector would see a 0.5 times increase (see Fig. 4.1). It should be noted, however, that industrial restructuring, energy conservation and efficiency improvement, among other factors on the demand side are taken into consideration in calculating the end user power demand. A significant growth in power demand is still expected under the 2°C and the 1.5°C scenarios as more ambitious actions will be taken for end user energy consumption electrification.

Massive emission reduction and decarbonization of end users will imply tremendous changes and challenges for these sectors, for which improvement in energy efficiency and electrification will provide the most important solutions. Taking decarbonization in the industrial sector as an example, measures to achieve decarbonization in the whole production process include using zero-carbon energy to provide power and heating, altering technical processes, adopting CCS to tackle carbon emissions from fossil fuel combustion and production process, switching to alternative feedstocks, etc. The biggest obstacle in decarbonization in the building sector is heating and cooking as 100% electrification is already achieved in lighting, cooling and home appliances. For transport, it requires fully electrified road transport, but for aviation and shipping where zero carbon emission is not within reach, efforts must be done to enable electrification for short-distance travels and opt for new carbon-free materials to make long-distance travels possible [13]. This will depend on the cost reduction of existing technologies (such as CCS and hydrogen-fueled steelmaking) and the emergence of new technologies and processes (such as hydrogen power generation and vehicle to grid).

4.1.2 Scenario Setting and Research Methods

In scenario setting for the power sector, emission reduction pathways are defined based on two phases in line with the overarching guidelines of the project. The policy scenario and the reinforced policy scenario are structured for Phase I from 2020 to 2030, guided by the target of the first stage of China's modernization drive, i.e. basic realization of modernization, fundamental improvement of ecological environment

Table 4.1 Future power demand under each scenario (unit: trillion kWh)

	2020	2030				2050			
		Policy scenario	Reinforced policy scenario	2°C scenario	1.5°C scenario	Policy scenario	Reinforced policy scenario	2°C scenario	1.5°C scenario
Industrial	4.59	5.66	5.87	6.06	6.27	6.21	6.67	7.8	7.99
Building	1.87	2.6	2.56	2.51	2.59	4.06	3.87	3.68	3.92
Transport	0.22	0.29	0.37	0.42	0.56	0.32	0.55	0.79	1.59
Other sectors and losses	0.59	0.63	0.65	0.63	0.62	0.78	0.82	0.86	0.76
Total demand	7.27	9.18	9.45	9.61	10.04	11.38	11.91	13.13	14.27

Fig. 4.1 Composition of power consumption increase by end users under the 2°C scenario

and overall fulfilment of the goal of building a beautiful China. In addition, the low-carbon policies shall be strengthened, the implementation program and action plan as pledged in the nationally determined contributions shall be carried out and reinforced, following the timetable of carbon emission peaking by 2030 and meeting the target of peak emission control. For Phase II from 2030 to 2050, emission reduction scenarios consistent with global temperature goals are studied and determined, targeting global emission reduction paths for 2°C or 1.5°C as well as the goals of developing China into a great modern socialist country and building a beautiful China. The four scenarios are conceived as follows:

1. The policy scenario: extension of the current policy trend of the power sector based on China's nationally determined contributions (NDCs) submitted in 2015. By 2030, the share of non-fossil energy shall exceed 20% in total energy supply (equivalent to 42% of non-fossil energy in total power supply).
2. The reinforced policy scenario: enforced emission reduction based on China's NDCs submitted in 2015. By 2030, power produced from non-fossil energy shall surpass 50% of total power generation; by 2050, non-fossil energy as a percentage of total energy supply shall exceed 50% (equivalent to 80% of non-fossil energy in total power supply).
3. The 2°C scenario: emission scenario consistent with the global target of 2°C by 2050, followed by net zero emission between 2065 and 2070.
4. The 1.5°C scenario: emission scenario aligned with the global target of 1.5°C by 2050, when negative emission is supposed in the power sector.

In this paper, a mathematical model, named the Long-term Multi-regional Load-dispatch Grid-based (LoMLoG) model has been developed for a quantitative analysis of the power sector under various scenarios. The LoMLoG model is to determine the least-cost development pathway of power sector with the constraints of meeting future power demand and realizing low-carbon targets. The objective function of this model is minimizing the total system cost in the planning horizon (2018–2050 in this paper), which is composed of capital cost, operation and maintenance cost, fuel cost and power transmission cost.

LoMLoG model features its high spatial–temporal resolution and technical details. China is divided into 17 regions based on resource endowment and grid structure. Existing and proposed interregional power transmission lines are set as fixed parameters whilst the incremental transmission capacity in the future are set as variables to be optimized along with power generation capacity. The balance of power generation and demand is based on different regions instead of a single country. In the real power system, the fluctuation of power load requires load dispatch at every moment. Hourly power balance is reflected in the model in order to describe the seasonal and daily variability of power demand and renewable energy availability. Each year is divided into 4 seasons (spring, summer, autumn, winter) and a representative day is selected in each season. Therefore, each year has 96 time slices in total. As a result, this model is able to simulate daily and seasonal balance of power system, with considering the temporal fluctuation of renewable generations.

4.2 Low-Carbon Pathways for the Power Sector

Power transformation pathways in each country are closely linked to its resources endowment, power source mix, economic development, and other factors. Regarding technological pathways, a consensus of various studies points to the necessity to vigorously tap into non-fossil power and develop technologies to enhance grid stability and flexibility. But these studies vary in the level of development for different non-fossil power sources. Among the available studies, a maximum of 630 GW [9], 540 GW and 510 GW [7] are forecasted for hydropower, nuclear power and intermittent renewable energy (wind and solar) respectively by 2050.

4.2.1 Installed Capacity and Power Generation of Different Emission Reduction Pathways

In this study, the nuclear power is supposed to maintain a steady growth to reach approximately 327 GW by 2050. For gas-fired power, despite its lack of price competitiveness, a size of 200 GW is set for 2050 considering its role in system balance and grid stability. According to the review of China hydropower survey in 2003, the theoretical reserves of water resources of all river basins are 694 GW, of which 542 GW are technically exploitable and 402 GW are economically exploitable. Therefore, the hydropower capacity by 2050 is set at 410 GW in this study. Raw materials of biomass industry in China mainly consists of agricultural and forestry residues, organic waste and energy crop/plants, with total available resources reaching nearly 600 million tce by 2050. Biomass liquid fuel technologies will be commercialized after 2030, biomass power generation will stabilize at around 106 million tce around 2030 [14], which is the basis for power generation from biomass and BECCS in this study. The installation and power generation from other sources are selected by the LoMLoG model for minimal social cost. As the cost of intermittent renewables will continue to fall in various forecasts, renewables will be the preferred option in models aiming at minimal cost.

Power installation in the future will be dominated by non-fossil energy, the share of intermittent energy (wind and solar) will see a remarkable increase, and coal-fired CCS and BECCS will be applied under the 2°C and the 1.5°C scenarios. As is illustrated in Table 4.2, total installed capacity of the power system will hit 3619 GW, 4291 GW, 5686 GW and 6284 GW respectively under the policy scenario, the reinforced policy scenario, the 2°C scenario and the 1.5°C scenario. The weight of non-fossil energy and intermittent renewables in the installed capacity is set to steadily rise under all four scenarios, where the proportion of non-fossil energy will respectively stand at 73.1%, 81.8%, 93.1% and 93.9% and that of intermittent renewables is forecasted to reach 54.1%, 64.5%, 79.4% and 81.3% by 2050. Coal-fired power with CCS and BECCS would reach an installed capacity of 68 GW and 32 GW respectively under the 2°C scenario, and would expand to 149 GW and 48 GW under the 1.5°C scenario.

Changes in power generation mix suggest that newly emerged demand is mainly met by power from non-fossil sources, which, in the long-term, will further replace coal-fired power stock to meet low-carbon emission goals. As is illustrated in Table 4.3, with other energy sources increasingly replaced by electric power for end users, total power demand in 2050 will reach 11.38 trillion kWh, 11.91 trillion kWh, 13.1 trillion kWh and 14.27 trillion kWh respectively under the policy scenario, reinforced policy scenario, the 2°C scenario and the 1.5°C scenario, while power generated from non-fossil energy will comprise 65.0%, 74.8%, 90.5% and 91.1% respectively. Increase in the share of intermittent renewable energy will also be seen in 2050 under the policy scenario, the reinforced policy scenario, the 2°C scenario and the

Table 4.2 Generation fleet in 2050 under different scenarios (unit: GW)

	Policy scenario	Reinforced policy scenario	2°C scenario	1.5°C scenario
Coal	773	583	123	32
Coal with CCS	0	0	68	149
Gas	200	200	200	200
Nuclear	280	327	327	327
Hydro	410	412	414	416
Wind	1063	1387	2312	2740
Solar	893	1380	2205	2367
Biomass	0	2	6	5
BECCS	0	0	32	48
Total	3619	4291	5686	6284

Table 4.3 Generation by technology in 2050 under different scenarios (unit: trillion kWh)

	Policy scenario	Reinforced policy scenario	2°C scenario	1.5°C scenario
Coal	3.64	2.62	0.45	0.11
Coal with CCS	0.00	0.00	0.40	0.79
Gas	0.34	0.37	0.39	0.38
Nuclear	1.97	2.38	2.35	2.34
Hydro	1.44	1.48	1.47	1.48
Wind	2.51	3.06	4.87	5.75
Solar	1.48	1.98	2.96	3.11
Biomass	0.00	0.01	0.04	0.03
BECCS	0.00	0.00	0.19	0.29
Total	11.38	11.91	13.13	14.27

1.5°C scenario, reaching 35.1%, 42.3%, 59.6% and 62.1% of total power generation. Under the 2°C and the 1.5°C scenarios, the share of intermittent renewable energy is predicted to surge to around 60% by 2050, presenting a greater challenge for system balance and grid flexibility. Compared to the 2°C scenario, power generation from coal-fired CCS and BECCS will soar under the 1.5°C scenario.

Under the reinforced policy scenario, non-fossil fuels will take up 25% of primary energy consumption by 2030, while 50% of total power generation will come from non-fossil energy, whose proportion in primary energy consumption will hit 50% in 2050. As is shown in Tables 4.4 and 4.5, the installed capacity and power generation of coal-fired power will continue to shrink, with a drop of nearly 50% in 2050 as compared to 2020. The installed capacity of natural gas power will reach 200 GW by 2050. But its share in total power production is insignificant due to its primary role

Table 4.4 Installed capacity in selected years under the reinforced policy scenario (unit: GW)

	2020	2030	2035	2040	2050
Coal	1100	992	914	772	583
Coal with CCS	0	0	0	0	0
Gas	110	140	155	170	200
Nuclear	53	136	184	232	327
Hydro	380	412	412	412	412
Onshore wind	205	585	752	1005	1315
Offshore wind	5	5	5	5	72
Centralized PV	164	362	434	502	934
Distributed PV	51	228	362	382	446
Biomass	18	30	29	17	2
BECCS	0	0	0	0	0
Total installed capacity	2086	2890	3247	3497	4291

Table 4.5 Power generation in selected years under the reinforced policy scenario (unit: trillion kWh)

	2020	2030	2035	2040	2050
Coal	4.63	4.52	4.09	3.59	2.62
Coal with CCS	0	0	0	0	0
Gas	0.19	0.24	0.27	0.30	0.37
Nuclear	0.39	1.00	1.37	1.74	2.38
Hydro	1.31	1.49	1.52	1.53	1.48
Onshore wind	0.42	1.27	1.65	2.21	2.78
Offshore wind	0.02	0.02	0.02	0.02	0.28
Centralized PV	0.21	0.51	0.62	0.76	1.39
Distributed PV	0.05	0.32	0.51	0.55	0.59
Biomass	0.09	0.20	0.20	0.12	0.01
BECCS	0	0	0	0	0
Total power generation	7.3	9.4	10.1	10.8	11.9

in peak regulation. Speedy growth and ambitious scale-up are needed for nuclear, wind and solar power, with the three combined taking up 72.1% of total installed capacity and 62.3% of total power generation by 2050.

Under 2°C scenario where net zero emission is to be achieved around 2050 in the power sector, the emission reduction pathway before 2030 is mostly the same as that under the reinforced policy scenario, while more ambitious actions are essential after 2030. In addition, CCS and BECCS are vital to the 2°C scenario. Compared to the reinforced policy scenario, coal-fired power will see a steep reduction under 2°C scenario, with an installed capacity of merely 123 GW in 2050, apart from 68 GW of coal-fired CCS and 32 GW of coal-fired BECCS in the installed capacity (see Table 4.6). Moreover, the 2°C scenario calls for redoubled efforts in wind and solar power, whose installed capacity would overtake that of the reinforced policy scenario by 67% and 60% respectively, and the power generation would growth by 59% and 50% (see Tables 4.6 and 4.7).

4.2.2 Application of CCS and BECCS Technologies

The future capacity of coal-fired and gas-fired units to be retained in China depends, in large measure, on CCS and its combination with bio-energy, i.e. BECCS technologies. Capturing 90% of carbon emissions, coal-fired power plants with CCS will make it a relatively low-carbon power generation technology. BECCS is a negative emission technology that can offset the residual emissions from the power sector. The penetration of CCS depends on its cost reduction in the future, whereas both cost reduction and biomass resources availability are indispensable factors for the scale-up of BECCS. Some studies argue that 400–700 GW of coal-fired power should

Table 4.6 Installed capacity in selected years under the 2°C scenario (in GW)

	2020	2030	2035	2040	2050
Coal	1100	1022	925	643	123
Coal with CCS	0	0	5	16	68
Gas	110	140	155	170	200
Nuclear	53	136	184	232	327
Hydro	380	414	414	414	414
Onshore wind	205	656	889	1295	2221
Offshore wind	5	5	5	25	91
Centralized PV	164	385	504	671	1518
Distributed PV	51	230	324	423	686
Biomass	18	37	36	24	6
BECCS	0	0	0	3	32
Total installed capacity	2086	3025	3441	3916	5686

Table 4.7 Power generation in selected years under the 2°C scenario (in PWh)

	2020	2030	2035	2040	2050
Coal	4.62	4.35	3.88	2.93	0.45
Coal with CCS	0	0	0.04	0.11	0.40
Gas	0.19	0.24	0.27	0.40	0.39
Nuclear	0.39	1.01	1.38	1.73	2.34
Hydro	1.31	1.51	1.53	1.52	1.47
Onshore wind	0.43	1.39	1.95	2.82	4.52
Offshore wind	0.02	0.02	0.02	0.10	0.34
Centralized PV	0.21	0.52	0.73	1.02	2.14
Distributed PV	0.05	0.32	0.45	0.58	0.82
Biomass	0.09	0.25	0.25	0.17	0.04
BECCS	0	0	0	0.02	0.19
Total power generation	7.3	9.6	10.5	11.4	13.1

be reserved by 2050 for basic load, peak load regulation and heating purposes, yet the existing units should be retrofitted for higher flexibility and combined heat and power generation (CHP). Our results are basically consistent with these studies if gas-fired generator sets are included.

CCS and BECCS technologies will play important roles in the 2°C and the 1.5°C scenarios. Under the 2°C scenario, CCS is expected to scale up in coal-fired power plants by 2035 and hit an installed capacity of 68 GW by 2050 with 320 million tonnes of CO_2 captured; BECCS, on the other hand, will be used in large scale by 2040 and reach a capacity of 32 GW by 2050 with 190 million tonnes of CO_2 captured. Under the 1.5°C scenario, the scale-up of CCS is moved up to 2030, with

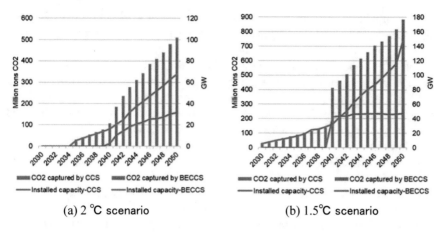

(a) 2 °C scenario   (b) 1.5°C scenario

Fig. 4.2 Installed capacity and carbon storage capacity of CCS and BECCS

a capacity of 149 GW by 2050 and 600 million tonnes of CO_2 captured in the same year, and an aggregate capture of 4.13 gigatonnes of CO_2 from 2040 to 2050. BECCS' deployment will soar from 2040 onward, reaching 48 GW by 2050 with 280 million tonnes of CO_2 captured (see Fig. 4.2).

4.2.3   Carbon Emission Trajectories of the Power Sector Under Different Scenarios

The carbon emission trajectories of the power sector under the four scenarios are illustrated in Fig. 4.3. Under the policy scenario, carbon emissions from the power

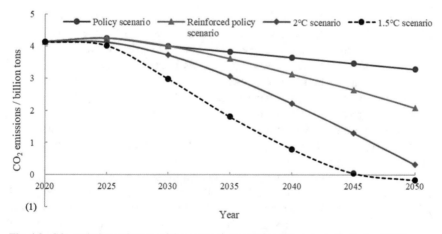

Fig. 4.3 CO_2 emission trajectory of the power sector under each scenario (including CCS)

sector continue to rise slowly until it peaks in 2025 at 4.25 billion tons CO_2 and then downward to 3.29 billion tons in 2050. Under the reinforced policy scenario, the emissions prior to 2030 stay on the same trajectory with policy scenario but fall rapidly after 2030 and to 2.09 gigatonnes in 2050. In this study, the target of "non-fossil energy comprising 20% of primary energy consumption by 2030" in the 2015 version NDCs is integrated into the policy scenario, and the target of "non-fossil energy representing 50% of primary energy consumption by 2050" is incorporated into the reinforced policy scenario, which is also characterized by a scale-up of renewables before 2030 and the rise of non-fossil energy to 25% of primary energy consumption by 2030. The carbon trajectories of both scenarios point to a failure in attaining the 2°C goal, i.e. China's current mid- to long-term energy and power policies and targets fall short of the 2°C target set in the Paris Agreement.

In this study, strengthened efforts in emission reduction are seen in the 2°C and the 1.5°C scenarios compared to the policy and reinforced policy scenarios prior to 2030, and the efforts are sped up after 2030 for the sake of the 2°C and the 1.5°C targets. Under the 2°C scenario, carbon emissions from the power sector hit the peak in 2023 at 4.21 billion tons and take a nosedive after 2030, falling to merely 320 million tonnes by 2050. Under the 1.5°C scenario, the power sector sees its emissions peak at 4.21 billion tons in 2023, followed by a steep drop after 2030, reaching a negative emission of 160 million tonnes by 2050 with the aid of BECCS. As is illustrated in the above figure, the carbon trajectory features a sharp fall after 2030 under the 1.5°C scenario, which presents daunting challenges in the short term to the industry, technology, market and policy, and puts enormous pressure on emission reduction in later phase with the massive retirement of coal-fired power units after 2030.

4.2.4 Demand for Emission Reduction Technologies Under Different Scenarios

Among carbon emission technologies under the four scenarios, nuclear, renewables, CCS and BECCS hold the key to deep emission reduction in the power sector. Relative to the policy scenario, the reinforced policy scenario witnesses the contribution of hydro, nuclear and renewables to emission reduction, to which wind and nuclear power are the biggest contributors. Compared to the policy scenario, the 2°C scenario sees hydro, nuclear, renewables, CCS and BECCS aiding with emission reduction, coupled with intensified efforts after 2030, and wind and solar power are the biggest contributors. Under the 1.5°C scenario, though no extra emission reduction technologies are adopted, the penetration of existing technologies and emission reduction efforts increase significantly (see Fig. 4.4).

Table 4.8 shows the technology demand of different emission reduction trajectories in two timeframes (2020–2030, 2030–2050). The difference in the growth of installed capacity is observed between 2020 and 2030 due to cost changes of technologies and different schedules for deep emission reduction. Between 2030

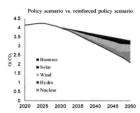 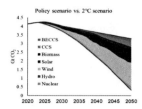 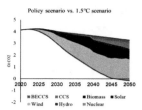

Fig. 4.4 Carbon emission reduction by different technologies

Table 4.8 Technology demand under different scenarios

		Policy scenario	Reinforced policy scenario	2°C scenario	1.5°C scenario
Growth of wind power capacity (GW/yr)	2020–2030	24	38	45	71
	2030–2050	31	40	83	91
Growth of solar power capacity (GW/yr)	2020–2030	14	38	40	42
	2030–2050	27	40	79	87
Growth of hydropower capacity (GW/yr)	2020–2030	3	3	3	3
	2030–2050	0	0	0	0
Growth of nuclear power capacity (GW/yr)	2020–2030	8	8	8	8
	2030–2050	7	10	10	10
Growth of biomass capacity (GW/yr)	2020–2030	0	1	2	1
	2030–2050	– 1	– 1	0	1

and 2050, deep emission reduction emerges in all pathways, where key emission reduction technologies (wind, solar, energy storage and cross-regional power transmission) all scale up to a similar high level. To be specific, the scale of wind and solar power installation is equally considerable, but due to variance in technological pathways and schedule of deep emission reduction, the size of power installation differs greatly between 2020 and 2030. Due to restrictions of cost and raw material utilization, impede biomass power generation can't play the major role for emission reduction.

4.3   Crucial Problems and Solutions for the Low-Carbon Transition of the Power Sector

4.3.1   Operation Security of the Power Grid

China is about to enter a phase that features high penetration of renewables with large-scale clustered integration and high-penetrated distributed integration. In that phase, it will provoke drastic changes for the power system. With the integration of high penetrated renewables, highly uncertain and volatile wind and solar power will be evolved from the supplementary to the primary power sources, and the operation and planning mechanism of power generation systems will be fundamentally changed. The coaction of high volatility in the generation side and the massive distributed sources in the demand side will then fundamentally change the operation and planning mechanism of power transmission and distribution systems. Besides, with the widely application of power electronic interfaces in renewable generation technologies, transmission facilities and load devices, the whole power system is changing towards a power electronics dominated trend. The features of low inertia, weak immunity, and multi-timescale response in power electronic devices will lead to a fundamental change in the stability mechanism and control methods of power electronics dominated power systems. Therefore, there are two key scientific difficulties required to be addressed: (1) the scarcity of power system flexibility and the structure evolution of power transmission and distribution networks due to the strong uncertainties from both generation and demand sides; (2) multi-timescale coupled stabilization mechanism and optimal operation of power electronics dominated power systems.

To tackle these challenges raised by the renewables integration, the combination of multiple technologies with varying costs is necessary for the power system (as shown in Fig. 4.5). When the share of renewable generation is not high (<30%), lower cost technologies, such as improved prediction of renewable generation, flexibility retrofit of thermal power plants, and advanced flexible scheduling, can be utilized to address the problems including generation ramping, peak-load regulation and power reserve. More expensive technologies, including flexible generation, pumped storage, cross-regional transmission, renewable energy curtailment, and biomass power generation, can be adopted as auxiliary approaches. When the share of renewable generation is over 30%, renewables' uncertainty and volatility will make a significant impact on the power system planning and operation, complicating the power system operation behaviors. In that case, multiple energy system integration is recommended to leverage gas storage, heat storage, cold storage, and other low-cost storage devices for offsetting the fluctuation of renewable generation. Other advanced technologies such as demand response, CSP (concentrating solar power) and emerging energy storage solutions (compressed air energy storage, flow battery and super-capacitor) are also optional. But these technologies are still in the demonstration stage and far from commercial scaled application, and hence are relatively expensive. When the share of renewable generation is over 50%, these above

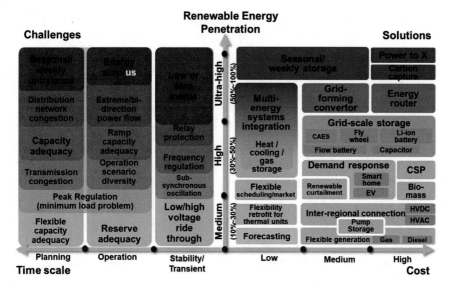

Fig. 4.5 Challenges and solutions for high penetration of intermittent renewable energy (*Note* CAES—compressed air energy storage; CSP—concentrating solar power; HVDC—high voltage direct current; HVAC—high voltage alternating current)

technologies will not be enough, due to new challenges including low inertia and seasonal supply–demand mismatch. In such a case, virtual synchronous generation (VSG) may need to be deployed to increase system synchronous inertia and thus strengthen the system stability. Seasonal supply–demand mismatch problem can be tackled either by seasonal energy storage (such as heat or cold storage) or more expensive solutions such as energy routers or power-to-gas devices (Fig. 4.5).

With the growing share of renewable generation, more flexible resources are needed to satisfy the peak-load regulating capacity requirement in the power system. The role of coal-fired power units is gradually shifted from satisfying basic load demand to providing flexibility and peak regulation services from 2030, while the corresponding installed capacity is also reducing. The intermittent renewable energy integration is supported by the flexibility retrofit of thermal power plants and the power grid interconnection before 2030. After then, the carbon emission reduction will accelerate, the penetration of renewables will further increase, and the generation capacity of flexible coal-fired power units will also reduce. Large-scale energy storage is necessary to accommodate renewable energy, especially in Inner Mongolia, Xinjiang, Shandong, and northwest China.

The cross-regional power grid interconnection can provide flexibility for renewable energy integration and enhance the optimization of resource allocation. The cross-regional power grid interconnection should be further strengthened and reaches 1.8–2.7 times of current cross-regional transmission capacity level by 2030 under different scenarios. After 2030, due to the limited availability of exploitable renewable energy resources in eastern and central China, northwestern China shall become

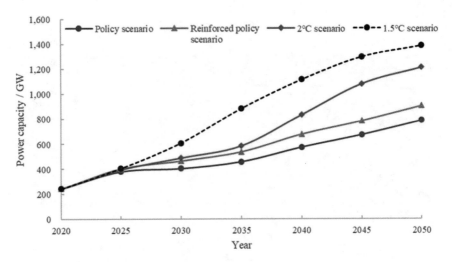

Fig. 4.6 Total capacity of inter-regional power distribution

a key area in renewable energy development to meet the demand for sustained emission reduction. Meanwhile, supporting facilities such as long-distance power transmission lines should be built to reach 3.5–6.2 times of current cross-regional transmission capacity level by 2050 under different scenarios (see Fig. 4.6). The research results show that the cross-regional transmission capacity requirement in 2050 respectively are 788 GW (policy scenario), 903 GW (reinforced policy scenario), 1212 GW (2°C scenario), and 1387 GW respectively (1.5°C scenario).

Energy storage could be categorized into short-term daily energy storage and long-term seasonal energy storage. Daily energy storage aims to mitigate renewables' volatility and address the daily power mismatch between renewable generation and load demand, while the seasonal energy storage enables seasonal energy transfer to address the monthly energy mismatch. In terms of energy storage requirement, the total capacity of 574 GWh, 677 GWh, 1168 GWh, and 1334 GWh will be needed respectively under policy scenario, reinforced policy scenario, 2°C scenario, and 1.5°C scenario. The energy storage capacity requirement in 2050 under 2 and 1.5°C scenarios are expected to be 2 times and 2.3 times of that under policy scenario (see Fig. 4.7).

4.3.2 Retirement of Coal-Fired Power Units

A country's choice for transition pathway is closely related to its resources endowment, power source mix and economic development level, etc. Duplicating the experience of developed countries in the retirement of coal-fired units would not be a

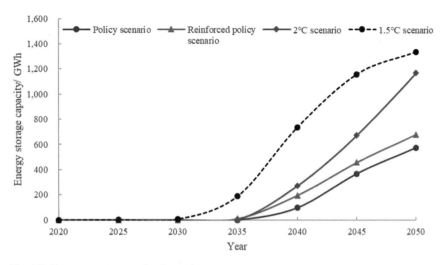

Fig. 4.7 Energy storage capacity demand

practical option for China. Despite the consensus on the necessity of decarbonization in current studies, opinions are highly different on the capacity of coal-fired
power that can be retained in the future. Some studies conclude that China is able
to reach the Paris climate targets with lower economic cost, phasing out coal-fired
power completely between 2050 and 2055 [13][15]; while others argue that 400–700
GW of coal power will be necessary in 2050 for basic load, peak load regulation and
heating, on condition that existing units are retrofitted in terms of flexibility and CHP
[6]. The retained capacity of coal power allowed for China will, in large measure,
hinge on the potential and progress of CCS and BECCS, where research is severely
insufficient [16].

The stranding of coal power assets is a matter of wide concern among various
stakeholders. In this research, the cost of stranded coal-fired power assets is defined
as the residual value of fixed assets in case of early retirement before reaching the
expected service life. Besides, early retirement refers to the fact of being put out of
economic service in the conventional sense, with some units still capable of offering
valuable services such as complementing the power load shortage at certain time
slot as the backup. The cost of stranded coal-fired power assets varies significantly
under different scenarios—707.9 billion RMB for the 1.5°C scenario, 60.9 billion for
the 2°C scenario, 3.1 billion in the policy scenario and 3.2 billion in the reinforced
policy scenario. Figure 4.8 illustrates the interannual variation of stranded coal power
assets under various scenarios. On the whole, under the 1.5°C scenario, most of the
stranded cost is incurred between 2031 and 2046, with a peak between 2038 and
2044 in particular. Under the 2°C scenario, the peak of stranding arrives between
2046 and 2048.

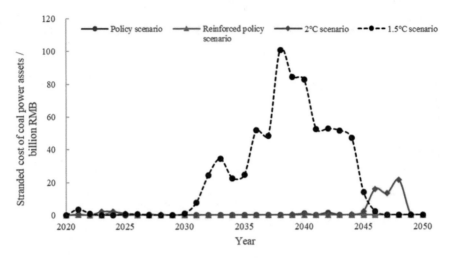

Fig. 4.8 Year-to-year stranded cost of coal power assets from 2018 to 2050

4.4   Conclusions and Policy Recommendations

This study shows that for the power sector, the emission reduction pathway based on the NDC of 2015 and related reinforced measures fall short of the 2°C and 1.5°C targets. To accomplish the emission reduction goals set in Paris Agreement, actions should be ramped up in the power sector to secure an earlier emission peak with faster and more ambition. Major measures include scaling up the development of non-fossil and renewable energy, accelerating the phasing-out of coal-fired power plants and massive deployment of CCS technologies. In the transition of the power sector, appropriate solutions must be explored for safe operation of power grid, retirement of coal-fired power fleet, investment as well as the research and deployment of CCS and BECCS technologies.

1.   **Development of non-fossil energy represents the ultimate approach for decarbonization of China's power sector, so measures must be taken to ensure the scale and pace of such development.**
     It's imperative to study the proper policy measures to boost renewable energy development, both in terms of the scale and speed. This Study found that China's power sector shall rely on non-fossil energy for its decarbonization, and on inter-mittent renewables for deep decarbonization in the long run. Under the 1.5°C scenario in particular, the development of various renewables has almost hit the ceiling of economically available resources with environment and ecolog-ical preservation taken into full account. Under the 2°C scenario, the annual installed capacity of wind power should stay above 45 GW between 2020 and 2030, and above 83 GW between 2030 and 2050; while for solar power, at least 40 GW between 2020 and 2030 and 79 GW between 2030 and 2050.

2. **The 14th Five-Year Plan must keep a tight rein on addition of coal-fired power units and make full use of the stock**

Some suspended or postponed coal-fired power projects have been re-approved in an effort to spur domestic demand and stabilize growth in the wake of COVID-19 and the subsequent economic downturn, resulting in a rebound of coal consumption. To reverse the situation, it is suggested that the number of new coal-fired power projects should be strictly controlled during the 14th Five-Year plan period and that the overwhelming majority of incremental power demand be met by renewable energy. More support should be provided for renewable energy in new infrastructure construction and remove the barriers and challenges in this connection. While avoiding increase in coal power capacity in principle, efforts should be made to enable a functional shift of coal-fired power and fully usage of existing coal-fired power. Major steps should be taken to enhance the flexibility retrofit of existing power units prior to 2030, after which the role of coal-fired power should shift from basic load provision to peak load regulation service. The 2 and 1.5°C scenarios require a transformation of coal-fired installations into peak-shaving units with prolonged shutdown for some to act as standby units. A large amount of social resources and investment cannot be recovered due to the sharp decline in the utilization rate of coal-fired power units. Therefore, it is necessary to establish a capacity cost recovery mechanism to properly handle the problems arising from the transformation of coal power units.

3. **Spare no effort to develop and deploy energy storage technologies, strengthen inter-regional power transmission channels, and advance power market reforms**

With high penetration of intermittent renewable power, higher flexibility is required to meet the need of hourly peak load regulation. The year 2030 will be a watershed, before which peak regulation depends on flexibility retrofit of coal-fired units and power grids interconnection and mutual support; whereas after this point, new construction of energy storage units is required to accommodate the surge in renewable power, especially in Inner Mongolia, Xinjiang, Shandong and northwest provinces. Therefore, it's important to proactively develop, demonstrate and utilize energy storage technologies, develop green financing tools to support the massive construction of inter-regional power transmission lines. Deep decarbonization of the power sector in future hinges, from a policy standpoint, on the power market mechanism, hence the need to advance power market reforms and minimize total social cost via market levers.

4. **Formulate roadmaps for the R&D, demonstration and application of grid technologies to secure a high penetration of intermittent renewable energy**

Compared to conventional flexible power sources such as coal-fired power, the volatility and seasonality of intermittent renewable energy can seriously affect the safety operation of grid. As the penetration of wind and solar power grows, the grid is increasingly featured by the randomization of power and energy balance, diversification of operation modes, scarcity of flexible resources, and complication of stabilization mechanism, etc. As challenges mount, it's essential to employ an array of technologies to ensure the normal working order of the grid.

When the share of renewable generation is less than 50%, extra flexibility resources would be required to offset the volatility and uncertainty of renewable generation for the timescale from minute to hour. This is mainly achieved through increasing frequency regulation capacity, peak-load regulation capacity and power reserve capacity. Major technical solutions include flexibility retrofit of coal-fired power plants, demand response management, multiple energy system integration, cross-regional power transmission, concentrating solar power and electrical energy storage. None of these solutions can address the renewable energy integration problem by itself alone. A reasonable and economic combination is necessary, and the optimal technology portfolio depends on the generation and transmission structure of power systems, as well as the maturity and economics of each technology.

When the share of renewable generation is more than 50% (up to 60% under the 2 and 1.5°C scenarios in this study), novel renewable energy integration challenges will arise, including low synchronous inertia, frequency stability issue, more periods with renewable generation higher than load demand, and the seasonal energy imbalance between renewable supply and load demand. The power system security operation requires to strengthen the system synchronous inertia, tackle the surplus renewables and the seasonal energy imbalance. The former can be addressed by equipping virtual synchronization in renewable power plants, while the latter calls for seasonal energy storage. Both technologies are currently under the development or the demonstration phase. Strengthened R&D and investment are supposed to promote technology maturity, cost reduction and large scale application through the supports from technology, policy, and markets.

Compared to conventional flexible thermal power plants, the strong volatility and uncertainty and seasonal mismatch feature of intermittent renewable energy bring significant impacts on power system secure and economic operation. As the penetration of wind and solar power grows, the power grid is increasingly featured by the randomization of power and energy balance, diversification of operation modes, scarcity of flexible resources, and complication of stabilization mechanism, etc. It is essential to employ an array of technologies to ensure the secure and economic operation of power grids.

5. **CCS should be highlighted as a technology of vital importance for more R&D and demonstration. It's particularly important to advance geological exploration for potential underground CO_2 reservoirs.**

Based on current understanding of renewable resources in China, this study shows that the zero carbon transformation would be impossible for the power sector without large-scale application of CCS and BECCS technologies, unless revolutionary technological breakthroughs take place. Therefore, CCS should be highlighted as an important technology, and the research and demonstration of the technology should be accelerated to achieve breakthroughs in capture technology, lay a solid foundation for the whole process technology integration and large-scale demonstration, strengthen the policy research on CCS industrialization, and strengthen international cooperation and technology transfer.

References

1. British Petroleum (BP). BP Statistical review of world energy (2020–06–17) [2020–06–22] http://www.bp.com/statisticalreview
2. International Energy Agency (IEA) (2018) World Energy Outlook 2018. IEA, Paris
3. International Renewable Energy Agency (IRENA) (2019) Global energy transformation: a roadmap to 2050. ARENA, Abu Dhabi
4. General Institute of Electric Power Planning and Design (2019) China power development report 2018. China Electric Power Press, Beijing
5. China Academy of Petroleum Technology and Economics (2018) World and China energy outlook in 2050. China Academy of Petroleum Technology and Economics, Beijing
6. Zhang N, Xing L, Lu G (2018) Prospects and challenges of China's medium—and long-term energy and electric power transformation. China Electr Power Enterp Manag 526(13): 60–65
7. Energy Research Institute of National Development and Reform Commission (2015) Research on the development scenarios and approaches of high-proportion renewable energy in China 2050. Energy Research Institute of National Development and Reform Commission, Beijing
8. Dai Y, Tian Z, Yang H et al (2017) Reshaping energy: China—roadmap for a revolution in energy consumption and production towards 2050 (Comprehensive volume). China Science and Technology Press, Beijing
9. Energy Research Institute of National Development and Reform Commission (2018) Research on China's energy emission scenario under the global 1.5°C target. Institute of Energy, National Development and Reform Commission, Beijing
10. National Center for Renewable Energy (2018) Energy economic ecosystem in beautiful China 2050. National Renewable Energy Center, Beijing
11. Zhou X Energy transformation in building a new generation of Chinese energy system (2019–02–03) [2019–05–21]. https://www.sohu.com/a/293084741_100006059
12. Liu Z Promoting the transformation and high-quality development of China's energy and power industry.(2019–04–19)[2019–06–07] http://www.slsdgc.com.cn/index.php?m=con tent&c=index&a=show&catid=12&id=2495
13. Energy Transitions Commission (ETC) and Rocky Mountain Institute (RMI) (2019) China 2050: a fully developed rich zero-carbon economy. ETC and RMI, Colorado
14. Qin S, Hu R (2015) China biomass energy industry development roadmap 2050. China Environmental Press, Beijing
15. Cui R, Hultman N, Jiang K, et al (2020) A high ambition coal phaseout in China: feasible strategies through a comprehensive plant-by-plant assessment. University of Maryland, Maryland
16. Chang S, Zheng D, Fu M (2019) Biomass Energy Combined with Carbon Capture and Storage (BECCS) under 2°C/1.5°C temperature control. Glob Energy Internet 2(03): 75–85

Chapter 5
Primary Energy Consumption and CO_2 Emissions

5.1 Analysis of China's Future Primary Energy Consumption

5.1.1 Total Primary Energy Consumption

Based on the previous analysis of energy demand of end users in Chapter 3 and power sector in Chapter 4, the primary energy consumption under policy scenario, reinforced policy scenario, 2 °C scenario and 1.5 °C scenario is shown in Fig. 5.1. The comprehensive analysis data of various scenarios is illustrated in Tables A.1–A.4 which can be found in the appendices of the report.

Under the policy scenario, China's total primary energy consumption increases rapidly between 2015 and 2040 from 4.94 billion tce in 2020 to 6.06 billion tce in 2030, and to 6.10 billion tce in 2040. With the achievement of socialist modernization, China's industrial structure dominated by energy-intensive sectors will gradually shift to one driven by service and high-tech sectors, with further improvement in energy efficiency, total primary energy consumption in 2050 will slowly reach 6.23 billion tce.

Under the reinforced policy scenario, from 2020 to 2040, China experiences continuous growth of total primary energy consumption, followed by a dip from 2040 to 2050, which is consistent with the strategic goal of China's social and economic development. With the tightening of carbon constraints and the improvement of energy efficiency, China's total primary energy consumption would see a decrease as a whole, peaking at 6.02 billion tce in 2040 before dropping to 5.63 billion tce in 2050, 600 million tce less than the policy scenario.

Under the 2 °C scenario, with strengthened carbon constraint, enhanced energy efficiency and increased energy saving on the demand side, China's total primary energy consumption peaks at 5.64 billion tce around 2030 before transitioning to 5.20 billion tce in 2050. As short-term drastic improvement is seen in the level of

Institute of Climate Change and Sustainable Development of Tsinghua University et al., *China's Long-Term Low-Carbon Development Strategies and Pathways*, https://doi.org/10.1007/978-981-16-2524-4_5

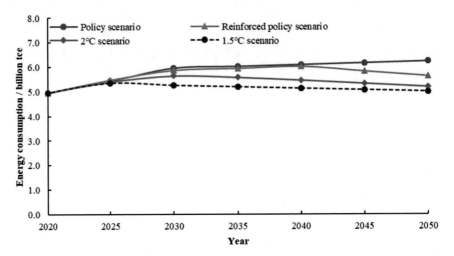

Fig. 5.1 Total primary energy consumption under the four scenarios

energy efficiency improvement and demand energy saving the 2 °C scenario, total primary energy consumption drops slightly under the 1.5 °C scenario compared to the 2 °C scenario between 2015 and 2030, hitting a peak of 5.34 billion tce in 2025 before a moderate decline to 5 billion tce in 2050.

Power consumption by end users is converted into primary energy equivalent, producing a breakdown of primary energy consumption by sector under various scenarios, as shown in Fig. 5.2. If primary energy consumption of the power sector is separately calculated and only the direct use of primary energy by end users is taken

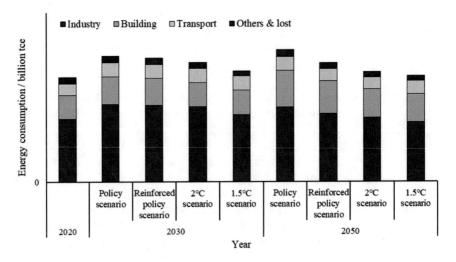

Fig. 5.2 Mix of primary energy demand under four scenarios

Table 5.1 Primary energy consumption under the reinforced policy scenario and the 2 °C scenario (Unit: billion tce)

		Reinforced policy scenario			2 °C scenario		
		2020	2030	2050	2020	2030	2050
Direct use of primary energy consumption by end users	Industry	1.62	1.88	1.23	1.62	1.73	0.69
	Building	0.55	0.51	0.37	0.55	0.41	0.26
	Transportation	0.49	0.55	0.40	0.49	0.53	0.30
	Other sectors and losses	0.12	0.22	0.5	0.12	0.09	0.04
Conversion of primary energy power generation		2.18	2.83	3.57	2.18	2.89	3.91
Total consumption of primary energy		4.94	5.98	5.63	4.94	5.64	5.20

into account, the reinforced policy scenario and 2 °C scenario would look differently in Table 5.1.

5.1.2 Mix of Primary Energy Consumption

Under the four scenarios, with the constraint of decarbonization, primary energy composition in China by 2050 features an overall trend of accelerated coal elimination and sharp rise in the share of non-fossil fuels. The composition of primary energy consumption in major years (2020, 2030 and 2050) under the four scenarios is shown in Fig. 5.3 and Table 5.2.

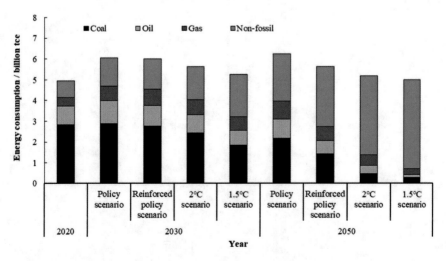

Fig. 5.3 Total primary energy consumption and its composition in major years under different scenarios

Table 5.2 Energy consumption composition by energy source in major years (Unit: %)

	2020	2030				2050			
		Policy scenario	Reinforced policy scenario	2 °C scenario	1.5 °C scenario	Policy scenario	Reinforced policy scenario	2 °C scenario	1.5 °C scenario
Coal	57.4	47.6	46.0	43.2	35.4	34.9	25.3	9.1	5.4
Oil	18.0	18.4	16.7	15.5	13.2	14.8	11.3	7.7	3.0
Gas	8.7	11.6	13.1	12.6	7	14.0	11.9	10.0	5.5
Non-fossil	15.9	22.4	24.3	28.6	38.7	36.3	51.5	73.2	86.1

- In the policy scenario, the total coal consumption decreases annually from 2.84 billion tce in 2020 to 2.18 billion tce by 2050, edging down 0.9% on an annual basis. Oil consumption rises from 890 million tce in 2020 to 1.1 billion tce in 2030 before falling to 920 million tce in 2050. Continuous growth is seen in natural gas consumption from 430 million tce in 2020 to 690 million tce in 2030 and 870 million tce in 2050. Non-fossil energy consumption experiences an annual growth of 3.6% from 790 million tce in 2020 to 2.26 billion tce in 2050. Under such scenario, by 2050, coal, oil, natural gas and non-fossil energy represents 34.9%, 14.8%, 14.0%, and 36.3% respectively in the energy mix.

- In the reinforced policy scenario, total coal consumption drops steadily since 2020 from 2.84 billion tce to 1.42 billion tce in 2050, an annual decline of 2.3%. Oil consumption grows from 890 million tce in 2020 to 980 million tce in 2030 before falling to 640 million tce in 2050. Natural gas consumption climbs from 430 million tce in 2020 to 770 million tce in 2030, then shows a downward trend and slides to 670 million tce in 2050. Non-fossil energy consumption jumps from 790 million tce in 2020 to 2.9 billion tce in 2050, averaging an annual growth of 4.2%. Under such scenario, by 2050, coal, oil, natural gas, and non-fossil energy each takes up 25.3%, 11.3%, 11.9%, and 51.5% respectively in the energy mix.

- In the 2 °C scenario, total coal consumption rapidly decreases from 2.84 billion tce to 470 million tce between 2020 and 2050, a 5.8% drop on an annual basis. Oil consumption falls from 890 million tce in 2020 to 400 million tce in 2050. Natural gas consumption, after a rise from 430 million tce in 2020 to 710 million tce in 2030, goes down to 520 million tce in 2050. Non-fossil energy consumption soars from 790 million tce in 2020 to 3.81 billion tce in 2050, growing at 5.4% annually. Under this scenario, by 2050, coal, oil, natural gas, and non-fossil energy comprises 9.1%, 7.7%, 10.0%, and 73.2% respectively in the energy mix.

- In the 1.5 °C scenario, total coal consumption foresees a steep fall from 2.84 billion tce in 2020 to 270 million tce in 2050, dipping 7.5% year by year. Oil consumption starts its drop since 2020 from 890 million tce to 150 million tce in 2050. In the wake of a climb to 660 million tce in 2030 from 430 million tce in 2020, natural gas consumption spirals down to 280 million tce in 2050. Non-fossil energy consumption, on the other hand, surges from 790 million tce in 2020 to 4.31 billion tce in 2050, an average annual growth of 5.8%. Under such scenario, by 2050, coal, oil, natural gas, and non-fossil energy makes up 5.4%, 3.0%, 5.5%, and 86.1% respectively in the energy mix.

Given the above analysis, in the transition from the policy scenario to 1.5 °C scenario, the decarbonization speed of the energy system continues to accelerate, non-fossil fuel energy becomes the main source in the primary energy consumption mix. However, to materialize these scenarios calls upfront strategic planning, as the earlier the transition, the better it is for the entire energy system.

5.1.3 CO_2 Emissions Under Different Scenarios

Analyzing from the macro perspective on the overall primary energy consumption and CO_2 emission from fossil fuel burning, in the transition from the policy scenario to 1.5 °C scenario, CO_2 emission from primary energy consumption would witness a steady decline, and carbon emissions from primary energy consumption under the four scenarios would fall from 10.25 billion tCO_2 in 2020 to 9.09, 6.18, 2.92, and 1.47 billion tCO_2 in 2050. CO_2 emissions from fossil fuel combustion under various scenarios are illustrated in Fig. 5.4 and Table 5.3.

The changes in CO_2 emissions from fossil fuel combustion under the four scenarios are mainly illustrated as follows:

- Non-fossil energy grows importance in primary energy consumption mix. By 2050, its consumption soars to 4.31 billion tce under the 1.5 °C scenario from the original 2.26 billion tce under the policy scenario, its share in primary energy consumption is 86.1% under the 1.5 °C scenario as compared to 36.3% under the policy scenario.
- There are striking differences in the year and level of CO_2 peak. In the policy scenario, energy-related CO_2 emissions will peak around 2030 at about 11.0 billion tCO_2. In the reinforced policy scenario, the peak occurs between 2025 and 2030 at around 10.6 billion tCO_2. In the 2 °C scenario, the peak will move up to 2020–2025, and reduced to about 10.3 billion tCO_2. To embark on the 1.5 °C pathway sooner, swift reduction in energy-related emissions is required around 2020.

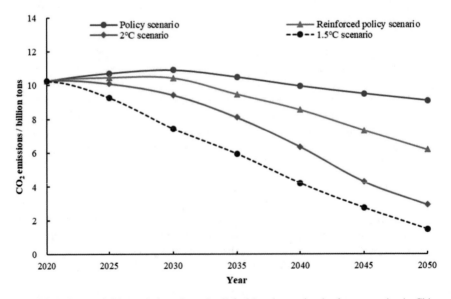

Fig. 5.4 Pathway of CO_2 emissions from fossil fuel burning under the four scenarios in China (excluding CCS)

Table 5.3 Primary energy consumption and CO_2 emissions in major years under four scenarios

	2020		2030		2050	
	Energy demand (billion tce)	CO_2 emissions (billion tCO_2)	Energy demand (billion tce)	CO_2 emissions (billion tCO_2)	Energy demand (billion tce)	CO_2 emissions (billion tCO_2)
Policy scenario	4.94	10.03	6.06	11.08	6.23	9.09
Reinforced policy scenario	4.94	10.03	5.98	10.61	5.63	6.18
2 °C scenario	4.94	10.03	5.64	9.42	5.20	2.92
1.5 °C scenario	4.94	10.03	5.25	7.44	5.00	1.47

- Great difference exists in the speed of total CO_2 emissions reduction. In the policy scenario and reinforced policy scenario, the average decrease of total carbon emissions after peaking in 2030 stands at 0.9% and 2.6% respectively, compared to 5.7% and 7.8% in the 2 °C and 1.5 °C scenarios.

The breakdown of carbon emissions by sector under the policy scenario, reinforced policy scenario, 2 °C scenario and 1.5 °C scenario is illustrated in Fig. 5.5. In all scenarios, carbon emissions from industrial sector makes up the largest share.

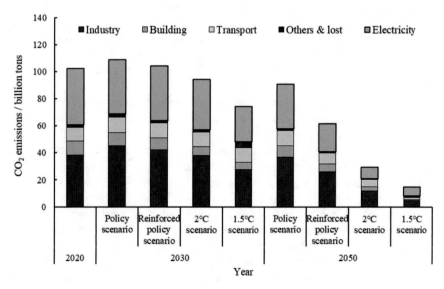

Fig. 5.5 Breakdown of CO_2 emissions under all scenarios

With the development of low-carbon electric power supply technology and negative emission technology, the power sector stands to be the biggest contributor to emissions reduction with a drop from 4.06 billion tCO_2 in 2020 to 3.29 billion, 2.09 billion, 490 million, and -150 million tCO_2 in 2050 under four scenarios respectively. Total CO_2 emissions from the building sector are curtailed, falling from 1.00 billion tCO_2 in 2020 to 830 million, 560 million, 310 million and 80 million tCO_2 in 2050 respectively. CO_2 emissions from transport experiences a minor growth before 2030, followed by an accelerated decline afterward to 1.11 billion, 800 million, 550 million and 170 million tCO_2 by 2050.

5.2   Economic Analysis of Emissions Reduction in the Energy System

5.2.1   Introduction of the Energy Economic Analysis Model

The Computable General Equilibrium (CGE) method can well characterize the correlation and interconnection between the economic and energy systems. This project employs the Global Energy Economy Computable General Equilibrium model (C-GEM) developed by the Institute of Energy, Environment and Economy of Tsinghua University to explore the socio-economic impact of China's low-carbon policy intervention from the perspective of global energy economy balance. As the world moves into a new phase of addressing climate change after the Paris Agreement, and as China embarks on supply-side structural reform amid the economic "new normal", the C-GEM team reintegrated sectors in the model, improved the power and energy-intensive industries, modified the energy substitution relationship, enhanced the automatic efficiency improvement factor, revised parameters of China's economic structure underlying the model, and updated and verified the data of global energy economy.

The C-GEM model can depict in detail the flow of elements, goods and services in economy and energy systems. Through factor input and intermediate product input, producers churn out domestic final products under certain technical conditions, some of which are supplied domestically to meet domestic market needs, while others are exported to the international market. Apart from domestic supply, total domestic demand for goods also comes from imports. Residents, governments, and other consumers, through production factor sales and taxation, obtain disposable income that is partly used for personal consumption and partly placed in saving, which then translates to investments, the combination of personal consumption and investment makes up the total domestic demand. Energy input (as part of the intermediate input) and carbon emissions allowance (as a scarce input) also enter the production process, thereby linking the economic system and the energy system. Given a perfectly competitive market, producers seek maximized profits through cost minimization with given production technologies. At a given income level, consumers maximizes their utility through preference selection. Given certain output level, domestic products and import and export commodities generate maximized sales revenue through the pricing mechanism. Production factors are best allocated for supply and demand in the production process.

C-GEM uses GTAP 10 as the basic database embedded in the model, with 2014 as the base year, putting together data from economy, energy, and bilateral trade in 65 industrial sectors across 141 countries and regions. A global computable general equilibrium model is then formulated, while accommodating the sectoral breakdown of official energy and economic statistics in China, capturing 17 regions and 21 sectors globally (19 production sectors and 2 consumption sectors). Furthermore, the policy scenario, reinforced policy scenario, 2 °C scenario, and 1.5 °C scenario are built within the model to diagnose carbon price changes and GDP losses in the low-carbon transition.

5.2.2 Carbon Price

Carbon price levels from 2020 to 2050 under the four scenarios are shown in Table 5.4. Noted that carbon price incentive mechanism in this study contains all policies for emission reduction adopted by the entire energy and economic system except the policies for renewable energy development and natural gas utilization. Carbon price, in this context, represents the marginal cost of abatement in the process of CO_2 emission reduction, thus reflecting the force of policy measures under different scenarios. Under policy scenario, reinforced policy scenario, 2 °C scenario, and 1.5 °C scenario, carbon price gradually climbs from RMB 46/ton in 2020 to RMB 265, 510, 1364, and 5,701/ton in 2050. By 2050, carbon price in the 2 °C and 1.5 °C scenarios is 5 times and 21 times of that in the policy scenario, and 3 times and 11 times of that in the reinforced policy scenario. This means that in order to fulfill the most stringent emission targets, the whole society would have to bear a very high cost.

Table 5.4 Carbon price under different scenarios (Unit: RMB/ton, at constant 2011 prices)

	2020	2025	2030	2035	2040	2045	2050
Policy scenario	46	73	99	139	192	245	265
Reinforced policy scenario	46	79	106	152	238	351	510
2 °C scenario	46	86	126	199	344	583	1364
1.5 °C scenario	46	106	166	285	622	1609	5701

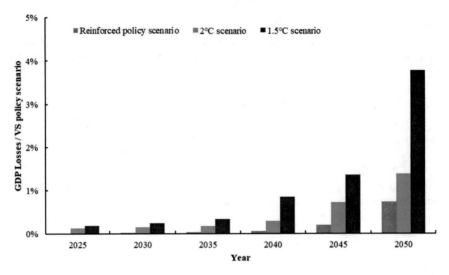

Fig. 5.6 GDP losses under different scenarios

5.2.3 GDP Losses

To promote deep decarbonization while securing sustainable economic and social development, it is essential to strengthen policy measures and ensure sufficient flow of investment. For the economic system, a cut in CO_2 emissions entails costs and sacrifices. Figure 5.6 provides the GDP losses of the reinforced policy scenario, 2 °C scenario and 1.5 °C scenario as opposed to the policy scenario. The reinforced policy scenario features relatively minor GDP losses of less than 1% by 2050, or approximately 0.7%. By contrast, the 2 °C and 1.5 °C scenarios are characterized by greater GDP losses, which grow to 1.4% and 3.8% by 2050, almost 2–5 times larger than in the reinforced policy scenario, reflecting to fairly high economic losses by China for the sake of the most stringent emission targets.

5.3 Synergy Between Deep Decarbonization and Environment Improvement

Cutting back on fossil energy consumption reduces CO_2 emissions while reducing SO_2, NOx, $PM_{2.5}$, and other conventional pollutants at the source, facilitating achievement of environmental standards and the "Beautiful China" target [1]. China's mid-long term environmental targets indicate that by 2035, all major cities and regions should meet the national standard of $PM_{2.5}$ concentration below no more than 35 $\mu g/m^3$ [2] and by 2050, $PM_{2.5}$ concentration below 15 $\mu g/m^3$ for all regions.

The Jing-jin-ji (Beijing-Tianjin-Hebei, respectively) Region and its periphery areas, Fenhe-Weihe Plain and Yangtze River Delta are the key regions for China's environment improvement programs. In 2015, the average concentration of $PM_{2.5}$ at all monitoring stations in China stood at 53.1 $\mu g/m^3$, and the three regions reported 81.3, 60.4, and 55.0 $\mu g/m^3$ respectively. Only 29.1% of Chinese cities fell within the standard [3].

In the policy scenario, the national $PM_{2.5}$ concentration falls to 19.2 $\mu g/m^3$ by 2030, down 63.8% from 2015. Jing-Jin-Ji Region and its periphery areas, Fenhe-Weihe Plain and Yangtze River Delta witness a decrease estimated at 59.8%, 65.4%, and 65.5% respectively. By then, air quality will see a dramatic shift across the country, with the Fenhe-Weihe Plain and Yangtze River Delta no longer being major $PM_{2.5}$ pollution areas. 96.6% of Chinese cities would have been met the Class two of environment quality standard of 35 $\mu g/m^3$ on an annual average. Though one third of cities would have reported a concentration level less than 25 $\mu g/m^3$, over 30% of cities in Beijing-Tianjin-Hebei Region and its periphery areas would fall short of the target. The level of $PM_{2.5}$ pollution in 2050 remains at par with that in 2030, with most regions far behind the 15 $\mu g/m^3$ target and some places in Jing-Jin-Ji Region and its periphery areas still struggling with $PM_{2.5}$ pollution. By 2050, the national $PM_{2.5}$ concentration is reduced to 14.0 $\mu g/m^3$, and that in Jing-Jin-Ji Region and its periphery areas, Fenhe-Weihe Plain and Yangtze River Delta drops to 24.9, 14.6 and 14.0 $\mu g/m^3$respectively. Only 76.0% of the cities in China and 4% of cities in Jing-Jin-Ji Region and its periphery areas measures up to the Class one standard of 15 $\mu g/m^3$. In general, the policy scenario foresees a great decline in national $PM_{2.5}$ concentration, but big gaps still exist if China is to achieve 35 $\mu g/m^3$ by 2030 and 15 $\mu g/m^3$ by 2050.

The reinforced policy scenario sees notable reduction in $PM_{2.5}$ concentration as compared to the policy scenario. By 2030, $PM_{2.5}$ pollution is somewhat improved in key areas such as Jing-Jin-Ji Region and its periphery areas compared to the policy scenario. China as a whole, Jing-Jin-Ji Region and its periphery areas, Fenhe-Weihe Plain, and the Yangtze River Delta is down to 17.3, 29.7, 19.8 and 16.1 $\mu g/m^3$ by 2030 respectively. But the 35 $\mu g/m^3$ target remains a challenge for 3.6% of cities in China and 9.6% of cities in the Jing-Jin-Ji Region and its periphery areas. By 2050, $PM_{2.5}$ pollution in China under the reinforced policy scenario is markedly alleviated compared to the policy scenario, with that in China as a whole, Jing-Jin-Ji Region and its periphery areas, Fenhe-Weihe Plain and the Yangtze River Delta falling to 10.1, 19.3, 11.4 and 9.3 $\mu g/m^3$, with some regions achieving 15 $\mu g/m^3$ in principle. However, 10% of Chinese cities and over 65% of cities in Jing-Jin-Ji Region and its periphery areas fall short of the 15 $\mu g/m^3$ Class one standard, presenting a health hazard that shouldn't be neglected.

In the 2 °C scenario, $PM_{2.5}$ concentration in China, Jing-Jin-Ji Region and its periphery areas, Fenhe-Weihe Plain and Yangtze River Delta is reduced to 16.2, 28.0, 8.2, and 16.0 $\mu g/m^3$ by 2030. All regions meet 25 $\mu g/m^3$ except for Jing-Jin-Ji Region. By 2050, concentration further declines to 8.3, 14.1, 9.6, and 7.5 $\mu g/m^3$, with no region in China classified as $PM_{2.5}$ high pollution zone, with majority of the regions realizing the 15 $\mu g/m^3$ standard. Though the 10 $\mu g/m^3$ goal is achievable in 81.7% of Chinese cities, but only 28.0% of cities in Jing-Jin-Ji Region meet this standard. In comparison, under the 1.5 °C scenario, all Chinese cities meet the 15 μg/m^3 target, with 88.8% reaching the 10 μg/m^3 target. But over 40% of cities in Jing-Jin-Ji Region and its periphery areas still fall short under this target.

In summary, achieving the deep decarbonization target under the 2 °C or even 1.5 °C significantly improves the environment quality. Even under the policy scenario, China's air quality sees considerable reduction in SO$_2$, NOx, PM$_{2.5}$ concentration, but it remains a challenge to meet the target of 35 μg/m^3 by 2030 and 15 μg/m^3 by 2050 in key regions. The 2 °C scenario goes a step further to realize the 35 μg/m^3 by 2030 and 15 μg/m^3 by 2050. And 1.5 °C scenario trumps the previous two scenarios to bring about fulfillment of 10 μg/m^3 target by 2050. In other words, deep decarbonization goal will drive substantial improvement of the environment and secure the success of building a "beautiful China".

Figure 5.7 shows the number of premature deaths from PM$_{2.5}$ exposures by 2030 and 2050 under various scenarios. Deep decarbonization will speed up the PM$_{2.5}$ reduction while ensuring better public health. In 2015, 1.379 million people died prematurely due to PM$_{2.5}$ exposures in China [4]. In the policy scenario, with continuous high PM$_{2.5}$ levels and the aging society, the number of premature deaths is 1.629 million and 2.424 million by 2030 and 2050 respectively, up 18.2% and 75.8% from the 1.379 million in 2015. In the reinforced policy scenario, the number is brought down to 1.541 million by 2030, an 88,000 people reduction than that in the policy scenario, which highlights the contribution of emissions reduction on public health, and by 2050, the health benefits are magnified with 545,000 fewer deaths. Both policy scenario and reinforced policy scenario see increased premature deaths

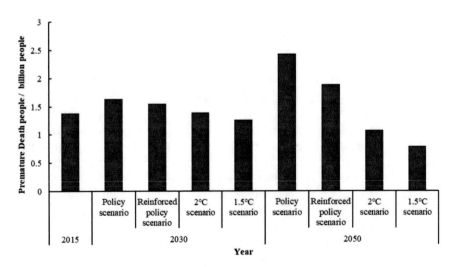

Fig. 5.7 Number of premature deaths from PM$_{2.5}$ exposure by 2030 and 2050 under various scenarios

by 2030 and 2050, suggesting that the health benefits of improved air quality under such scenarios are insufficient to offset the health losses from the ageing population.

Under the 2 °C and 1.5 °C scenarios, the number of premature deaths from $PM_{2.5}$ exposures by 2030 is 1.387 million and 1.26 million respectively, an 8,000 people increase and a 118,000 decrease compared to that of in 2015 respectively, equivalent to a 14.8% and 22.6% reduction than that of in the policy scenario. The number of such deaths falls to 1.065 million and 782,000 by 2050, a 23.2% and 37.9% reduction compared with 2030. Concludingthat with the profound transformation of energy mix and tightening of control measures, the health benefits brought by air quality improvement can offset most of the health losses from population ageing and changes in demographic distribution.

5.4 Policy Suggestions

1. Implement coordinated measures for win–win outcome in economic development, environmental protection, and carbon emissions reduction

China actively now pursues a new development philosophy under the new normal of its economy, working to achieve innovative, coordinated, green, open and shared development by fostering new growth drivers through innovation, while shifting its growth model through green development. Green development, in essence, seeks to boost the harmonious coexistence between human and nature while pursuing green and low-carbon development, thereby promoting the coordinated and sustainable development of economy, society, resources, and environment. This echoes the path of climate-friendly low-carbon economic development advocated in the Paris Agreement. China's enormous strides in energy conservation, carbon reduction, and economic transformation are also attributed to the combined efforts in addressing climate change and promoting domestic sustainable development. China has become a major contributor and leader in driving energy transition and low-carbon economic transformation in the world.

The key to green and low-carbon development lies in promoting the reform of the energy system and transformation of economic development model. With the ongoing industrialization and urbanization in China, the economic takeoff is hampered by resource and environmental constraints. It's important to facilitate revolution in energy production and consumption, save energy, enhance energy efficiency, cap fossil fuel consumption, and vigorously develop new and renewable energy, promote the low-carbonization of the energy mix in a bid to curb conventional pollutants and CO_2 emissions. Domestic efforts in conserving environment and building ecological civilization are consistent with the goals and measures of addressing climate change and protecting ecological safety of the earth, and extensive synergies can be harnessed in this regard. It's essential to devise holistic plans for coordinated progress, aim for sustainable domestic development, and ramp up goal-driven initiatives for long-term low-carbon development and emission reduction. Currently, given the enhanced prevention and treatment of environment pollution, tightened emission standards, and control measures of conventional pollutants in the use of coal, petroleum, and other fossil fuels, more focus should be placed on cutting and replacing coal and oil in the end-use consumption, increasing the share of electricity in this mix, pacing up the development of renewable energy, and increasing the share of renewable power generation in primary energy consumption in an effort to set the stage for faster growth of renewables amid a slowdown in total energy demand. The substitution of non-fossil power generation for coal and oil in the end consumption could enable the dual effect of environment improvement as well as CO_2 emission reduction.

2. Optimizing energy mix is major task and long-term goal for achieving low-emission development

Under the policy scenario, non-fossil energy makes up 22.4% in 2030, further rises to 36.3% in 2050 in the energy mix; coal consumption reaches peak around 2030, and its proportion reduces to 34.9% by 2050. Under the reinforced policy scenario, energy mix is improved in comparison to the policy scenario, with coal and oil dropping by 1.6 and 1.7 percent respectively, and natural gas and non-fossil increasing by 1.5 and 1.8 percent by 2030 respectively. Coal consumption meets its peak around 2020 at 2.84 billion tce, and its proportion drops to 25.3% in the energy consumption mix by 2050. In the same year, non-fossil energy in primary energy consumption increases to 51.5% as a dominant energy source. Under the 2 °C scenario, power generation from coal, natural gas, and biomass with CCS would have evolved around 2040, with non-fossil energy and fossil energy with CCS (carbon-free energy in short) taking over the energy mix. Non-fossil energy as a percentage of primary energy consumption is 28.6% in 2030 before jumping to 73.2% in 2050. Under the 1.5 °C scenario, power generation from coal, natural gas and biomass with CCS would have been developed around 2035, with non-fossil energy in primary energy consumption standing at 38.7% in 2030 before hitting 86.1% in 2050.

As things stand, the installation of coal-fired power units would reach approximately 1.1 billion kilowatts at the end of the 13th Five-Year Plan period, with adequate capacity to support future growth of electricity demand. But in the long run, such capacity will inevitably diminish, and extra caution should be exercised for new installations. In principle, no new coal power units should be added to avoid waste. Furthermore, a shift in coal power usage should be considered for the full use of existing units. Major steps should be taken to make existing coal power plants more flexible prior to 2030, and these plants should gradually undertake more peak-shaving task than the provision of base load. Under the 2 °C and 1.5 °C scenarios, some coal power units need to retire before the end of their service life, which causes tremendous waste of social resources and investment. Therefore, an orderly withdrawal mechanism for existing coal power units should be created for the appropriate management of retiring units.

3. Industrial structure upgrade and technological innovation underpin and safeguard the energy revolution

On its way to a wholistic modernization, China can't afford to replicate the path and model of developed countries, new modes of production and new drivers of modernization should be fostered. The traditional pattern of China's economic development resulted in inefficient resource allocation with mounting economic risks.

The adjustment and upgrading of industrial structure can dramatically mitigate the economic impact from the low-carbon transformation. Therefore, accelerating the transformation of the development pattern is a crucial step along the way to achieving China's economic modernization. Meanwhile, the global energy transformation will prompt major changes in the mode of economic and social development worldwide, thus affecting the dynamics of international economic and technological competition. One important trigger and strategic goal of major countries in mitigating climate change is to increase its competitiveness in advanced energy technologies. Only with more breakthroughs and massive application of key low-carbon technologies for technological and institutional innovations can low emission pathway be made much more accessible.

To start, multiple technological means should be adopted to ensure a high uptake of renewable energy. As more wind and solar power is fed into the grid, grid itself is increasingly challenged by the balance of power, diverse operation pattern, scarce flexible resources, and complicated stability mechanism. The mounting challenges of grid operation calls for various technical solutions to ensure the smooth continuous operation of the grid. More peak-frequency modulation and reserve capacity are often used in case intermittent renewable energy comprises less than 50% of power production, which entails a sound portfolio of technologies, which in turn hinges on the structure of power sources and grid as well as the maturity and economy of various technologies. If intermittent renewable energy reaches over 50%, virtual synchronization and seasonal energy storage technologies should be adopted for the renewable power generators. These two technologies are still under development or demonstration, thus more R&D and investment are crucial to enhance their maturity and cost reduction through a mixture of technology, policy, and market means in order to prepare for their massive application.

Secondly, efforts should be made to actively promote the R&D and application of energy storage technology, strengthen the construction of trans-regional power exchange channels, and press ahead with power market reform. As the grid operates with more intermittent renewable energy, power systems need to provide better flexibility to meet hourly peak demand. Prior to 2030, peak regulation through flexibility upgrading and grid interconnection could well accommodate the integration of intermittent renewables. After 2030, with carbon emissions accelerating, the share of renewables further increased, and the installed capacity of flexibility-supporting coal power on the decline, new energy storage units need to be built for intermittent renewables, especially in Inner Mongolia, Xinjiang, Shandong, and northwest China. Therefore, proactive steps should be taken to develop, demonstrate and apply energy storage technology, create green financing tools to provide funding for building massive cross-regional power exchange channels. Deep decarbonization of the power industry in the future, from the policy standpoint, rests on the improvement of power market mechanism. This provides the rationale for continued reform of the power market and the use of market means for minimum total social cost.

Lastly, step up the R&D and industrialization of advanced energy technologies. Alongside renewable energy technologies such as solar and wind, nuclear technology and hydrogen technology, energy storage, and smart grid technologies, China should place particular attention to CCS and advanced nuclear technology with more focus on R&D and demonstration. Studies have found that the power sector won't succeed in its migrating to the 1.5 °C carbon-free pathway without revolutionary technological breakthroughs. With this in mind, CCS should be deemed as a key technology with intensive R&D and demonstration to set the scene for large-scale application. In addition, with coal long taking the center stage in China's primary energy consumption—nearly 60% by 2020—and urgent global emission reduction targets and high carbon prices, apart from efforts in clean coal and its efficient utilization, CCS will be a major alternative technology by 2040. CCS may sequester hundreds of millions of tons to one billion tons of carbon, playing a pivotal role in emission reduction in the long run. Given the difficulty of replacing fossil fuels in chemical and cement industries, BECCS technology, as a relatively predictable negative emission technology right now, will be instrumental to achieving ultra-low emission targets in the future.

4. Market mechanism and leverage pricing signals to encourage low-carbon green development should be improved

Studies show that China's low-carbon energy transition must be accompanied by forceful policies and legislations, including laws and regulations, targeted policies for renewable energy development and natural gas utilization, market mechanisms like carbon trading, and institutional mechanism reform, etc. By translating all policies other than the targeted policies for renewable energy development and natural gas utilization into carbon price incentives, under the policy scenario, the marginal cost of CO_2 emissions reduction is approximately CNY100/ton in 2030, and CNY 265/ton in 2050. Under the reinforced policy scenario, the cost is around CNY106/ton in 2030 and CNY510/ton in 2050. As for the 2 °C and 1.5 °C scenarios, the cost in 2050 soars to CNY1,364 and CNY 5,701 /ton respectively.

Low-carbon development entails clear policy guidance and, more importantly, long-term market signals. It's essential to stay rooted in reality, deepen reform in energy prices, ensure equitable access to quality energy services for low-income families to boost harmonious social development, alongside efforts in curbing irrational consumption and promoting energy conservation. In particular, institutional reform of the energy market should be strengthened to restore the attributes of energy as a commodity so as to create a just, fair, and effective market structure and system.

Carbon market will be a major institutional safeguard for the early peak of CO_2 emissions in China, and part and parcel of ecological civilization in the country. Now, on top of the carbon trading pilots in five cities and two provinces, China has launched a single national carbon market, which will be continuously expanded and improved. Carbon market is a crucial policy tool that seeks to minimize the cost of emission reduction for the whole society and helps attain national target of emission reduction through government-led market mechanism. It quantifies the value of carbon emission quotas and environmental capacity as scarce public resources and production factors and internalizes the social cost of resource and environmental losses, thereby encouraging the conservation of fossil energy, the development of new and renewable energy, and the transformation of energy mix. The accounting, monitoring, reporting, and verification systems built for the unified national carbon market also constitute the main institutional guarantee for building a green, low carbon, and circular economic system in China. It's important to create holistic plans for the varied indicators and assessment of energy conservation and emissions reduction, for instance, to coordinate the implementation of energy use right of enterprises and emission allowance systems. Compared with the right to energy use, the emission allowance system stresses not only energy saving, but more importantly, the development of alternative energy sources by encouraging enterprises to tap into distributed renewables and focus on the curtailment of coal, oil and other fossil fuels, which is more aligned, in a comprehensive manner, with the vision of energy revolution featuring a clean, low-carbon, safe, and efficient system of energy supply and consumption.

References

1. Xinhua News Agency (2018) Opinions of the Central Committee of the Communist Party of China and the State Council on comprehensively strengthening ecological environmental protection and resolutely fighting the battle against pollution, 24 June 2018. http://www.gov.cn/zhengce/2018-06/24/content_5300953.htm
2. Ministry of Environmental Protection, Environmental Air Quality Standard (GB3095-2012). 2012
3. Institute of Climate Change and Sustainable Development, Tsinghua University (2019) Coordinated governance of environment and climate—successful cases in China and beyond. Tsinghua University, Beijing
4. Liu J, Han Y, Tang X et al (2016) Estimating adult mortality attributable to $PM_{2.5}$ exposure in China with assimilated $PM_{2.5}$ concentrations based on a ground monitoring network. Sci Total Environ 568:1253–1262

Chapter 6
Non-CO$_2$ GHG Emissions

6.1 Scope and Status of Non-CO$_2$ GHG Emissions

There is still a substantial gap between the Intended Nationally Determined Contributions (INDCs) submitted by countries to the UNFCCC secretariat and the lower levels of emissions needed to hold warming below 2 °C and 1.5 °C. If the temperature rise is to be limited to 2 °C, total emissions in 2030 must be reduced by another 15 billion tons of CO$_2$e on top of the existing INDCs; and to enable 1.5 °C, 32 billion tons of CO$_2$e must be cut. This means that from 2020 to 2030, an annual emission reduction of 2.7% is required to achieve the 2 °C goal, and emissions must be cut by 7.6% on average annually to cap warming below 1.5 °C [1]. CO$_2$ emissions which accounts for the largest share of greenhouse gases, are a major cause of global climate change. However, a great many studies also highlight the unneglectable impact of non-CO$_2$ GHG emissions. The Fifth Assessment Report of the United Nations Intergovernmental Panel on Climate Change shows that non-CO$_2$ emissions have consistently hovered around 25% of the global GHG emissions. A host of international agencies, including the U.S. Environmental Protection Agency (EPA), anticipate that non-CO$_2$ emissions will continue to rise in the future, and without effective control, could offset efforts in carbon reduction. Studies have demonstrated that the reduction of non-CO$_2$ GHG emissions, as a vital alternative emissions reduction plan, features the edge of being low-cost, highly flexible and responsive, and synergistic, hence its extensive application in developed countries. To narrow the emissions gap, the reduction of non-CO$_2$ GHG emissions has gained more prominence and interest in the global response to climate change.

To effectively mitigating GHG emissions in tackling global climate change, Articles 4.1 and 12 of the UNFCCC stipulate that national communications—the most critical component of which is national inventories of anthropogenic emissions and removals by sinks of all GHGs—are an obligation of Parties to the Climate Change Convention. The 16th and 17th sessions of the Conference of the Parties to the UNFCCC held in 2010 and 2011 adopted decisions 1/CP.16 [2] and 2/CP.17, which

© China Environment Publishing Group Co., Ltd. 2022
Institute of Climate Change and Sustainable Development of Tsinghua University et al.,
China's Long-Term Low-Carbon Development Strategies and Pathways,
https://doi.org/10.1007/978-981-16-2524-4_6

mandated that beginning in 2014, non-Annex I Parties shall submit a biennial update report covering national GHG inventories, mitigation actions, finance, technology and capacity-building needs and support received that is consistent with the Party's level of international support, and that the report should be subject to international consultation and analysis.

The non-CO$_2$ GHGs covered in this report are primarily methane, nitrous oxide, and F-gases. The main emission sources and sectors involved are presented in Table 6.1 (CO$_2$ emissions generated from industrial production processes are analyzed in Chapter 3).

This chapter summarizes non-CO$_2$ data from official emissions inventories such as the Initial National Communication on Climate Change of the People's Republic of China [3] (hereinafter referred to as the Initial Communication), the Second National Communication on Climate Change of the People's Republic of China [4] (hereinafter referred to as the Second Communication), the PRC's First Biennial Update Report on Climate Change [5] (hereinafter referred to as the Biennial Update Report), and the Second Biennial Update Report on Climate Change. The specific results are shown in Fig. 6.1.

Table 6.1 The scope of non-CO$_2$ GHG emissions defined by this report

		Methane (CH$_4$)	Nitrous oxide (N$_2$O)	Fluorinated greenhouse gases (F-gas)
Energy	Coal mining	●		
	Oil and gas fugitive emissions	●		
	Transportation	●	●	
	Biomass burning	●	●	
Industrial production	Adipic acid		●	
	Nitric acid		●	
	Electrolytic aluminum			●
	HFC			●
	SF$_6$			●
Agriculture	Rice planting	●		
	Agricultural land		●	
	Fermentation in animals' gastrointestinal tracts and manure	●	●	
Waste disposal	Urban waste	●		
	Domestic sewage	●	●	
	Industrial waste	●		

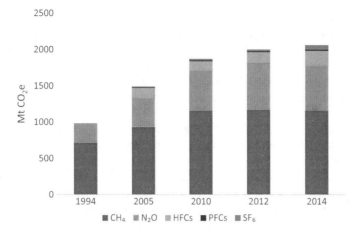

Fig. 6.1 Non-CO₂ GHG emissions in China

In 1994, China's total non-CO₂ GHG emissions stood at 980 million tCO₂e. That number rose to roughly 1.49 billion tCO₂e in 2005, and more than 2.06 billion tCO₂e in 2014. In 2014, non-CO₂ GHGs comprised 16% of China's total emissions, as shown in Fig. 6.2. Methane emissions, which stemmed mainly from energy and agriculture activities, made up 56% of total non-CO₂ emissions and were approximately three times the level in 1994.

Total GHG emissions from waste treatment amounted to 195 million tCO₂e, of which 100 million tCO₂e arose from the treatment of solid waste (landfill, incineration, and biological treatment), accounting for 52.4% of total emissions, and 91 million tCO₂e resulted from wastewater treatment, or 47.6% of the total. HFCs

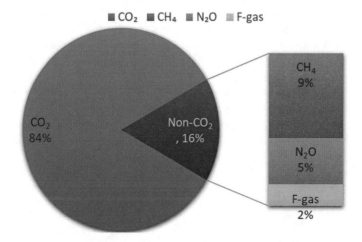

Fig. 6.2 Breakdown of GHG emissions by gas in 2014

were first introduced into the market by developed countries in the late 1980s and were mostly used as raw materials to produce refrigerants, foaming agents, fire extinguishing agents, aerosols, and chemical products in a wide array of industrial sectors. SF6 emissions principally involve four sectors: power transmission and distribution equipment (electric equipment for short), magnesium smelting, semiconductor production, and SF$_6$ production. Currently, China has, by and large, ceased the use of SF$_6$ [6, 7] in the production of semiconductors and magnesium, and electric equipment is now the primary source of SF$_6$ emissions. Studies have shown that the increase in global SF$_6$ emissions mainly stems from Non-Annex I countries, and in particular China [8]. The global warming potential (GWP) of perfluorocarbons (PFCs) is 6,500–9,200 times that of CO$_2$, and its main emission sources are the production of electrolytic aluminum, which contributes over 95% of the emissions, and semiconductors [9]. Therefore, the emission of non-CO$_2$ GHGs from industrial processes constitutes an essential part of China's overall strategy on climate change.

6.2   Scenarios and Results of Non-CO$_2$ GHG Emissions

Four scenarios are outlined in this chapter to examine non-CO$_2$ emissions and reduction potential. 2015 is chosen as the base year, and 2050 the target year. The policy scenario is used as the basis for comparison, and a reinforced policy scenario is built on of that to intensify the reduction in non-CO$_2$ GHGs under the current policy. The 2 °C scenario is also built to accommodate the trend of non-CO$_2$ GHGs under foreseeable emission reduction policies and technological potential. Last but not least, the 1.5 °C scenario is created to examine various emission reduction policies and the potential of applying all possible emission reduction technologies available.

The differences between the various scenarios lie, on the one hand, in the changes in activity levels, and on the other hand, in the degree of diffusion of emission reduction technologies. Changes in the activity levels of industries are, in large measure, aligned with changes in the level of economic development. Take methane emissions from coal mining as an example. With the continuous advancement of China's policy on curbing coal-fired power, coal production has seen a steady decline, which has in turn played a critical role in reining in methane emissions from mining. Emission reduction pathways for various industries mainly include the complete reduction of demand, replacement of demand, end-of-pipe recycling, disposal and decomposition measures. The latter scenario always represents more aggressive emission reduction targets and higher levels of technology adoption than the previous one. With the progression of scenarios, each is assigned a more intensified position in such aspects as the level of technology diffusion and demand reduction than the one before.

6.2.1 Non-CO$_2$ GHG Emissions Under the Policy Scenario

Under the policy scenario, non-CO$_2$ GHG emissions climb from 2.17 billion tCO$_2$e in 2015 to 3.17 billion tCO$_2$e in 2050, up 46.2% with an average annual growth of roughly 1.1% (Fig. 6.3 and Table 6.2). The growth of emissions primarily arises from HFC refrigerants and the agricultural sector during the 2015–2050 period. Non-CO$_2$ GHG emissions from the waste sector largely remain stable, and energy-related methane emissions no longer increase due to the peaking of coal consumption.

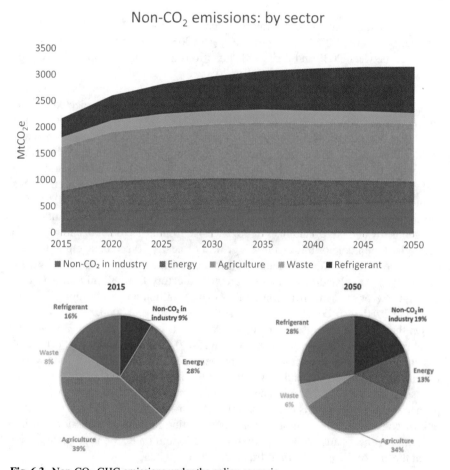

Fig. 6.3 Non-CO$_2$ GHG emissions under the policy scenario

Table 6.2 Non-CO$_2$ emissions under various scenarios (excluding LULUCF) (GtCO$_2$e)

	2020	2030	2040	2050
Policy scenario	2.44	2.97	3.13	3.17
Reinforced policy scenario	2.44	2.78	2.63	2.37
2 °C scenario	2.44	2.49	2.15	1.76
1.5 °C scenario	2.44	1.94	1.59	1.27

6.2.1.1 Methane

Under the policy scenario, CH$_4$ emissions amount to roughly 1.22 billion tCO$_2$e in 2015 and grow by 14.2% to 1.39 billion tCO$_2$e in 2030, up nearly 0.9% year-on-year. However, between 2030 and 2050, CH$_4$ emissions are slated to drop by an average of 0.7% annually, to 1.2 billion tCO$_2$e in 2050.

In terms of the distribution of emissions by sector, coal mining is the largest source of methane emissions. CH$_4$ emissions from coal mines are projected to slide slowly from 540 million tCO$_2$e in 2015 to 530 million tCO$_2$e in 2020 before falling steadily to 310 million tCO$_2$e in 2050. Between 2015 and 2030, the share of methane emitted during coal mining ranges between 38.8 and 44.2%, and that number drops incrementally to 25.5% by 2050. Enteric fermentation in animals and rice cultivation in the agricultural sector are also major sources of CH$_4$ emissions. Methane emissions from these sources are estimated to rise from nearly 470 million tCO$_2$e in 2015 to 580 million tCO$_2$e in 2030, and to 630 million tCO$_2$e in 2050. Their share in total CH$_4$ emissions, by and large, stabilizes at 38.5–42.4% from 2015 to 2030, and climbs to 52.8% by the middle of this century, a growth of nearly 10 percentage points from the base year. Landfilling waste also contributes to CH$_4$ emissions. Emissions from this category are slated to maintain a growth trajectory from 80 million tCO$_2$e in 2015—6.8% of total methane emissions—to 104 million and 80 million tCO$_2$e in 2030 and 2050, respectively, with the share rising to 7.5% in 2030.

On the makeup of emission increase, the growth in CH$_4$ emissions amounts to roughly 167 million tCO$_2$e between 2015 and 2030. The most significant drivers of growth are enteric fermentation, oil and gas leaks, and the landfilling of solid waste, which resulted in 120, 30, and 20 million tCO$_2$e, respectively. The increase in CH$_4$ emissions is reduced by 180 million tCO$_2$e between 2030 and 2050, during which the primary source of growth is intestinal fermentation—40 million tCO$_2$e. The decline in coal consumption leads to a continuous reduction in methane emissions during the mining process after 2030. Approximately 180 million tCO$_2$e is projected to be cut at the end of the model's time-frame compared to 2030.

6.2.1.2 Nitrous Oxide

Under the policy scenario, N$_2$O emissions stands at roughly 490 million tCO$_2$e in 2015 and are projected to register a 44% growth and increase at an annual average of 2.3% to reach 700 million tCO$_2$e in 2030. The growth of N$_2$O emissions slows dramatically between 2030 and 2050 when they maintain at a level of less than 700 million tCO$_2$e.

On the sources of emissions, the fertilization of agricultural soils using nitrogen fertilizers and manure is the largest source of nitrous oxide emissions. Under the policy scenario, N$_2$O emissions from nitrogen fertilizers are slated to grow steadily from 360 million tCO$_2$e in 2015 to peak at 450 million tCO$_2$e in 2020, and then witness gradual growth in the ensuing plateau period. This projection largely aligns with China's commitment to peak the use of nitrogen fertilizers in 2020 as proposed in its NDC. The proportion of N$_2$O emissions from using nitrogen fertilizers and manure is set to decline gradually from 74.0% in 2015 to 66.9% of total N$_2$O emissions. The sources of N$_2$O emissions are diverse. Apart from the primary sources—agricultural soils, the production of nitric acid and adipic acid in industrial processes, waste-water treatment, and animal manure management in the agricultural sector all contribute to N$_2$O emissions, with adipic acid representing the largest share. Under the policy scenario, N$_2$O emissions from adipic acid production peak at 210 million tCO$_2$e in 2025, with the share growing from 14.1% in 2015 to 27.7% in 2030 before trending slightly downward to 24.2%.

From 2015 to 2030, N$_2$O emissions grow by roughly 220 million tCO$_2$e, of which the biggest drivers are adipic acid, nitrogen fertilizers, manure management and the use of manure as a fertilizer, contributing 130 and 80 million tCO$_2$e, respectively. Between 2030 and 2050, the increase in N$_2$O emissions is significantly reduced to 20 million tCO$_2$e. The major sources of growth during this stage are manure management, energy activities, and the use of manure as a fertilizer. As emissions from other sources are slated to decrease, all growth stems from the above-mentioned sources.

6.2.1.3 F-Gases

Under the policy scenario, F-gas emissions soar by 93.8% from roughly 450 million tCO$_2$e in 2015 to 870 million tCO$_2$e in 2030, averaging 4.5% growth annually. Between 2030 and 2050, the growth sees a notable drop to around 1.9%. Emissions in 2050 are up 181% relative to the base year, reaching 1.27 billion tCO$_2$e.

On the distribution of emissions by sector, HCFC-22 production and air conditioning are the biggest contributors to F-gas emissions. Emissions from HCFC-22 production are projected to grow steadily from 110 million tCO$_2$e to the peak at 160 million tCO$_2$e in 2025 before entering a plateau. In the meantime, emissions from indoor, automobile and commercial air conditioning witness exponential rise at an average of 4.5% annually from 220 million tCO$_2$e in 2015 to 430 million tCO$_2$e in 2030, double the level relative to the base year. Between 2030 and 2050, the growth slows to around 2.3%. Total emissions hit 690 million tCO$_2$e, up 213% from the base year. Resulting from the substantial increase in emissions from air conditioning, the share of F-gas emissions from HCFC-22 production in the total F-gas emissions is projected to plunge from 38.8% in 2015 to 23.9% and 16.9% in 2030 and 2050, respectively. The share of F-gas emissions from air conditioning jumps from 50.9% in the base year to 66.4% in 2030 and 76.4% in 2050.

F-gas emissions are expected to climb by 420 million tCO$_2$e from 2015 to 2030, the biggest drivers of which are air conditioning and HCFC-22 production, which contribute 290 and 40 million tCO$_2$e, respectively. Together they account for 89.2% of the total increase, of which the contribution of air conditioning stands at 77.8%. From 2030 to 2050, the increase in F-gas emissions falls to 390 million tCO$_2$e. At this stage, emissions attributed to air conditioning, the chief source of increase rise by 260 million tCO$_2$e, comprising over 95% of the total increase. Air conditioning and indoor cooling are the largest growth drivers, accounting for approximately 77.5% of the total increase.

6.2.1.4   Carbon Sink

Forest carbon sinks and grassland carbon sinks/emissions are vital parts of the global carbon cycle in terrestrial ecosystems. The cycle mainly consists of photosynthesis, respiration, the burning and decaying of biomass, and the decomposition of soil and other organic matter. Forests store large amounts of carbon. Forest biomass is most closely linked to the stage of the life cycle, and the accumulation rate of carbon is the largest in middle-aged forests while mature and over-mature forests—whose biomass largely ceases to grow—are in equilibrium with a net carbon balance. This report presents three scenarios for evaluating changes in forest carbon sinks. The policy and reinforced scenarios are similar and thus merged into one. Under the three scenarios, China's forest stock keeps continuous growth through to 2050, and the changes in carbon uptake and storage by forests are precisely in sync with that of forest stock. However, carbon sink capacity reflects the rate of change in forest carbon storage during a given time interval (one year or multiple years) and the ability to cleanse carbon accumulations in different periods.

Under the policy/reinforced policy scenarios, the primary consideration is to appropriately implement measures such as creating new forest areas, increasing the share of young and middle-aged forests, and harvesting mature forests, so that by 2030, the forest stock jumps by more than 20% compared with 2015, while the carbon sequestration capacity increases by less than 1% to reach 412 million tCO$_2$, and Chinese forests' capacity to absorb and store carbon grows gradually to 424 million tCO$_2$ in 2050, almost double the amount in 2015. Under the 2 °C scenario, more sophisticated and intensified measures are adopted, including afforestation, and the judicious selection of plant species, growth rate and age for afforestation. The carbon sequestration capacity is projected to hit 654 million tCO$_2$ in 2030 before a slight decrease to 502 million tCO$_2$ by 2050. Under the 1.5 °C scenario, the capacity exceeds 900 million tCO$_2$ in 2030, and in 2050, the forest stock surges by 90% over the 2015 level, and the carbon sequestration capacity increases by 1.7 times to reach 576 million tCO$_2$.

Carbon emissions and sequestration in grasslands are part and parcel of the ecosystem carbon cycle. The majority of carbon sequestered in grasslands is stored in vegetation and soil organic matter. There are numerous types of grasslands in China with great discrepancies in carbon storage capacities. Climate and vegetation zonality exert a profound impact on the carbon storage capacity of grasslands. Changes in the carbon storage volume and area of grasslands are closely associated with grassland management policies and measures after 2020. Wetlands, forests, and grasslands are collectively known as the three major terrestrial ecosystems. Wetlands sequester carbon primarily through soils and plant biomass, with the former accounting for over 90% of total carbon storage in wetlands. An analysis of the scenario in this paper shows that the capacity of China's grasslands and wetlands to sequester carbon from the atmosphere remains at roughly 200 million tCO_2 in 2050.

It can be established from the scenario analysis in conjunction with related research that the capacity of China's forests, grasslands, and wetlands to store carbon stands at 700–800 million tCO_2 in 2050.

6.2.2 The Emission Reduction Potential of Non-CO₂ GHGs

Under the reinforced policy scenario, non-CO_2 GHG emissions are estimated to peak around 2030 before entering a plateau and experiencing a gradual decline. The peak is at roughly 2.78 billion tCO_2e with an average annual growth of around 1.7% during this period. From 2030 to 2050, emissions slowly drop to 2.37 billion tCO_2e, down 0.8% year-on-year. Emissions are reduced by roughly 25% compared with the policy scenario. Detailed data are shown in Table 6.3 and Fig. 6.4. The emissions of non-

Table 6.3 Non-CO₂ GHG emissions in 2030 and 2050 under the reinforced policy scenario (MtCO₂e)

		CO₂	CH₄	N₂O	F-gas	Subtotal
2030	Energy activities		583.9			584
	Industrial processes			149.7	205.4	355
	Agriculture		580.0	438.0		1018
	Waste management		212.1	38.1		250
	Refrigerants				573.4	573
	LULUCF carbon sinks	−612				−612
	Total emissions	−612	1376	626	779	2169
2050	Energy activities		400.4			400
	Industrial processes			52.0	284.4	336
	Agriculture		622.0	456.0		1078
	Waste management		168.3	33.7		202
	Refrigerants				350.56	351
	LULUCF carbon sinks	−624				−624
	Total emissions	−624	1191	542	635	1743

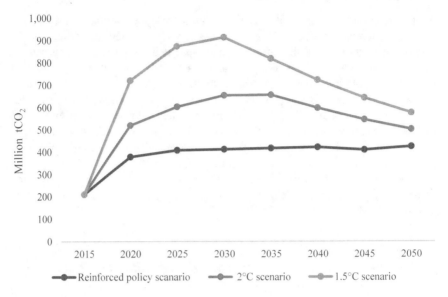

Fig. 6.4 Forest carbon sink scenarios

CO$_2$ GHGs are expected to peak simultaneously with CO$_2$ emissions in 2030 under this scenario. Due consideration should be given to establishing non-CO$_2$ peaking targets in China's future NDC ambitions, which will underline the importance of non-CO$_2$ GHGs to the country's overall climate strategy. Currently, nearly one-third of the emission reductions in developed countries is achieved through the abatement of non-CO$_2$ GHGs, which indeed warrants wider attention.

Under the 2 °C scenario, non-CO$_2$ GHG emissions are estimated to peak in 2025 at approximately 2.51 billion tCO$_2$e, with an average annual growth of 1.5% during this period, falling to 1.76 billion tCO$_2$e in 2050, down 1.4% annually. Compared with the reinforced policy scenario, non-CO$_2$ emissions are slashed under the 2 °C scenario to 1.76 billion tCO$_2$e by 2050, roughly 44% lower than the policy scenario. Under the 1.5 °C scenario, non-CO$_2$ emissions peak in 2020 at 2.38 billion tCO$_2$e before dropping to 1.2 billion tCO$_2$e in 2050, around 60% lower than the policy scenario and half of its peak. The details are presented in Table 6.4 and Fig. 6.5, which show that even when all emission reduction technologies are adopted, the most radical efforts toward cutting non-CO$_2$ GHG emissions would fail to deliver near-zero emission by 2050. Moreover, the costs of technologies for near-zero emissions are exceedingly high. It is estimated that there would be a gap of at least 1.2 billion tCO$_2$e. Therefore, early opportunities must be seized upon when it comes to the emission reduction of non-CO$_2$ GHGs, and the aggressive implementation of negative emission technologies holds the key in the latter phases.

In 2015, non-CO$_2$ GHG emissions amounted to roughly 2.17 billion tCO$_2$e, of which methane, nitrous oxide, and F-gases took up 56.3%, 22.8%, and 20.9%, respectively. Under the policy scenario, emissions of various GHGs increase in 2050, albeit

Table 6.4 Non-CO$_2$ GHG emissions in 2030 and 2050 under the 2 °C scenario (MtCO$_2$e)

		CO$_2$	CH$_4$	N$_2$O	F-gas	Subtotal
2030	Energy activities		476.9			477
	Industrial process			123.6	190.5	314
	Agriculture		515.0	415.0		930
	Waste management		196.6	37.6		234
	Refrigerants				532	532
	LULUCF carbon sinks	−854				−854
	Total emissions	−854	1188	576	722	1633
2050	Energy activities		201.1			201
	Industrial process			19.7	213.7	233
	Agriculture		466.0	406.0		872
	Waste management		134.4	28.9		163
	Refrigerants				291.32	291
	LULUCF carbon sinks	−700				−700
	Total emissions	−700	801	455	505	1061

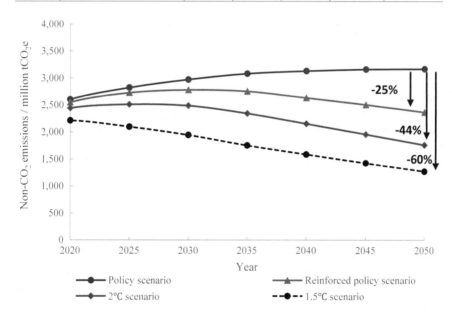

Fig. 6.5 Non-CO$_2$ emission reduction in different scenarios

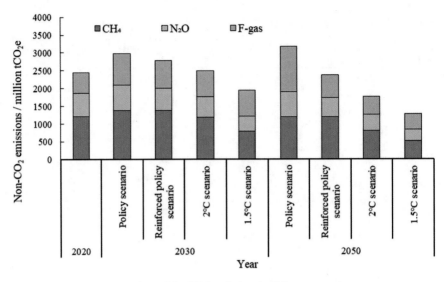

Fig. 6.6 Volume and share of non-CO$_2$ GHG emissions in different scenarios

at different rates. F-gas emissions spike to account for 40.1% of total emissions while the share of methane emissions is down to 37.9% and that of nitrous oxide emissions remains broadly unchanged. The details are presented in Fig. 6.6. Under the reinforced policy scenario and the 2 °C scenario, methane still represents the highest share in 2050 at about 50.3% and 45.5%, respectively. The sharp rise of F-gas emissions pushes up their shares under the two scenarios to 26.8% and 28.7%, both higher than the levels in 2015. Under the 1.5 °C scenario, the share of methane emissions is 5 percentage points higher than that of F-gases, standing at 40.5% of the total, which means methane remains the largest source of non-CO$_2$ GHG emissions. The proportion of nitrous oxide emissions sees moderate growth to 25%. However, the gap in the share of emissions among the three is be narrowed.

Based on comparisons of the various scenarios, it is anticipated that as socio-economic development and energy transition accelerate, CO$_2$ emissions will take a nosedive, and the share of non-CO$_2$ GHG emissions will grow year by year. In 2015, non-CO$_2$ gases comprised roughly 18% of all GHG emissions (national net GHG emissions being 11.2 billion tCO$_2$e). In 2050, the share of non-CO$_2$ GHG emissions in the reinforced policy scenario is projected to rise to 37%, and it further grows to 58% under the 2 °C and 95% under the 1.5 °C. Therefore, prompt efforts to curtail non-CO$_2$ emissions are vital to fast-track China's low-carbon transition and development.

6.2.2.1 Methane

CH$_4$ emissions are set to drop by a certain extent under the reinforced policy scenario compared with the policy scenario. CH$_4$ emissions peak in 2030 at 1.37 billion tCO$_2$e under the reinforced policy scenario; under the 2 °C scenario, its peaking occurs earlier in 2020 at a reduced level of 1.22 billion tCO$_2$e; and under the 1.5 °C scenario, the peak is moved up further to the time around 2015 at 1.22 billion tCO$_2$e; in 2030 under the reinforced, 2 °C and 1.5 °C scenarios, CH$_4$ emissions amount to 1.38 billion, 1.18 billion, and 790 million tCO$_2$e, respectively, down 0.7%, 14.2%, and 43.0% from the levels in policy scenario.

In terms of the distribution of emissions reduction by sector, under the 2 °C scenario, total CH$_4$ emission reduction is 200 million tCO$_2$e in 2030. The sectors contributing the most to the reduction are coal mining, solid waste treatment, rice cultivation, enteric fermentation, and animal manure management, with 140, 20, 20, and 20 million tCO$_2$e cut, respectively.

Under the 2 °C scenario, the reduction of CH$_4$ emissions grows to 400 million tCO$_2$e in 2050, which is primarily attributed to coal mining, animal intestinal fermentation, rice cultivation, and solid waste management, which witness a cut of 190, 170, and 10 million tCO$_2$e respectively. More ambitious efforts on non-CO$_2$ reduction are seen in the 1.5 °C scenario, with emission reduction reaching 590 million tCO$_2$e in 2030 and 680 million tCO$_2$e in 2050. A comparison of the scenarios reveals that part of the increase in emissions reduction in 2030 comes from the coal mining process, while the growth in 2050 is chiefly due to efforts coal mining, animal intestinal fermentation, and solid waste treatment.

6.2.2.2 Nitrous Oxide

N$_2$O emissions under reinforced policy, 2 °C and 1.5 °C scenarios witness a downward trajectory relative to the policy scenario, with the overall decrease exceeding that of CH$_4$ emissions. Under the reinforced policy scenario, N$_2$O emissions trend slowly downward after peaking in 2020, and maintain at 560–650 million tCO$_2$e from 2030 to 2050. At 2 °C scenario, the peak of N$_2$O emissions arrives in 2020 at 650 million tCO$_2$e. At 1.5 °C scenario, the peak remains in 2020 at a slightly reduced level of 580 million tCO$_2$e. Under these three scenarios, N$_2$O emissions in 2030 amount to 630, 570 and 420 million tCO$_2$e, which are 11.9%, 18.9% and 40.6% lower than the policy scenario.

In terms of the distribution of emission reduction by sector, under the 2 °C scenario, N$_2$O emissions are cut by 240 million tCO$_2$e in 2030, and the sectors with the highest contribution are nitrogen fertilizer application, animal manure management and manure as a fertilizer, which help reduce 77, 23 and 20 million tCO$_2$e respectively. In 2050, N$_2$O emissions are down by 150 million tCO$_2$e, mostly from energy activities such as nitrogen fertilizer application and biomass combustion and mobile sources, which contribute 80 and 64 million tCO$_2$e respectively. The 1.5 °C scenario sees stronger efforts in N$_2$O emission reduction, which rises to 280 and 380 million tCO$_2$e in 2030 and 2050 respectively. The incremental part was mainly achieved through the emission reduction measures during the production of adipic acid and nitrogen fertilizer application.

6.2.2.3   F-Gases

The emissions of F-gases start to decline slowly after 2035 under the reinforced policy scenario, while the growth of F-gases in the 2 °C scenario and the 1.5 °C scenario largely slows down after 2030 after peaking at 730 million tCO$_2$e, which is 16.3% lower relative to policy scenario. In 2050, F-gas emissions are 510 million and 440 million tCO$_2$e, down 60.3% and 65.5% compared to policy scenario.

The 2 °C scenario is characterized by a F-gas reduction of 140 million tCO$_2$e compared with the policy scenario, the biggest contributors being household air conditioners, HCFC-22 production and automobile air conditioners, with an emission reduction of 80, 30 and 30 million tCO$_2$e. By 2050, emission reduction rises to 760 million tCO$_2$e, mostly from refrigerant substitution in household and automobile air conditioners, which reduce 470 million and 290 million tCO$_2$e respectively. The 1.5 °C scenario features rising emission reduction to 150 and 830 million tCO$_2$e in 2030 and 2050. It is found that the increase of F-gas emission reduction in 2030 is largely contributed by the production process of HCFC-22 (120 million tCO$_2$e), while the contribution in 2050 mainly comes from the production process of HCFC-22 (20%), household air conditioners (19%) and automobile air conditioners (19%), etc.

6.2.2.4   Carbon Sink Analysis of Land Use and Land Use Changes

Land use is responsible for 25% of global greenhouse gas emissions, with agricultural production directly responsible for 10–14%; and the other 12–17% are caused by vegetation changes, including forest degradation [10]. The greenhouse gas inventory report of China's land use, land use changes and forestry in 2014 covers greenhouse gas emissions and carbon sequestration from six types of land use, including woodland, agricultural land, grassland, wetland, construction land and others. The enormous potential of land use in cutting greenhouse gas emissions has long been neglected. The sustainable management of land use and regulation of land carbon cycle not only facilitate carbon emission reduction per unit of land area, but also carbon storage of soil.

Stopping deforestation and massive tree planting represent one of the most effective approaches, as healthy forests support local economic development by ensuring water security, improving soil health, and helping regulate the climate. Under the reinforced policy scenario, China's forestry carbon sink in 2030 can reduce 412 million tons of CO$_2$, which not only mitigates climate risks, but also enhances adaptation to climate change. And the 2 °C scenario sees even higher contribution of forest—a reduction of 654 million tons of CO$_2$—a potential that should be reckoned with.

Compared with forest ecosystem, grassland carbon sink/carbon absorption is more susceptible to natural factors. Temperature, precipitation and evaporation are the main climatic factors affecting the carbon cycle of grassland ecosystem. Yet human measures for grassland preservation and restoration are also crucial for enhancing its potential of carbon sequestration. For the existing grasslands in China, the carbon sequestration potential of the three measures of enclosure, grass planting and conversion of cultivated land to grassland reaches $39.06 Tg \cdot a^{-1}$ [11], equivalent to the carbon sequestration capacity of 140 million tCO_2.

Forestry, grassland and wetland are major components of China's emission reduction efforts, and the international community is also turning greater attention to the role of forestry and grassland in global emission reduction. Forestry and grassland are unique in coping with climate change, as they can be either "carbon sources" or "carbon sinks", depending on the actions and measures of humanity.

6.3 Important Technologies and Measures to Reduce Non-CO$_2$ GHG Emissions

6.3.1 Key Technologies for Non-CO$_2$ Emission Reduction

6.3.1.1 Industrial Process

Greenhouse gas emissions from industrial production are mostly attributed to the process of physical and (or) chemical changes, which confines available emission reduction pathways to raw material substitution or the choice of solvent and reducing agent, with only a few technological options to choose from, hence the need to leverage other technologies from downstream production process to maximize energy conservation and emissions reduction in the industry.

N_2O emission reduction technology in industrial production process can be grouped into source control technology, process control technology and end treatment technology according to the stage of its action or workshop section.

- Cyclohexane production is the mainstream technology in China for producing adipic acid, whose capacity in this regard comprises over 95% of the total output of adipic acid in the country, and the N_2O source control mainly takes place through substituting other raw materials or methods for the cyclohexane adipic acid capacity. First applied by Japan-based Asahi Kasei Corporation, the cyclohexanol method enables N_2O emission reduction thanks to the dramatically reduced consumption of hydrogen and nitric acid compared to the cyclohexane method despite similar raw materials. Not only that, literature reveals other N_2O-free pathway for adipic acid production [12], which features 30% hydrogen peroxide under the effect of sodium tungstate and catalyst for a direct oxidation of cyclohexene into adipic acid, with a 90% yield. Yet this technology—yet to be industrialized—is not attractive unless hydrogen peroxide is cheap enough and the policy framework puts a tight leash on N$_2$O emissions. Devoid of nitric acid oxidation, butadiene [13] and ozone plus ultraviolet irradiation [14] are also free from N$_2$O emissions, and are currently under laboratory research.

- No process control technology is available in adipic acid industry to reduce N$_2$O emissions. But the end control technology of catalytic decomposition is available, whose mechanism is similar to the end control of nitric acid industry. Still, it's not a viable option given its apparent drawback in high catalyst cost and short lifespan. What is more commonly used in China is thermal decomposition [15] for terminal treatment, a method that directly sends N$_2$O tail gas into the incinerator and decomposes it into N$_2$, O$_2$ and NO at high temperature without catalyst, and recycles part of the nitric acid and waste heat. But its flaw lies in the consumption of fuel, whose combustion produces additional carbon dioxide emissions. Other literature outlines the direct purification of N$_2$O in chemical production process before being recycled to produce phenol and medical nitrous oxide anaesthetic, etc. These methods are closely associated with the demand-side pricing factor of downstream products and have not been commercialized yet.

As the substitute for the Ozone Depleting Substances (ODS) of CFCs and HCFCs, HFCs have witnessed notable growth in its global consumption and emissions along the journey to phase out ODS and are mostly used as refrigerants, foaming agents, fire extinguishing agents, aerosols and raw materials for chemical products, involving a range of industrial sectors.

1. In the refrigeration and air conditioning industry, fluorine-containing gases, as refrigerants, exchange energy with the outside through changes in their thermal state for the purpose of refrigeration.

   - For small car air conditioning, the refrigerant alternatives mainly include HFO-1234yf, CO$_2$, HFC-152a and Mexichem AC6. A slightly flammable refrigerant, HFO-1234yf has got the nod for United States SNAP, Europe REACH and China new material registration, and has been widely commercialized, with more than 60 million vehicles using the refrigerants globally in 2018, including some Chinese brands with car exports to Europe and the United States. CO$_2$, a natural working medium, has been used successfully as a refrigerant. For small and light systems such as automotive and mobile air conditioning, CO$_2$ systems represent a promising option. HFC-152A enjoys the potential of fuel-saving, low refrigerant price, higher system efficiency and available deceleration for refrigeration, but lacks a full-fledged SAE (Society of Automotive Engineers) standard.
   - The refrigerant alternatives for indoor air conditioners mainly include HFC-32 and R290, the former being a slightly flammable refrigerant whose cooling performance is comparable to HFC-410A with less refrigerant charge under the same refrigerating capacity. The technology has been adopted in the

markets of Japan, Europe and the United States. R290, with fairly close physical parameters to HCFC-22, is a very close direct substitute, plus its GWP is less than 20 and is 15% more efficient than conventional systems. However, its flammability raised concern from the United States, which outlawed hydrocarbons as an alternative to HFCs. Several Chinese companies have established production lines for R290 air conditioners. With enhanced safety, heightened safety awareness and the enactment and improvement of regulations, R290 is poised to gain more penetration in air conditioning.

- As products vary greatly in the refrigeration for industrial and commercial air conditioners, a wider portfolio of alternatives is available. Apart from the CO$_2$ and HCs mentioned above, natural working medium ammonia, with zero ODP and GWP, represents an environmentally friendly refrigerant with outstanding thermal and physical properties, large cooling capacity per unit volume, small viscosity, low price and high operating efficiency. Ammonia refrigerant is still being used in many large industrial systems, including mid-and-large freezers in China. Nevertheless, its toxicity and flammability have always been a cause for concern.

2. Cyclopentane demonstrates the most practical value out of HCs foaming agents. Compared with other HCs foaming agents, it features the lowest thermal conductivity, which explains its best thermal insulation performance. A relatively mature technology, hydrocarbon is already widely used in PU foam industry. Water is arguably a universal foaming agent for PU foam; it's environmentally friendly, safe and requires no changes of foaming equipment compared to other alternatives; besides, it's the most economically viable option considering the cost of equipment and plant renovation and changes in product production. HCFO-1233zd and HFO-1336mzz, with their non-flammability, low GWP value, high energy efficiency and superior insulation performance, represent extremely promising new generation foaming agents, which have obtained certification of US SNAP and the EU REACH. Honeywell and Arkema have registered the patent for HCFO-1233zd foaming agent, while Dupont owns the patent for HFO-1336mzz, but the massive worldwide penetration is yet to begin.

3. There are three main approaches for cutting HFC-23 emissions:

- Incineration of HFC-23. A mainstream solution to the treatment of HFC-23 in businesses at home and abroad, the method includes thermal oxygen decomposition and plasma digestion, etc. As China is involved in CDM projects implementation, HCFC-22 producers mostly opt for incineration and decomposition to dispose HFC-23, and the technologies are mature.
- Reducing HFC-23 by-product rate. By improving the production process of HFC-22, developed countries have managed to keep the by-product rate of HFC-23 at 1.5 ~ 3%, and below 2% on average. At present, due to process, equipment or management obstacles, the by-product rate of HFC-23 hovers at a high level of 3% in most domestic enterprises.

- HFC-23 recycling, that is, converting HFC-23 into fluoride compound with economic value through chemical reaction, and achieving the effective utilization of fluorine resources.

6.3.1.2 Agricultural Production

The sources of agricultural GHG emissions mainly include animal intestinal fermentation, animal manure management, rice cultivation and agricultural land, which account for 24%, 16%, 19% and 40% of GHG emissions respectively.

1. The emission reduction technologies in rice fields can be divided into five categories: variety selection, water management, fertilizer management, tillage management and new technologies [16–18], as shown in Table 6.5.

The study shows that if nitrogen fertilizer application in the rice field is reduced from 225–450 kg N ha^{-1} to 90–200 kg N ha^{-1}, N$_2$O emissions from rice field would be down by 42%. A 10–70% drop in nitrogen fertilizer would make for an 8–57% reduction in N$_2$O emissions without significant impact on CH$_4$ emissions from rice fields and SOC fixation [19]. Water-saving irrigation technologies, such as shallow wet irrigation, controlled irrigation, intermittent irrigation and film-covered dry farming, have been made widely available. Such water-saving irrigation technologies not only saves large quantities of water for rice, but also lowers greenhouse gas emissions. Other studies suggest that controlled irrigation can effectively reduce the global warming potential, and the total amount of CH$_4$ discharged from paddy fields under controlled irrigation would be down by more than 80% [20]. Under intermittent irrigation, CH$_4$ emissions of the paddy field was 5.4% of that under continuous flooding. Despite a 6.5-fold increase in N$_2$O emissions, the comprehensive greenhouse effect under intermittent irrigation is reduced by 90% [21].

Table 6.5 Methane emission reduction technology system in rice field

Category of emission reduction technologies	Specific technology
Variety selection	Low emission variety
Water management	Control irrigation Shallow wet irrigation Intermittent irrigation Combination of intermittent drainage and baking during growing period
Fertilizer management	Reducing nitrogen fertilizer application Biogas residue replacing farm organic manure
Tillage management	Paddy-upland rotation of rice oil and rice wheat
New technologies	Coated controlled release fertilizer Methane inhibitor Biochar

Table 6.6 Technology system for reducing nitrous oxide emissions from upland fields

Category of emission reduction technology	Specific technology
Variety selection	Low emission variety
Fertilizer management	Reducing chemical nitrogen fertilizer application
	Mixing organic fertilizer (farm organic fertilizer, biogas residue, straw, etc.) and chemical fertilizer
	Test soil formula fertilization
	Deep placement of nitrogen fertilizer
Tillage technology	Conservation tillage (less or no till, wheel tillage, straw mulching or returning to paddy field, etc.)
New preparations	Nitrification inhibitor
	SRFs and CRFs

2. The emission reduction technology system of nitrous oxide in upland can be grouped into four categories: variety selection, fertilizer management, tillage technology and new preparations (see Table 6.6).

Nitrogen fertilizer reduction is the most direct approach for reducing dryland. Studies show that if nitrogen fertilizer is reduced by 10–30%, N$_2$O emissions from wheat, corn and vegetable crops would be down by 11–22%, 17–30% and 27–45%, respectively [19]. Analysis indicates that for different types of farmland, including wheat, corn, and vegetable field, adding nitrification inhibitors serves to bring down N$_2$O emission coefficient by around 40%. SRFs and CRFs, which enable nitrogen to be slowly released for the sustained crop absorption and growth, is the best alternative to traditional organic fertilizer and the preferred choice for enhancing utilization of fertilizer and reducing its environmental impact.

3. The GHG gases in livestock and poultry breeding mainly stem from CH$_4$ in livestock and poultry intestines and CH$_4$ and N$_2$O produced in the process of excrement management. At present, GHG emission reduction technologies in livestock and poultry industry include four categories [16, 17, 22, 23], namely grain ration management, improved breeding, feces collection and storage, and feces treatment (as shown in Table 6.7).

Studies find that feed conversion can reduce CH$_4$ emissions by 4–6%, and improving the management of feed or energy intake can facilitate a 11% reduction in CH$_4$ emissions (equivalent to reducing 142.7 kg CO$_2$e of CH$_4$ emissions per head of beef cattle per year). Revising the feed mix of ruminants, especially adding lipid additives, can slash CH$_4$ by about 15% (equivalent to reducing 142.7 kg CO$_2$e of CH$_4$ emissions per head of beef cattle per year) [19].

4. The past 30 years have witnessed an overall increase in China's surface soil organic carbon (SOC) pool, which have served as a carbon sink (see Table 6.8). The annual carbon sequestration of soil at a depth of 20 cm in farmland ranged from 9.6 to 25.5 Tg, and the depth of 30 cm from 11 to 36.5 Tg. Carbon sequestration rate per unit of cultivated land is 74–184 kg C/ha per year at 20 cm depth and 85–281 kg C/ha per year at 30 cm depth [24].

Category of emission reduction technology	Specific technology
Breeding model	Combination of farming and breeding Low-carbon eco-breeding Fermentation bed breeding
Ration management	Straw silage and ammonification New feed additive Proper mix of concentrated/coarse feed TMR
Improved varieties	Cultivate high-yield varieties
Feces collection and storage	Dry feces collection Separation of feces and urine Feces drying treatment Centralized feces collection and storage Adjust fecal carbon—nitrogen ratio
Feces treatment	Integrated nutrient management Anaerobic digester Aerobic compost directly returned to the field Biogas project

Category of carbon sequestration technology	Specific technology
Tillage	No or less tillage
Soil management	Degraded soil remediation Land reclamation
Fertilizer management	Organic fertilizer use
Pasture plant management	Improved grassland variety/grassland component
Pasture animal management	Proper grazing density (grazing capacity)

Adjusting grazing density proves to be a major contributor to avoiding soil SOC loss and promoting grazing for soil carbon sequestration. High grazing intensity significantly reduces soil SOC content. Studies find that by lowering grazing density from high to medium or low level, the soil SOC content would increase by 0.77 tons of CO$_2$ ha^{-1} yr^{-1}. Rest-rotation grazing could enable a 1.48% improvement in the soil SOC, or 1.06 tons of CO$_2$ ha^{-1} yr^{-1}. Remediation of degraded soil could increase the soil carbon sequestration by 4.22 tons of CO$_2$ ha^{-1} yr^{-1}.

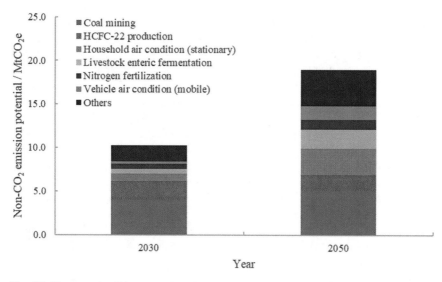

Fig. 6.7 Key areas for China's non-CO$_2$ GHG emission reduction (MtCO$_2$e)

Figure 6.7 illustrates the aggregate non-CO$_2$ gas emission reduction in selected sectors in 2030 and 2050 under the 1.5 °C scenario. It's found that the six sectors of coal mining, HCFC-22 production, household air conditioning, animal intestinal fermentation, nitrogen fertilizer application and automobile air conditioning contributed to the largest reduction of total non-CO$_2$ gases, standing at 410, 200, 90, 60, 60 and 30 million tCO$_2$e respectively in 2030, which adds up to 82.6% of total reduction. In 2050, the figure is 520, 170, 300, 220, 110 and 160 million tCO$_2$e respectively, with a minor drop of the cumulative contribution to 78.3%.

6.3.1.3 Summary

Given the characteristics of non-CO$_2$ emission reduction technologies occurring in different stages during implementation, this report groups them into six technology categories, namely complete reduction of demand, replacement of original demand, improvement of production mode and utilization efficiency, end-of-pipe recovery, disposal and decomposition.

In 2030, non-CO$_2$ gas emissions in the 2 °C scenario are 490 million tons less than the policy scenario, which is primarily attributed to complete reduction in demand, replacement of original demand, and end-of-pipe recycling, disposal and decomposition—steps that account for 37.6%, 17.6%, 21.5% and 10.9% of total emission reduction. The emission reduction of non-CO$_2$ gases in the 1.5 °C scenario grows by 540 million tons relative to the 2 °C scenario, which derives from increased end-of-pipe recycling, disposal and decomposition, strengthened efforts in demand reduction and substitution in the early stage—measures that contribute 28.5%, 23.0%, 27.6% and 13.3% of emission reduction respectively.

In 2050, the composition of technology category for non-CO$_2$ gas emission reduction in the 2 °C scenario is similar to that in 2030, with complete reduction of demand, replacement of original demand, end-of-pipe recycling, disposal and decomposition contributing 55.2%, 19.9%, 11.6%, and 6.5% to emission reduction. In contrast to 2030, the incremental non-CO$_2$ emission reduction under 1.5 °C scenario is primarily due to early demand reduction and substitution, with much less contribution from end-of-pipe disposal and decomposition. Additional emission reduction from the above three measures is estimated at 150, 210, and 160 million tCO$_2$e, or 47.1%, 26.6% and 14.8%, respectively.

Research and practice suggest that in the near and medium term, the spread and use of end-of-pipe recycling and disposal measures, such as the recovery of coalbed methane and the catalytic decomposition and treatment of industrial exhaust gases, should be prioritized for the reduction of non-CO$_2$ gas emissions. In the long run, more contribution is expected from the steady progress of upfront demand reduction or substitution, such as the encouragement of a healthier diet structure and substitution of new refrigerant with low GWP.

6.3.2 Key Measures for Non-CO$_2$ Emission Reduction

This report classifies non-CO$_2$ gas emission reduction technologies into three stages—early stage, intermediate stage and behind stage, based on the phases and modes of emission reduction behaviors (as shown in Table 6.9).

The early stage—prior to the release of non-CO$_2$ gases—includes two groups of emission reduction measures, namely complete reduction of demand and replacement of original demand. The so-called demand reduction represents the curb and reduction of activity levels of the emission sources, including transforming the energy system to cut coal consumption and subsequently dampen activities in coal mining; encouraging a shift to healthier diet to reduce the consumption of red meat, which indirectly leads to the decrease of the stock of livestock breeds; and diverting solid waste from landfill to incineration to cut back on the amount of landfill at the source, thus achieving a decline in emissions before landfill gas is generated. For the replacement of original demand or activity level, typical examples can be found in automobile air conditioning, where HFC-134A with a higher GWP value is replaced by HFO-1234yf with a lower GWP value as a refrigerant, and the use of R290 as a working medium replaces R410a in indoor air conditioning.

The intermediate stage refers to the process of non-CO$_2$ gas generation, and the emission reduction measures comprise improvement of production mode and utilization efficiency. Typical measures of the former include: promoting the combination of wet irrigation and intermittent irrigation in rice planting process, and of intermittent drainage and baked field during growing period; rationing the concentrated/coarse feed ratio for livestock raising; substituting automatic quenching anodic effect and non-effect aluminum electrolysis process for the original production line with obvious anodic effect in the aluminum smelting industry. The emission reduction

Table 6.9 Grouping of non-CO$_2$ emission reduction measures

Stage	Type	Emission source	Measure
Early stage	Complete reduction of demand	Coal mining	Energy system transformation
		Animal intestinal fermentation	Improvement and propagation of high-productivity livestock breeds and upgrading diet
		Animal excrement management	Improvement and propagation of high-productivity livestock breeds and upgrading diet
		Agricultural land	Lowering the use of chemical nitrogen fertilizer
		Solid waste	Incineration
	Replacement of original need	Power system	SF$_6$ + N$_2$ mixture substitution
		Semi-conductor production	COF$_2$ substitution
		Car air conditioning	HFO-1234yf substitution
		Indoor air conditioning	R290 substitution
Intermediate stage	Improvement of production mode	Rice planting	Promote the combination of wet and intermittent irrigation, and of intermittent drainage and baked field during growing period
		Animal intestinal fermentation	Proper rationing of concentrated/coarse feed ratio
		Aluminum smelting	Automatic quenching anodic effect, no effect aluminum electrolysis process
	Efficiency improvement	Agricultural land	Test soil formula fertilizer (formula fertilizer)
Behind stage	End-of-pipe recycling	Coal mining	Recovery for power generation
		Animal excrement management	Composting methane

(continued)

Table 6.9 (continued)

Stage	Type	Emission source	Measure
		Solid waste	Recycling and recovery for power generation
		Waste water treatment	Recovery for power generation
	End-of-pipe disposal and decomposition	Coal mining	Closed combustion, oxidation treatment
		Solid waste	Combustion emptying
		HCFC-22 production	Thermal decomposition
		Nitric acid production Adipic acid production	Secondary treatment, tertiary treatment
			Catalytic decomposition

measure through enhancing utilization efficiency mainly include the popularization of soil test formula fertilization in the application of nitrogen fertilizer to improve the efficiency of nitrogen fertilizer.

The behind stage, meaning post non-CO$_2$ release period, primarily features end-of-pipe recycling, disposal and decomposition for non-CO$_2$ emission reduction. To illustrate, CH$_4$ emissions from coal mining and solid waste landfill can be recycled to generate electricity; and emissions from HCFC-22, nitric acid and adipic acid production require heat treatment or catalytic processing to be eliminated.

6.3.3 Emission Reduction Cost of Non-CO$_2$ GHG Emissions

Research shows the marginal cost curve of non-CO$_2$ emission reduction features a flat pattern in the first half but a pretty steep trajectory in the latter half (as shown in Fig. 6.8), which indicates that despite the physical constraints of emission reduction technologies, non-CO$_2$ emissions can be further reduced through a proper level of carbon tax. However, the marginal cost of emission reduction sees a surge when certain emission reduction threshold is crossed. Modeling suggests the threshold ranges from 40 to 50% of emission reduction in 2030, and from 50 to 70% in 2050. Moreover, owning to technology limitation, zero emission of non-CO$_2$ gases is hard to come by—around 40% of such emissions remain stubborn under the 1.5 °C scenario.

With a carbon price of CNY 20/tCO$_2$e, N$_2$O, F-gas, CH$_4$ would be down by 31.2%, 12.9% and 14.5% respectively in 2030 and by 50%, 21.5% and 42.5% in 2050—equivalent to a reduction of 53 million tons and 130 million tCO$_2$e by 2030 and 2050. Under the reinforced policy scenario, the cost of emission reduction in 2030 stands at CNY 3.85 billion and CNY 16.05 billion in 2050; whereas the 2 °C

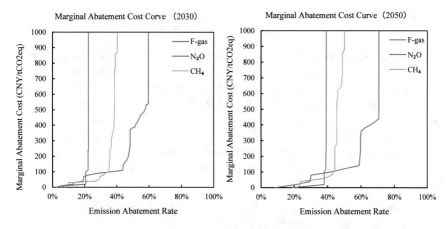

Fig. 6.8 Cost curve of Non-CO$_2$ emission reduction (2030 and 2050)

scenario sees the cost soaring to CNY 10.58 billion in 2030 and CNY 30.62 billion in 2050. Furthermore, the maximum economically achievable emission reduction is limited by the physical constraints of technologies. Once carbon is priced above CNY 200/tCO$_2$e (around $30/tCO$_2$e), the level of emission reduction would not increase in proportion to price rises. Assuming the cost of technology remains constant, the cost for non-CO$_2$ emission reduction in the 1.5 °C scenario would multiply that in the 2 °C scenario by a minimum of 3–5 times. Future R&D and technological advances may result in a reduction in technology outlay. Yet due to the scarcity of data on the cost reduction of non-CO$_2$ emission reduction technologies, future cost reduction prompted by technological progress is not taken into account in this study.

The conclusion is that, first, such non-CO$_2$ gases as methane is an aggressive greenhouse gas, whose greenhouse effect is equivalent to 80 times of CO$_2$ in 20 years. So it's essential to take non-CO$_2$ emission reduction seriously by creating a strategy that targets non-CO$_2$ gases under the umbrella of the national strategy for long-term low-carbon development. Second, the potential for cutting non-CO$_2$ gases can be fulfilled in the early period through low cost technologies and carbon pricing. But when certain threshold is reached, deeper emission reduction can only be enabled through behavior change to reduce the activities, which renders emission reduction more challenging. Third, the cost and difficulty of non-CO$_2$ emission reduction demonstrate non-linear rises after the threshold, prompting the use of negative emission technology for the remaining emissions in order to achieve net zero greenhouse gas emission by 2050.

6.3.4   Nature-Based Solutions

The notion of Nature-based Solutions (NbS) first appeared in the World Bank's report entitled *Biodiversity, Climate Change, and Adaptation: Nature-Based Solutions from the World Bank Portfolio World Bank (2008–09)*, which underscored the importance of biodiversity conservation for climate change mitigation and adaptation. "Nature-based Solutions" is a nature-inspired solution that supports and utilizes nature to address social challenges with effective and appropriate means, enhancing social resilience, and bringing economic, social, and environmental benefits.

"Nature-based solutions" fall into three categories based on different ways of using ecosystems:

- The direct use of ecosystems with no or minimum intervention, such as natural ventilation for air purification, rain flood management through urban green space;
- The repair of ecosystems such as soil management and restoration of soil ecosystem function;
- The imitation of nature to create ecosystems, such as new green roofs and green wall façade to adjust microclimate.

Nature-based climate solutions produce both economic benefits and efficiency perks, and showcase fully the importance of trees, soil and other natural elements in tackling climate change. The proposition of nature-based climate solutions provides a path toward the harmonious development of man and nature, and is expected to inspire more ambitious and effective actions and reasonable use of various resources by Contracting Parties on climate change issues.

Forestry carbon sink is a practical embodiment of "nature-based solutions" for managing climate change in the forestry and grassland sectors. It taps into the carbon storage service of forests through afforestation, strengthened forest management, reduced deforestation, preservation and restoration of forest vegetation, etc. Under the policy and reinforced policy scenarios, China's forest stock is on track to reach 20.8 billion cubic meters in 2030 and 26.2 billion cubic meters in 2050, and the forestry carbon sink is stabilized at a level of 400 million tons per year. The 2 °C scenario would push up the stock to 22.7 billion and 30 billion cubic meters in 2030 and 2050 respectively with a carbon sink of 650 million tons in 2030 and 500 million tons in 2050.

Considering the analysis and conclusions of related research, plus around 200 million tons of carbon sink in grassland, wetland and forest products, it is generally estimated the total carbon sink of China would reach at 700 to 800 million tons by 2050. As non-CO$_2$ gas emission sources are relatively dispersed, and the cost for reducing them would experience a non-linear surge after reaching a certain reduction rate, hence enormous challenge in driving emissions down on this front. This necessitates the efforts to tap into carbon sink potential, which is one of the important avenues to near-zero emissions. It is advised to coordinate agroforestry and rural development strategies to unleash the full potential of blending technology-based and nature-based solutions, and make better use of the sinks from agroforestry and land use change in order to offset technically difficult emission reductions in other sectors.

References

1. UNEP. The Emissions Gap Report 2019. United Nations Environment Programme (UNEP). https://www.unenvironment.org/resources/emissions-gap-report-201
2. UNFCCC. Decision 1/CP.16, the Cancun Agreements. Bonn, Germany; 2010. [2018-02-26]. http://unfccc.int/resource/docs/2010/cop16/eng/07a01.pdf
3. National Development and Reform Commission (2004) Climate change initial state information of the People's Republic of China. Beijing. http://unfccc.int/resource/docs/natc/chnnc1c.pdf
4. National Development and Reform Commission (2014) Second National Communication on Climate Change of the People's Republic of China. http://www.ccchina.org.cn/archiver/ccchin acn/UpFile/Files/Default/201302181420/20138656.pdf
5. Ministry of Ecology and Environment. The People's Republic of China's first biennial update on climate change. http://www.mee.gov.cn/ywgz/ydqhbh/wsqtkz/201904/P02019041952273 5276116.pdf
6. Xuekun Fang, Xia Hu, Greet Janssens-Maenhout et al (2013) Sulfur hexafluoride (SF$_6$) emission estimates for China: an inventory for 1990–2010 and a projection to 2020. Environ Sci Technol 47:3848–3855
7. Gao Feng, Cao Yancui, Liu Yu, Gong Xianzheng, Wang Zhihong (2016) Analysis of factors influencing greenhouse gas emissions from raw magnesium production in China. Environ Sci Technol 39(05):195–199
8. Emission Database for Global Atmospheric Research (EDGAR) (2011) Emission Database for Global Atmospheric Research (EDGAR), release version 4.2. European Commission, Joint Research Centre (JRC)/Netherlands Environmental Assessment Agency (PBL). http://edgar. jrc.ec.europa.eu (accessed April 9, 2018)
9. Xiang Zhen (2011) Mitigation of perfluorocarbons in response to global climate change. Environ Ecol Three Gorges 33(04):15–18
10. Paustian K, Lehmann J, Ogle S et al (2016) Climate-smart soils. Nat 532(7597):49
11. Guo Ran, Wang Xiaoke, Lu Fei et al (2008) Current status and potential of grassland and soil ecosystems in China. Ecologica Sinica 28(2)
12. Kang Li (1999) A "green" route for the production of adipic acid. Pet Nat Gas Chem Ind (02):118
13. Hao Jingquan, Hua Weiqi, Cha Zhiwei, Li Shikun (2012) Development of adipic acid production technology and market analysis. Mod Chem Ind 32(08):1–4
14. Shi Huaxin (2015) New technology for synthesis of adipic acid without N$_2$O. Energy Conserv Emiss Reduct Pet Petrochem 5(02):40
15. JJiang Yu, Xu Yekun and Ai Xiaoxin (2008) Review of N$_2$O emission reduction technology in adipic acid production. Chem Des Commun 44(09):56–57
16. Dong Hongmin, Li Yue, Tao Xiuping, Peng Xiaopei, Li Na, Zhu Zhiping (2008) China agricultural greenhouse gas emissions and emission reduction technologies. J Agric Eng 10:269–273
17. Mi Songhua, Huang Zuhui (2012) Screening of applicability of greenhouse gas emission reduction technologies and management measures for agricultural sources. Chin J Agric Sci 45(21):4517–4527
18. Wang Xiaomeng, Sun Yu, Wang Qi, Song Qiulai, Zeng Xiannan, Feng Yanjiang (2018) Research progress on greenhouse gas emission and emission reduction in paddy fields. Heilongjiang Agric Sci 07:149–154
19. Nayak D, Saetnan E., Cheng K et al (2015) Managing opportunities to mitigate greenhouse gas emissions from Chinese agriculture. Agr Ecosyst Environ 209:108–124

20. Peng SZ, Yang SH, Xu JZ et al (2011) Field experiments on greenhouse gas emissions and nitrogen and phosphorus losses from rice paddy with efficient irrigation and drainage management. Sci China Technol Sci 54(6):1581
21. Li Xianglan, Ma Jing, Xu Hua, Cao Jin-liu, Cai Zucong, Yagi Kazuyuki (2008) Effects of water management on seasonal changes of CH$_4$ and N$_2$O emissions in rice growing period. J Agric Environ Sci (02):535–541
22. Guo Jiao, Qi Desheng, Zhang Niya, Sun Lvhui, Hu Ronggui (2017) Current situation and peak forecast of greenhouse gas emission from livestock industry in China. J Agric Environ Sci 36(10):2106–2113
23. Li Shunjiang, Liu Jing, Du Lianfeng, Ma Maoting, Bi Xiaoqing (2018) Technology model for reducing pollution from large-scale livestock and poultry breeding. World Environ (03):56–58
24. Wang Chong, Yu Dongsheng, Zhang Haidong, Zhao Yongxiang, Shi Xuezheng, Wang Ning (2014) Study on the significant change characteristics of soil carbon pool and its influencing factors in farmland in typical black soil region. J Soil Sci 51(04):845–852

Chapter 7
Technical Support for Long-Term Deep Decarbonization

According to the Intergovernmental Panel on Climate Change (IPCC) Special Report on Global Warming of 1.5 °C released in October 2018, achieving the 1.5 °C target requires net-zero CO_2 emission by the middle of this century. And this necessitates the innovations and applications of a wide array of key revolutionary technologies.

Long-term deep decarbonization development of the world and China entails a variety of medium- and long-term emission reduction technologies. To enable the 1.5 °C target, the analysis on potential and cost of deep decarbonization technologies is essential.

Comprehensive cost–benefit analysis of deep decarbonization technologies underpins technology selection and emissions strategy. This chapter begins by analyzing the comprehensive impact of multiple energy efficiency and low-carbon technologies in terms of technology maturity, economic impact, social impact, environmental impact and ecological impact, followed by a roadmap for the development of major technologies.

In this chapter, over ten deep decarbonization technologies were diagnosed and evaluated, including research and development progress, cost effectiveness, development potential and policy requirements.

7.1 Comprehensive Cost–Benefit Analysis of Long-Term Deep Decarbonization Technologies

7.1.1 Research Background

China's decarbonization technological advances are playing a leading role. The emission reduction potential of traditional technologies is limited, and the contribution of rational demand and structural adjustment and optimization has been on the rise.

© China Environment Publishing Group Co., Ltd. 2022 177
Institute of Climate Change and Sustainable Development of Tsinghua University et al.,
China's Long-Term Low-Carbon Development Strategies and Pathways,
https://doi.org/10.1007/978-981-16-2524-4_7

China still has tremendous potential for energy conservation and carbon reduction by using negative and low-cost technologies, which may grow with technological progress and structural upgrade. Deep emissions reduction in end-use sectors calls for the development of low-carbon or zero-emission technologies, processes and products, collaborative innovation of disruptive energy technologies, new materials and information intelligence, and the deep integration of advanced technologies with the shift of green consumption concepts and behavior patterns.

A spate of domestic and international studies has identified carbon capture and storage (CCS), advanced nuclear energy, hydrogen energy, and geoengineering technologies, as well as technologies in the building, transportation, industrial, and energy storage sectors as crucial medium- and long-term emission reduction technologies (see Fig. 7.1). Most of these studies develop roadmaps for the development and deployment of emission reduction technologies based on industry demand and technical economics, without comprehensive cost–benefit analysis (i.e. impacts on environment and ecology). Thus, further research is needed.

As shown in Fig. 7.2, existing research mainly focuses on the research and development of new technologies and cost reduction of existing emission reduction technologies. Domestic research primarily concentrates on technical details and the improvement and dissemination of existing technologies, instead of focusing on the new technologies and other related as the international studies did. The current mainstream perception of medium and long-term emission reduction strategy is more of qualitative understanding, with more interest on improvement of available technologies and systems, cost reduction and proliferation, and less attention to potential new technologies and quantitative assessment of their emission reduction potential. Furthermore, most of the existing research is based on technical and economic costs estimation, lacking goal-driven mid- and long-term technological

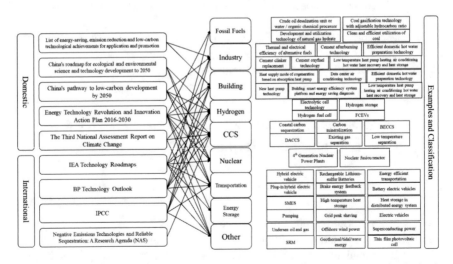

Fig. 7.1 Examples and classification of mid- and long-term emission reduction technologies in domestic and international research

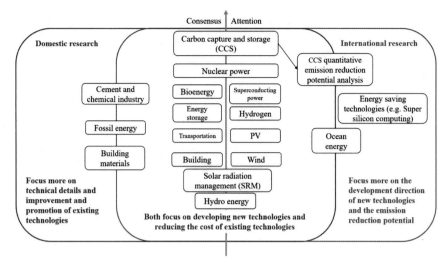

Fig. 7.2 Comparison and analyze of mid- and long-term emission reduction technologies in domestic and international research

strategies or roadmaps that accommodate ecological and social impact after the massive penetration of such technologies.

7.1.2 Dimensions of Analysis

As shown in Table 7.1, this study selects renewables (wind and solar), biomass, negative emission technology (CCS), hydrogen, nuclear as well as other technologies, such as demand-side management and energy efficiency improvement technologies, and synthesizes the conclusions on impact assessment in existing studies.

Among these technologies, opinions are divided on the technology maturity of demand side management. Some studies suggest that this kind of technology is fairly mature as it's widely used in daily life. But there is still much room for improvement from a climate perspective, for example, to what extent car-sharing will reduce the demand for cars. There is still big gap in the study of demand influenced by technology breakthroughs, while the impact of economic, social, environmental and ecological on the demand is relatively certain.

As shown in Fig. 7.3, this study rises above traditional technical and economic analysis, and sorts out the diverse categories of technology maturity, employment, environment, ecology, population health, public acceptance and other factors. Comprehensive cost–benefit analysis of wind, solar, biomass, carbon capture and storage, hydrogen, and other technologies is provided in detail, and the conclusions can be reference for holistic evaluation of medium and long-term emission reduction technologies.

Table 7.1 Conclusion of comprehensive impact of mid- and long-term emission reduction technologies made in existing studies

	Technological maturity	Economic impact	Social impact	Environmental impact	Ecological impact
Demand side management	Immature	Relatively certain	Relatively certain	Relatively certain	Relatively certain
Energy efficiency improvement	Mature	Relatively certain	Relatively certain	Relatively certain	Relatively certain
Wind and solar	Mature	Relatively certain	Relatively certain	Relatively certain	Uncertain
Biomass	Relatively mature	Uncertain	Uncertain	Relatively certain	Uncertain
Hydrogen	Immature	Relatively certain	Relatively certain	Relatively certain	Relatively certain
Nuclear	Relatively mature	Relatively certain	Uncertain	Relatively certain	Uncertain
CCS	Immature	Uncertain	Uncertain	Relatively certain	Uncertain

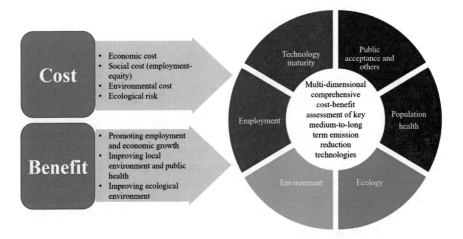

Fig. 7.3 Comprehensive analysis framework of key technologies for deep decarbonization

7.1.3   Horizontal Comparison of Key Technologies and Priorities

Table 7.2 illustrates the horizontal comparison of seven key technologies in terms of their comprehensive cost–benefit from six dimensions. A quasi-quantitative analysis is conducted on the whole despite possible controversies.

Table 7.2 Horizontal comparison of key technologies in terms of comprehensive cost–benefit

	Technological maturity	Employment impact	Local environmental impact	Ecological impact	Population health	Public acceptance
Demand side management	±	+	++	++	++	++
Energy efficiency improvement	++	+	++	++	++	++
Renewables (such as wind and solar)	++	++	++	−	++	++
Biomass	+	+	±	−	±·	++
Hydrogen	−	+	+	±	+	+
Nuclear	+	+	+	−	+	−
CCS	−	+	−	±	−	−
Legend	++: mature +: relatively mature −: not yet fully mature	++: great boost to employment +: boost to employment	++: great boost to local environmental improvement +: boost to local environmental improvement ±: uncertain impact to local environment −: potential damage to local environment	±: uncertain impact on ecology −: potential risk for ecology	++: great boost to population health +: boost to population health ±: uncertain impact on health −: potential adverse impact on population health	++: high public acceptance −: low public acceptance for now

Note + and − signify subjective qualitative judgment of technology impact, of which + denotes positive impact, and − refers to negative impact, and the number of + suggests level of impact

Opinions are divergent on the technology maturity of demand side management, however, there is rising consensus that demand side management produces positive impact on environment, ecology and population health. In general, demand side management and energy efficiency technologies feature universal benefits and represent the key priority areas for future low-carbon development. Yet their potentials and scale are subject to uncertainties in terms of future technological progress and breakthroughs as well as policy implementation.

Renewables such as wind and solar power generation have positive employment and health benefits, with relatively high technology maturity and a high degree of public acceptance. Thus, these renewables could be the priority for developing long-term emission reduction technologies. But their exponential growth should accommodate the eco-friendly spatial layout. For instance, installation of such facilities in high-risk areas could potentially pose local ecological damage, hence precautions are needed.

The amount of biomass is relatively small compared with other renewables, but the maturity of technology is higher. However its impact on the environment and people health cannot be generally described due to its wide variety and various ways of utilization. For example, biogas power generation helps reduce environment pollution, but indoor biomass combustion can be detrimental to human health and air quality. From an ecological viewpoint, the rapid development of biomass means more consumption of land and water resources, and more land was inappropriately explored with abundant artificial irrigation generates mounting risks for the ecological system. The utilization of biomass should be integrated with other sectors such as transport and power generation, and should be intensified to avoid negative impact on health.

Experts pointed out that China's hydrogen technology relies on policy support, and technological breakthroughs. While the key barriers for nuclear power are ecological risks and low public acceptance.

Zero emissions technologies, in particular CCS, are not mature. If future power system still highly dependent on coal-fired generation, the large-scale deployment of CCS and other zero and negative emission technologies would bring negative impacts on environment, not to mention ecological and commercial risks or public acceptance obstacles.

This study assesses the ecological and health impact of the five key technologies on emission reduction cost (see Fig. 7.4). It's observed that improvement in energy efficiency brings all aspects of benefits; the development of wind, solar and other

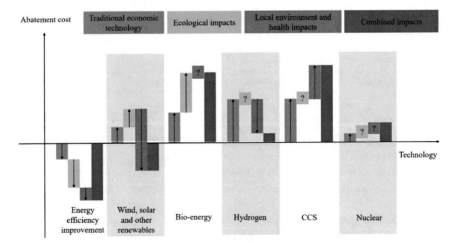

Fig. 7.4 Schematic diagram of abatement cost changes of China's key mid- and long-term emission reduction technologies considering ecological and health impacts

renewables, with the benefits for local environment and population health, features certain overall benefits; biomass, hydrogen and CCS involve high cost of emission reduction because of their potential ecological risks or environmental impact.

It should be noted that, with a comprehensive evaluation, the cost of emission reduction from of energy efficiency improvement, renewables such as wind and solar, biomass, hydrogen, and CCS are increasing in turn. It is expected that with the decarbonization of the power industry, the cost–benefit analysis and comparison of these technologies will be subject to more changes.

7.1.4 The Strategic Importance of Advanced Low-Carbon Technologies to Deep Decarbonization

As shown in Fig. 7.5, power supply system, energy system and strategic technologies in the technology innovation system are mutually reinforcing. Deep decarbonization entails the development of advanced low-carbon technologies as a strategic support. With close interplay between the technology innovation system and the energy system, the selection, management, and breakthrough of technologies all coincide with the development of energy system.

In general, the development of low-carbon technologies in the energy system will be complemented by technological innovation and revolution in the whole society. Specifically, the low-carbon technologies of the power supply system are interrelated with the energy Internet, energy big data, energy and artificial intelligence

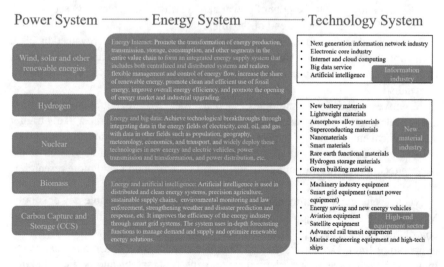

Fig. 7.5 Mutual support of strategic technologies in power system, energy system, and technological innovation system

technologies of the energy system, and are further correlated with the development of industries such as information technology, new materials and high-end equipment.

7.1.5 Conclusions and Suggestions

A holistic assessment of such dimensions of low-carbon technologies as employment, environment, ecology, health impacts and public acceptance, would improve the assessment on technology potential, cost effectiveness and spatial layout, which would also help to promote the synergies between carbon emission reduction and sustainable development.

Technologies of demand side management and energy efficiency have positive effects on sustainable development, and represent the key priority areas for the future of low-carbon development. Yet their potentials and scale are subject to uncertainties in terms of future technological progress and breakthroughs.

Renewables such as wind and solar power could be the priority of mid-to long-term emission reduction technologies. But their exponential growth should accommodate eco-friendly spatial distribution and forestall potential ecological risks.

Resting only on marginal land and rainwater irrigation, energy crop farming cannot meet the demand of deep emission reduction in China. Near-term development of energy crops primarily serves the transport sector. Yet given its potential risks of taking up land resources and straining water resources, it should be deemed as a transitional technology for the low or even zero carbon transportation before electrification takes hold in the sector. The long-term development of energy crops should concentrate on the supply of power and heating of the electricity sector, with the installation of CCS at the end of the pipe.

Hydrogen is projected to serve a high percentage of end-use energy need by 2050 with considerable potential for emission reduction. Hydrogen market in China has huge potential. Its development in the future hinges on the level of cost reduction as demand soars.

Negative emissions technologies, such as BECCS, are not mature for now. Large scale deployment of negative emission technologies would most likely trigger adverse local environmental impact, not to mention their ecological and commercial risks as well as public acceptance challenge. Therefore, it should be viewed as the backup technology for mid-and long-term emission reduction, and more research is needed.

Nuclear energy is instrumental in securing clean, safe and reliable power supply. Meanwhile, it can boost jobs along the industry chain, and take the role of a major contributor to deep decarbonization of the power system in line with the targets of temperature limit. Despite its great potential, nuclear power in China is hampered by such challenges as supply chain development, cost, safety, and political factors. Thus, stronger policy support is essential for its future development.

For future development, this chapter puts forward the following suggestions.

- Strengthen multi-dimensional and systematic research on the cost–benefit of emission reduction technologies, and identify the spatial variation pattern of diverse technologies under varied future scenarios and varied constraints.
- Speed up efforts in building a market-oriented system of green technology innovation, and harness the role of various market in promoting the development of emission reduction technologies.
- Make systematic and continuous efforts to establish an innovation system for demand side management and energy efficiency technologies; employ a mixture of levers such as policy design, R&D of science and technology, education and publicity to raise public awareness, boost penetration of new products and technologies, foster innovation of the consumption pattern and improve the demand structure.
- Ensure the strategic security of green technology industrial supply chain, and speed up the breakthrough of bottleneck technologies.
- Explore the pathway of strengthening intellectual property protection and enhancing the balance of interests of intellectual property to prompt technological development and diffusion.
- Create a diverse investment and financing guarantee mechanism for the R&D of strategic technologies and a risk sharing mechanism for backup technologies.
- Deeply engage in global cooperation and trade in green technologies via the Belt and Road Initiative and the South-South Cooperation Framework, and facilitate global cooperation on emission reduction technologies and build a shared future for the technological innovation community.

7.2 Evaluation of Deep Decarbonization Technologies

This section analyzes and evaluates 12 deep decarbonization technologies under four categories in terms of their R&D progress, cost effectiveness, development potential, and policy requirements, etc.

7.2.1 Deep Decarbonization Technologies for Energy and Power Storage Systems

With a high proportion of intermittent renewable power fed to the grid in the future, large-scale energy storage, smart grid and distributed renewable energy network technologies are essential for the safe and stable operation of the power system.

1. Electrochemical Energy Storage Technology

As of the end of 2018, the cumulative installed capacity of energy storage projects worldwide had reached 181.0 GW, of which, pumped storage was the largest, standing at 170.7 GW, or 94.3% of the total. Electrochemical energy storage takes second spot with 6,625.4 MW, of which lithium ion battery held the largest capacity at 5,714.5 MW.

Table 7.3 compares key operating and economic parameters of energy storage technologies. Recent years have seen extensive research on energy storage technologies, especially electrochemical energy storage, with commercial demonstration and applications in massive grid connection with renewables, demand-side storage, frequency regulation, etc. [1]. But large-scale deployment is yet to be commercialized. Despite continuous cost reduction of lithium battery in recent years, there is still a big gap compared to the cost of pumped storage [2, 3], and challenges in technology, cost, and business model need to be overcome in order to gain penetration. Superconducting magnetic energy storage, supercapacitor, flywheel, and other energy storage technologies feature low cost per unit output power, but their unit energy costs are high, hence mostly applied for high-load and short-time occasions [4, 5] without massive adoption. Continuous revenue is the driver for the development of energy storage technologies, most of which, due to the cost barrier, have not been extensively used. The good news is the cost of electrochemical energy storage is rapidly falling, as studies found, and is on track to reach a level comparable to pumped storage.

With the improvement of safety, lifespan, and energy conversion efficiency, the market of lithium ion battery for electric energy storage has seen steady growth. Nowadays, domestic lithium battery technologies primarily consist of three mainstream routes: lithium iron phosphate (LiFePO$_4$) system, NCM system, and lithium titanate system. Considering the return of investment, lithium iron phosphate battery is being commercialized at a faster pace. The safety and stability of NCM lithium battery are mediocre. Lithium titanate battery has relative low energy performance compared with the other two, but it has excellent low temperature performance and can be used over ten thousand times under the condition of high charging and discharging rate. Battery energy storage technologies has flourished in the world

Table 7.3 Comparison of energy storage technologies by key metrics and economics

	Pumped storage	Electrochemical energy storage	Superconducting magnetic energy storage	Flywheel	Supercapacitor
Efficiency (%)	75	85	90	90	90
Energy (W·h/kg)	–	10 ~ 200	–	5 ~ 100	5 ~ 30
Power (W/kg)	Low	Low	High	High	High
Cost ($/kg)	60 ~ 2,000	300 ~ 2,500	300	350	300
Lifespan	Very long	Mid	Long	Long	Long
Length of discharging	Several hours	Several hours	Several seconds	Several seconds	Several seconds

in recent years, and represent a major route for the development of energy storage technologies.

The barriers for lithium battery energy storage technology lie in recycling and safety. The internal temperature of NCM battery rises to 200 ~ 300 degrees when it is subjected to collision, acupuncture, overcharging, or short circuit, which triggers reactions of anode materials and the subsequent swelling and explosion, among other safety hazards. Plus, control measures are not readily available when accidents occur.

2. Hydrogen Energy Storage

Hydrogen production through renewable energy electrolysis attracts great interests as hydrogen can be used as a cross-seasonal and cross-regional means of energy storage.

Different from traditional battery energy storage, the technology of hydrogen energy storage utilizes water electrolysis to produce hydrogen and store energy in the form of gaseous fuel, which can be used in chemical industry, hydrogen fuel cell vehicles, and gas stations, etc. Such technology is not only good for local application, but also long-distance transport with the aid of natural gas pipelines. Hydrogen energy storage technology is a major branch of power-to-gas (PtG) technology—the process of converting electrical energy into gas, usually hydrogen, which is injected into a natural gas pipeline or converted to methane by methanation.

The core concept of power-to-gas emerged as early as the nineteenth century. But it's not until 2009 that the first power-to-gas equipment was launched. So far, the practical application of both technologies has been rather limited. Relatively speaking, Power-to-gas is used more often and is mainly concentrated in Germany and other European countries. In the aforementioned projects, the transformed gas is either fed into the natural gas pipeline or directly used for transportation and power generation.

The German government is pinning its hope on power-to-gas, which is deemed in some studies as the silver bullet for the energy transformation of the country. The northern part of Germany has the largest concentration of wind farms, whereas electricity is primarily consumed in the south. Thus, Germany chose to convert electricity into hydrogen for storage and replace direct power transmission.[1]

Relevant studies show that, since hydrogen storage technology is still at the stage of R&D and demonstration, which is hard to evaluate future cost precisely. Plus, it entails more energy conversion than electrochemical energy storage, which makes for greater energy loss and equipment investment, hence greater difficulty in cost reduction.

At present, the overall efficiency of operational power-to-gas demonstration projects is roughly 60%. The conversion efficiency of each project depends on the specific power-to-gas technology. As shown in Table 7.4, the investment cost and life span of varied power-to-gas technologies are different—the larger the power-to-gas capacity, the smaller the investment per unit of capacity. The AWE (Alkaline water electrolysers) represents a better choice for large-scale application than PEME (Proton Exchange Membrane electrolysers); whereas the technical constraint

---

[1] http://www.juda.cn/news/19716.html.

Table 7.4 Technical and economic parameters of varied power-to-gas technologies

	Type of technology	Conversion efficiency (%)	Life span (Year)	Fixed investment cost (USD/kW)	Energy consumption (kWh/Nm^3)^a
1	AWE (Alkaline water electrolysers)	70–80	20–30	800–1,500	5–7.5
2	PEME (Proton Exchange Membrane electrolysers)	75–90	3–5	1,500–3,000	5.8–6.3
3	CM (Chemical Methanol)	70–85	3–8	500–1,500	
4	BM (Bio-Methanol)	95–100		500–800	

Note 1 Nm^3/h = 0.0899 kg/h。
^{a}This set of data is related to the rated power. Varied power makes for data variance

of BM (Bio-Methanol) makes it suitable for small power-to-gas projects despite its higher efficiency and lower cost, and CM (Chemical Methanol) should be the preferred option for methanation for large power-to-gas projects. Annual operating/maintenance costs of power-to-gas stations are generally considered to be around 5% of fixed investment costs. Studies estimate that the average daily operating cost of AWE and CM is approximately $3500–5300, and that of PEME and BM combination is roughly $1400–2300.

China also attaches great importance to the research of power-to-gas technology. The R&D and demonstration project of 70 MPa hydrogen stations based on renewable energy/hydrogen storage was launched during the 12th Five-Year Plan period, with a focus on the application of power-to-gas in hydrogen stations for fuel cell vehicles. The research and demonstration project for direct hydrogen production through wind power and fuel cell power generation system was also deployed, targeting key technologies for wind power hydrogen and fuel cell integrated system. However, compared with the strategic planning and rapid development of hydrogen storage technology in advanced countries, there is still a big gap in key technologies and applications for China.

Currently, the cost of hydrogen storage in power-to-gas technology remains persistently high for two main reasons. Firstly, the electrolytic hydrogen production device is expensive. Economical operation can only be ensured with high utilization and a long running time throughout the year. Secondly, the energy conversion process incurs heavy loss of energy. Therefore, the key to the development of hydrogen energy storage technology is cost reduction and efficiency improvement. Only focusing on technical issues would not be enough to enhance the integrated energy application of hydrogen. Newer and more applications should also be developed to make new business models possible.

3. Nuclear Hydrogen Production Technology

Nuclear hydrogen production uses the heat generated by nuclear reactors as the main energy source, to make the large amount of hydrogen produced from hydrogen-containing substances (e.g. water or fossil fuels) in an efficient and carbon-free manner.

The pathway of developing nuclear hydrogen production technology should take into account the following factors: technical features (including capacity, purity of hydrogen, end-user, and waste management), cost (i.e. price of hydrogen, technical economics evaluation, and R&D cost, etc.) and risk (technology development, maturity, and R&D), etc. The process heat provided by the reactor must be utilized to achieve the efficient conversion of nuclear energy to hydrogen. The mainstream technologies include thermochemical cycle (sulfur-iodine cycle and mixed sulfur cycle) and high-temperature steam electrolysis. The former mainly involves chemical technology with a more complex process yet easier scale-up, so is suitable for large-scale hydrogen production. The latter is mainly determined by material technology, and its process is simple, which applies to small-scale hydrogen production.

Currently, the pressurized water reactor (PWR), which is widely used for power generation, is mostly utilized for power generation due to its relatively low outlet temperature. High temperature gas-cooled reactor is considered as one of the most suitable types for hydrogen production on account of its high outlet temperature, inherent safety and proper power.

The past decades have witnessed continued worldwide research efforts on hydrogen production via nuclear energy.

1. Japan: from the 1980s to now, Japan Atomic Power Agency (JAEA) has been conducting research on high temperature gas-cooled reactor and hydrogen production through iodide-sulfur recycling. The outlet temperature of its proprietory 30 MW high temperature gas-cooled test reactor (HTTR) was further raised to 950 °C in 2004. Combined with its commercial reactor (GTHTR300C), the co-generation of heat and power based on sulfur-iodine circulating hydrogen is being developed, and basic designs including cost estimates have been completed.

2. United States: the main research has been focusing on sulfur iodine cycle, mixed sulfur cycle and high temperature solid oxide electrolysis hydrogen production. In 2003, a preliminary evaluation and cost estimate of HTGR combined with sulfur-iodine circulating hydrogen production was conducted, which showed that the most important factors affecting the cost were nuclear thermal cost and electricity price of hydrogen production unit.

3. Canada: The supercritical water cooling reactor (SCWR) was developed, but the maximum output did not meet the demand for iodide-sulfur cycle. Then, a new cycle—copper-chlorine cycle—was proposed and developed with around 400–500 °C reaction temperature only, and supports with a net output of hydrogen and oxygen from water. The feasibility of the cycle has been verified by experiments, and a modification scheme for the model has been proposed to solve potential problems.

4. Argentina: At the end of last century, a recycling hydrogen production scheme from iron chlorine, cobalt chlorine, vanadium chlorine and vanadium chlorine was proposed. As estimated that the energy efficiency of the four methods are roughly 26%, 25%, 27% and 71%, respectively.

5. China: Institute of Nuclear and New Energy Technology of Tsinghua University developed and designed a sulfur-iodine cycle and high-temperature steam electrolysis system, and built an integrated laboratory-scale bench with hydrogen production capacity of 100 NL/h. At the end of 2013, the facility verification of the I-S closed cycle was achieved, and the production rate of hydrogen reached to 60 NL/h. The schedule is to complete the research on key equipment and technologies of high-temperature gas-cooled hydrogen reactor by 2020, validate the pilot project of high-temperature gas-cooled hydrogen reactor by 2025, and carry out the project demonstration of ultra-high-temperature reactor—nuclear energy hydrogen–hydrogen production and hydrometallurgy by 2030.

The International Atomic Energy Agency (IAEA) established the Coordinated Research Program (CPR) in 2012, in order to evaluate the technical economics of nuclear hydrogen production. The CPR intended to assess the technical and economic potential of nuclear hydrogen production by comparing alternative technologies, and through information sharing among member countries. Except for China, Members from 11 countries are currently involved in the project, such as US, Germany and Japan. Nuclear power routes identified for hydrogen production could be: supercritical water-cooled reactor (SCWR), pebble bed modular reactor (PBMR), high temperature gas-cooled reactor (HTGR) and prismatic gas-cooled reactor (PRISM). Possible technologies for hydrogen production could be: sulfur-iodine cycle, copper-chlorine cycle, mixed sulfur cycle, steam methane reforming and high temperature electrolysis.

Based on the data produced by 2014, IAEA estimated the cost of four available technologies using HEEP software: Canada CANDU pressurized-water reactor, China INET HTR-PM, Germany HTR-Modul and Japan GTHTR300C ultra-high temperature reactor, corresponding to CASE I, II, III and IV. CRP estimated the average cost of hydrogen under four scenarios (excluding costs of storage and transportation) at \$2–3/kg, with about 40% coming from hydrogen production and 60% from nuclear heat production. Considering the unique economic feature in different regions, the fluctuation would be around 20%.

As shown in relevant studies, the cost of nuclear hydrogen production would decrease in the future, depending on the large-scale industrialization of technologies. Around 2030, the cost may drop to a level of competitive with conventional processes.

4. Deep Decarbonization Technologies for Power Sector

Low-carbon power generation, smart grid, energy storage, demand-side response, and energy Internet techonologies will be the key to support net zero emission of the power system.

Smart grid and energy Internet are crucial deep decarbonization technologies to the power sector with five key technologies. As combined with renewable power generation technologies, a zero emission power system with 100% clean energy could be achieved.

Key technology 1: Virtual synchronization technology. Virtual synchronous technology mainly simulates the characteristics of the synchronous generator, such as the ontology model, active frequency modulation and reactive voltage regulation, so that the grid-connected inverter could be comparable with the traditional synchronous generator considering operation mechanism and external characteristics. Distributed power sources, such as energy storage, wind power, photovoltaic, electric vehicles, etc., are mostly connected to the distribution network in the form of an inverter interface, which provides a well application scenario for virtual synchronous generators. Virtual synchronization technology can increase the stability of new energy access and improve the potential of power grid to absorb clean energy.

According to studies, the cost of virtual synchronous machine renovation for wind farm is around 100 RMB/kW. There is no additional cost of standard virtual synchronous machine for new wind farms. The renovation of virtual synchronous machine of photovoltaic power plant adds the cost to about 600 RMB/kW. The standard virtual synchronizer for new photovoltaic power plant adds the cost to 200 RMB/kW. The cost of centralized virtual synchronizer is 220 RMB/kW, when a continuous decrease of energy storage prices followed, the construction cost would fall to around 100 RMB/kW by 2050.

Key technology 2: Solar thermal power generation technology. Solar thermal power generation collects solar energy by reflecting sunlight to solar collector, and provides high-pressure superheated steam to drive steam turbine for power generation through heat exchanger. The heat storage medium heated by solar energy can be stored in huge containers that can still turn a turbine to generate electricity hours after sunset. As the working mode of wind power and PV is greatly affected by the weather, the green power from wind and PV is random and intermittent. Therefore, the grid has an upper limit to accept the green power from wind and PV. Solar thermal power generation technology can provide controllable and low-randomness clean and green power, and also improve the potential of power grid to absorb clean energy.

According to professional forecasts, the investment of solar thermal power generation in the future will decrease by 10% year-on-year. At present, the unit kilowatt cost of solar thermal power generation in China is about 12,000 ~ 14,000 RMB/kW, which will drop to 9,000 ~ 11,000 RMB/kW by 2050. The normalized power generation cost of solar thermal power plant will fall to 0.67 ~ 0.76 RMB/kWh by 2020, and 0.50 ~ 0.56 RMB/kWh by 2050.

Key technology 3: Grid-side energy storage technology. It is difficult to harness the systematic and overall benefits of energy storage from power generation side and demand-side, as it has small energy storage capacity, scattered installation, and unlikely to follow unified grid regulation. On the contrary, as grid-side energy storage is usually large in scale and connected to the dispatching center at all levels. It contributes to the operation of the power system including load regulation, frequency regulation, blocking relief, voltage support, reactive power control, emergency standby in case of failure, etc. Grid-side energy storage also can ensure the spatial–temporal transfer of renewable energy, adjust frequency with thermal power units, and participate in various auxiliary power services, which could greatly improve the potential of the grid to integrate clean energy.

As shown in the research, the cost of existing energy storage technology is about 0.62 ~ 0.82 RMB/kWh. In the future, with further improvement of cell structure and process and material utilization, the system could be used by up to over 7,000 times, and the electricity cost of power plant would be reduced to about 0.3 RMB/kWh by 2050, while meeting the need of large-scale commercial application of energy storage.

Key technology 4: Demand-side response technology. Demand-side response refers to users actively changing their inherent power consumption patterns to respond to power supply by reducing or shifting the power load in a certain period, so as to ensure the stability of the power grid. Demand-side response can be divided into two categories: price demand response and incentive demand response. It can make up for the poor flexibility and slow response of conventional units on the power-supply side, greatly enrich the dispatching resources of the power grid, and enhance the flexibility of power grid dispatching. This technology can regulate the load to a certain extent, bring about flexible adjustment of energy use on the demand-side, so as to improve the integration of new energy on the power supply side equivalently, and also increase the potential of the grid to absorb clean energy.

Research shows the cost of demand response technology is extremely low, standing at 0.001–0.004 RMB/kWh now due to the low investment required in hardware and software. In the future, with the tariff reform of power transmission and distribution, mature market mechanism, and cost reduction of communication technology, the cost of demand-side response can go down by another 20% by 2050.

7.2.2 Deep Decarbonization Technologies for Industrial Processes

CO_2 emissions in steel, cement and other industrial processes are difficult to reduce deeply, which requires the promotion of decarbonization technologies.

1. Hydrogen Steel Making Technology

Generally speaking, long-process steel production mainly consists of three processes. First, iron ore is reduced to iron (commonly known as crude iron or pig iron). Next, crude iron is further decarburized to create crude steel, which is finally formed into different types of steel. Since the greenhouse gas emissions of the steel industry mainly come from the crude iron and crude steel production processes, direct reduction technology based on hydrogen and CCS are two important options to reduce CO_2 emissions. Compared to the end-of-pipe solution of CCS, hydrogen steelmaking replaces coke and other fossil fuels for reducing iron ore to crude iron, which helps eradicate CO_2 emissions and becomes one of the best decarbonization technologies known to the steel industry.

If hydrogen itself is produced in a zero-carbon manner, either through water electrolysis or by applying carbon capture and storage (CCS) technology to methane steam reforming or coal chemical production of hydrogen, direct hydrogen reduction to iron can help enable carbon-free steel production.

Currently, there are three projects in Europe that are focusing on hydrogen steel technology: HYBRIT, SALCOS and H_2Future/Susteel. The first two are mainly based on the existing reduction technology, whereas the last one uses the plasma melt reduction technology, but none of which need CCS or CCUS. SSAB in Sweden has launched a hydrogen direct reduced iron (DRI) pilot plant, which aims to achieve zero carbon steel production by the early 2040s. Salzgitter Steel in Germany is also conducting a pilot project, and Arcelor Mittal, the world's biggest steel producer, is also brewing this technology. Japan, South Korea, the United States and China are all have hydrogen steel projects under planning or demonstration. In China, Baowu Group has collaborated with China National Nuclear Corporation and Tsinghua University on hydrogen steelmaking in 2019, and is planning to work with Rio Tinto group on low-carbon metallurgy innovations.

Studies showed the cost estimate of hydrogen steel is fuzzy. Researchers from Tsinghua University estimated the cost of nuclear steelmaking by referring to parameters of the direct reduction project of Japan JAEA. Although hydrogen reduction steelmaking has not been commercialized, it can be calculated from the price of direct reduction steelmaking. Using the 10-year average price of steel from 2000 to 2010 as a reference and setting \$670/ton steel for coke blast furnace and \$675/ton steel for natural gas reduction, then the estimated cost of nuclear hydrogen is \$2.45/kgH$_2$, and that of hydrogen steelmaking through nuclear is \$628/tons of steel. Some studies suggested that hydrogen steel will be well positioned to compete with conventional processes.

By the progress of major hydrogen steelmaking projects around the world, this technology has great potential in the future, but its implementation cost is relatively high (see Table 7.5). Chances are predicted to be slim for its massive scale-up by 2030, and its potential is set to be released around 2050.

In 2018, the global steel output stood at 1.8 billion tons. If taking CO_2 intensity per unit of steel product as 1.8-tons/ton steel, the carbon emission of the steel industry was about 3.2 billion tons. If the current direct reduction technology can achieve a 50% reduction in emissions, 1.6 billion tons of CO_2 emissions can be saved in the steel industry. By adopting the technological combination of EAF and zero carbon electricity, an overall reduction of 80% could be achieved (2.5 billion tons of CO_2).

On hydrogen energy demand, in 2018, China produced around 900 million tons of crude steel (including roughly 800 million tons of long-process steel). If 50 kg of hydrogen is required for reducing each ton of iron, with hydrogen-based technology utilized in the long-process steelmaking, then approximately 40 million tons of hydrogen is needed.

2. Low-Carbon Chemical Technology based on Electricity and Hydrogen

CO_2 emissions in the petrochemical industry mainly come from atmospheric pressure reduction, catalytic reforming, catalytic cracking, hydrogen production, ethylene, ammonia synthesis, and other installations and power engineering [6].

Table 7.5 Major global hydrogen steelmaking projects

NO	Project	Technology	Emission reduction	Investment cost	Notes
1	HYBRIT (Sweden)	Hydrogen reduction steelmaking	HYBRIT is on track to cut 10% of CO_2 emission in Sweden[a] and lower CO_2 emissions by 7% in Finland[b]	The pilot phase investment will cost SKr 1.4 billion (almost half from Swedish Energy Agency)	The production cost of HRBRIT is about 20–30% higher than traditional steelmaking processes, and the gap is expected to narrow as the cost of renewable energy falls and the cost of carbon dioxide emissions rises. HYBRIT requires lots of cheap renewable electricity, and what's special about the HYBRIT process is that all hydrogen is produced by electrolyzing water. Although the process is energy-intensive, the carbon emissions from the whole process would be negligible if the electricity needed could be renewable.
2	SALCOS (Germany)	Hydrogen reduction steelmaking	Down by 50–85%[c]		
3	Voestalpine	The development of groundbreaking green hydrogen to replace coke smelting technology	Ultimately reduce 80% of CO_2 emissions by 2050	18 million Euro	The project lasts for 4.5 years.

(continued)

Table 7.5   (continued)

NO	Project	Technology	Emission reduction	Investment cost	Notes
4	Projects of big four steelmakers in Japan	Test blast furnace operation with hydrogen	Down nearly 10% compared to average blast furnace operation[d]		Assuming the same scale as the fine pulverized coal injection equipment for the existing technology, it is estimated that each blast furnace would require an investment of several billion yen[e]
5	South Korean project	Successful development of hydrogen reduction steelmaking technique	Down by over 15%[f]	R&D costs 915 million RMB; each blast furnace costs around 250 million RMB	
6	MIDREX H$_2$® (US)	hydrogen reduction steelmaking	Emission reduction by 80%		

[a]http://www.worldmetals.com.cn/viscms/bianjituijianxinwen1277/20180906/245527.html
[b]http://www.sohu.com/a/293905313_313737
[c]https://salcos.salzgitter-ag.com/en/index.html?no_cache=1
[d]http://www.worldmetals.com.cn/viscms/bianjituijianxinwen1277/20170112/240018.html
[e]http://www.sohu.com/a/231443680_313737
[f]https://mp.weixin.qq.com/s/SXiB_i95fhcH8Rl5zTEYXA

Carbon emissions from heat and electricity consumption account for more than 60% while emissions from fossil energy only made up less than 40% [7]. CO$_2$ emissions of this industry are different from any other manufacturing sector. The contribution of fossil fuels is not only about the energy released by combustion, but also from the coupling process of chemical reaction and energy conversion, in which a large amount of carbon is fed into the products through chemical reaction.

In the future, there would be three major ways to reduce carbon emission in the petrochemical industry. The first and foremost is the adjustment of energy supply and raw materials, including the source of heat and electricity, raw material, and corresponding process update, i.e. reducing the source of carbon emissions from the process and techniques. The second is the improvement of energy efficiency and fundamentally reduce energy consumption. Finally, some processes in petroleum and petrochemical production are characterized by high CO$_2$ emission concentration, such as coal hydrogen production, ethylene glycol production, acrylonitrile production and methanol production, etc. Then, CCUS technology can be used to collect and store or utilize the carbon dioxide emitted.

In a word, developing low carbon chemical technology based on power and hydrogen, implementing technical processes to produce methanol, olefins, hydrocarbons, synthetic ammonia, refined oil, and other petrochemical products based on electricity and hydrogen, and reducing or replacing coal, petroleum, or other fossil energy as chemical raw materials, are of crucial importance to achieve deep decarbonization.

2.1. Technical Route of Producing Methanol by Electric Hydrogen

This technology is to use hydrogen and pure CO_2 to produce methanol. Table 7.6 shows the comparison of such technical process with the traditional one in terms of the technical parameters of energy consumption and emission. As is shown below, methanol produced by electric hydrogen has significant advantages in carbon emission reduction.

Compared with the traditional petrochemical route, the economic cost of this technology is relatively high. The main challenge lies with the fixed cost of electrolytic cells, which accounts for about 75% of the total fixed cost. Depending on different electrolysis technologies, the current methanol synthesis cost is approximately 6,000–13,000 RMB/ton, far higher than the current traditional technological pathways. The cost is expected to drop to 4,400 RMB per ton by 2050.

2.2. Olefin MTO/MTP Technology Based on Methanol Production from Hydrogen and Pure CO_2

Methanol to Olefins (MTO) and Methanol to Propylene (MTP) are two important new chemical processes. At present, MTO and MTP technologies in China are quite mature. In the current MTO/MTO processes, the yield of low carbon alkenes such as ethylene and propylene can reach more than 80%.

The attainment of the zero carbon target for this technological route primarily hinges on methanol synthesis. If the previous methanol synthesis route is considered, the olefin technological route can also achieve negative emission.

However, similar to the technological route of producing methanol based on hydrogen and pure CO_2, its economic cost remains high, and the key is to reduce the cost of electrolytic tank. Considering the price fluctuation of coal and oil, the current cost of coal to methanol olefin is about 4,000–5,000 RMB/ton, while steam cracking of traditional mixed oil costs approximately 5,000–8,000 RMB/ton. The olefin cost of methanol production based on hydrogen and pure CO_2 will be more

Table 7.6 Comparison of methanol production through electric hydrogen with traditional petrochemical technological pathway

Methanol per ton	Petrochemical route	Electric hydrogen route
Total energy consumption (GJ)	37.5	39.7
Carbon emissions induced by raw materials (t)	0.97	−0.79
Carbon emissions in process (t)	0.52	0.123
Total carbon emissions (t)	1.49	−0.67

than 10,000 RMB/ton. At the current methanol market price of 1600 RMB, the olefin cost of this technological route will still rise by more than 2000 RMB/ton even if future methanol cost based on hydrogen and pure CO_2 is reduced to 4400 RMB/ton.

2.3. Technical Route of Synthetic Ammonia by Electricity and Hydrogen

Synthetic ammonia is the product of nitrogen and hydrogen under the joint action of high temperature, high pressure, and catalyst. At the moment, the existing technology relies on coal or natural gas to produce the hydrogen needed for ammonia synthesis. The most important source of carbon emissions from synthetic ammonia is hydrogen production processes. This ammonia synthesis technology is based on electric hydrogen, instead of coal and natural gas as the raw material to produce hydrogen.

Table 7.7 shows the comparison of such technological route and the traditional petrochemical route by technical parameters of energy consumption and emissions. It can be seen that the synthetic ammonia of hydrogen produced by electric power has significant advantages in reducing carbon emissions.

The economic cost of this technology is relatively high, which is roughly twice or more than traditional technologies. Some studies suggest that the cost per ton of synthetic ammonia in Chile and Argentina is around $460–700, which is slightly higher than the $300–600 per ton of synthetic ammonia produced from the existing steam methane reforming technology imported to Chile. The cost of synthetic ammonia based on electric hydrogen in northern Europe is likely to be 431–528 euros per ton in 2050, which is close to the market price.

2.4 Oil Products Produced by Hydrogen and Pure CO_2 via Fischer-tropsch (FT) Synthesis

In both crude oil refining and coal-to-oil technologies, hydrogenation process is neccessary, because the hydrogen content in crude oil and coal is insufficient to meet the demand of subsequent light oil products. As one ton of hydrogen is produced by water gas conversion, one ton of carbon dioxide would be emitted.

For every one ton of naphtha and gasoline produced by a traditional crude oil refinery, about 0.4–0.5 ton of CO_2 is emitted, and around 0.17–0.2 ton of CO_2 is discharged for every one ton of diesel [8]. For coal-to-oil technology, its carbon

Table 7.7 Comparison of technology of synthetic ammonia based on electric hydrogen and traditional petrochemical technology

Methanol per ton	Traditional petrochemical technology	Electric hydrogen technology
Total energy consumption (GJ)	35.04	45.1
Carbon emissions induced by raw materials (ton)	1.33	–
Carbon emissions in processes (ton)	0.5	0.12
Total carbon emissions (ton)	1.83	0.12

emission per unit oil product is very high without CCS technology, at roughly 5.5–6.9 tons of CO_2 [9].

The technology of producing oil products from hydrogen and pure CO_2 via FT synthesis relys on electric hydrogen, then use hydrogen and CO_2 to prepare syngas through the reverse water gas conversion reaction in order to synthesize oil products. Since hydrogen is produced without water and gas conversion, carbon emissions can be greatly reduced. Given that hydrogen production accounts for more than 25% of the carbon emissions from traditional technologies, carbon emissions per unit product of the technology can be reduced by more than 25%.

Compared with the traditional petrochemical technology, the economic cost of this technology is relatively high. The main challenge lies in the cost of electrolysis. Taking diesel as an example, the cost estimates of oil product synthesis based on different electrolysis technologies vary greatly. Some studies show that current cost is 4100–7000 RMB/ton, which is close to the traditional technologies, but the main-stream studies argue that the current cost is between 11,000–22,500 RMB/ton, which is much higher than the level of about 4,000 RMB per ton of traditional technology. The cost is expected to drop by 2050, but the cost per ton would still exceed 10,000 RMB.

3. Low-Carbon Cement Technology of Raw Material Substitution

Carbon dioxide is released in the process of calcining raw material into clinker to produce cement. The material ground to a certain degree of fineness after mixing in proportion by calcareous, clayey and a small amount of adjustment raw materials (sometimes mineralization agent and crystal seed are added; coal is also added to shaft kiln production) is known as cement raw meal. The substitution of raw materials has become a vital technology for low-carbon cement production.

As long as each chemical component in raw meal is proportioned properly, a variety of cement clinker conforming to standards can be produced. The traditional cement raw meal mix is shown in Table 7.8.

Producing clinker from alternative raw materials helps save raw material costs, reduce emissions, alleviate environmental pollution from all varieties of industrial by-products and save a bundle on the cost. The avaliable alternative raw materials are shown in Table 7.9, including calcium carbide slag, iron tailings, steel slag, silicon slag, silica sludge, and paper sludge.

Table 7.8 Content of main chemical substances in each component of traditional cement raw meal and burning loss rate

	Content of chemical constituents (%)					Burning loss rate (%)
	SiO_2	Al_2O_3	Fe_2O_3	CaO	MgO	
Limestone	5.58	0.42	0.31	50.58	1.51	39.75
Slag	34.72	10.86	1.85	38.4	9.67	1.2
Sandstone	66.09	8.9	4.06	6.66	3.5	8.37
Sulfate slag	10.51	2.84	65.21	8.69	3.64	5.52
Bauxite	39.18	32.00	12.38	0.92	0.14	14.35

Table 7.9 Content of main chemical substances of available alternative materials and burning loss rate

	Content of chemical constituents (%)					Burning loss rate (%)
	SiO_2	Al_2O_3	Fe_2O_3	CaO	MgO	
Carbide slag	2.29	2.45	0.36	68.35	0.78	23.48
Iron tailings	37.67	4.11	30.67	7.72	2.99	9.07
Steel slag	12.00	2.83	23.57	42.16	10.48	–
Silicon-calcium slag	22.07	2.41	5.00	57.75	0.81	10.84
Quartz sludge	85.96	3.73	1.57	0.51	0.26	4.51
Paper mill sludge	28	20	2.7	40.8	5.4	49.64

The chemical composition and content of various alternative materials are suitable for substitution, and some of them are even better than the original ones in terms of adjustment material content. When using alternative materials, it is necessary to pay attentions to the proportion of various raw materials and make appropriate adjustments to the subsequent processes.

The following obstacles, nonetheless, impede the extensive and continuous use of alternative raw materials in the cement industry.

Raw material supply. The global reserve of alternative materials is a technical challenge that hampers the wide use of alternative materials in cement plants. For example, the global output of granular blast furnace slag and fly ash in 2008 and 2007 was approximately one billion tons per year, which was insufficient for the high utilization rate of the global and cement industries.

Properties of cement products. Compared with the ash produced by fossil fuels, some alternative materials have different composition and content. The clinker composition produced by these materials in the kiln features has tremendous fluctuations. If the phosphorus in clinker exceeds the limit, the early strength of cement produced will decrease and the setting time will also be longer. In addition, the use of alternative materials can affect the long-term strength of the cement.

Economic challenge. Adjust the cost of purchasing, handling, and transporting raw materials to cement plants may result in additional costs.

In short, tapping into low-carbon cement technology based on alternative raw material, developing and applying alternative materials, and optimizing the mix of the raw materials, are all serve to reduce carbon dioxide emissions. But the cost of these technologies constitutes a major barrier, which calls for more technological breakthrough.

7.2.3 Deep Decarbonization Technologies for Transport Sector

The use of electricity and hydrogen instead of fossil fuels in the transportation sector represents a crucial deep decarbonization technology. It is essential to accelerate the development and proliferation of technologies for electric vehicles and hydrogen fuel cell vehicles.

1. Electric Vehicle Technology

Electric vehicles mainly consist of battery electric vehicles (BEV) and plug-in hybrid electric vehicles (PHEV). The former is driven by electric motors with electricity as the power source, while the latter is driven by internal combustion engine and motor separately or simultaneously, and the power is generated from gasoline, diesel or electricity. Many countries are actively taking measures to promote the strategic transformation of traditional gasoline vehicles and vigorously develop electric vehicles.

Despite the rapid progress in technology and performance, electric vehicles are yet to be competitive in terms of comprehensive cost. Breakthroughs have been made in key components such as electric motors, electronic controls, and battery management systems, with continuous improvement in system integration, vehicle performance and comfort as well as power consumption. At present, the volumetric energy density of lithium-ion batteries for cars is 200 ~ 300 Wh/L, and the battery cycle can reach 1000 times. The light vehicle battery market is dominated by NCM lithium batteries with high energy density, whereas lithium iron phosphate with higher cycle life and safety performance is the mainstream for medium and heavy vehicles. On the whole, the total cost of electric vehicles is currently higher than that of traditional internal combustion engine vehicles, which is largely driven by the high battery cost. With the breakthrough of key battery material technology, the cost of battery is on the decline.

The technologies and comprehensive performance of electric vehicles will witness tremendous improvement in the future. The comprehensive efficiency of powertrain and energy efficiency of vehicle will see significant improvement, with a dramatic reduction in vehicle power consumption, and the driving range will be equivalent to that of gasoline vehicles. Car bodies and parts and components will be much lighter through steady expansion of aluminum-magnesium alloy, high-strength steel, and carbon fiber materials, etc. The energy density and safety of lithium-ion batteries will be greatly enhanced, and innovative batteries will be gradually applied and commercialized on a large scale. In addition, with breakthroughs of Internet, big data and artificial intelligence and their rapid penetration into the automotive industry, new technologies and models, such as telematics, vehicle-to-grid (V2G), autonomous driving, vehicle-road collaboration, and intelligent manufacturing, will get flourished.

In the future, the comprehensive cost of electric vehicles will keep decreasing, and narrowing the gap with gasoline vehicles. With more stringent global regulations for car emissions, traditional internal combustion engine (ICE) technology will be further complicated, prompting rising cost for gasoline vehicles. In comparison, with the improvement in energy density of electric vehicle batteries, performance enhancement, and lower cost of comprehensive research, development, and manufacturing, the cost of electric cars will keep falling. Therefore, the comprehensive cost of electric vehicles will be superior over traditional internal combustion engine vehicles by 2025, and the cost will be competitive over gasoline vehicles and PHEVs by 2030, and the best comprehensive value/performance will be achieved by 2050.

The charging infrastructure for electric vehicles (EV) is undergoing rapid development. With surging sales and ownership of electric vehicles, charging facilities also show a rising trajectory. Countries vary in their priorities in supporting charging infrastructure. For instance, the United States provides incentives primarily in the form of tax relief and direct investment; Japan concentrates on supporting technology R&D and innovation; while China employs a mixture of policy instruments, covering planning, finance, infrastructure, technology, electricity, etc. Countries also vary in their charging technologies and standards, with diverse models of commercial operation for charging facilities. The comprehensive scale-up of high-power charging technology is now underway.

With increasing ownership of EVs, the relationship between vehicle charging/discharging and power grid has become another point of interest. EV energy storage can meet the needs of load regulation, frequency modulation, and grid connection of renewables by participating in auxiliary services and demand-side response. Especially with the increasing integration of renewable power in the future electric power system, EV energy storage will become a major flexible resource for the power system.

In short, the rapid development of EV technology and industrialization, declining cost, booming development of charging infrastructure, and the increasingly competitive overall cost over traditional ICE vehicles, will underpin deep decarbonization of the transportation sector.

The following obstacles hinder the development of EVs which needs particular attention:

1. The EV market is still primarily driven by policy factors, and market forces should promptly come into play to drive the market growth. There is also a dilemma as to which one should take the precedence over the other: infrastructure or vehicles. And the foundation of the EV industry remains fragile.
2. With the growing EV penetration, the sustainable supply of lithium, nickel, cobalt and other key resources needed for batteries is facing severe challenges, and the environmental pollution from the recycling and disposal of batteries is another issue for concern. In China, for example, the lithium reserve is around 3.2 million tons, or roughly 20% of the world's total reserves. However, due to the poor resource endowments and lack of industrial competitiveness, 70% of lithium supply in China depends on import. Nickel and cobalt resources are extremely scarce, accounting for only 3.4% and 1% of global reserves respectively, with over 60% and 90% of import dependency. With the massive scale-up and usage of EVs in the future, the scarcity of key battery resources will be

exacerbated. Moreover, the production, scrapping and recycling of batteries may cause serious pollution to the environment, such as the discharge of waste water, waste gas and waste liquid from the production process, electrolyte leakage in the process of recycling, the spread of heavy metals, and dust from materials dismantling, etc. The current battery recycling technology and the market are underdeveloped, and disposal processes are immature. The efforts to establish the recycling system have just started, and the legal framework is far from enough.

2. Hydrogen Fuel Cell Vehicle Technology

Hydrogen fuel vehicle technology represents the key technology pathway for low-carbon transformation in the field of automobile transportation. It involves the production, storage, transportation, filling, and final use of hydrogen energy for vehicles, including hydrogen energy supply, fuel cell system, stack, and fuel cell vehicle related technologies.

Regarding hydrogen production, storage, transportation, refueling technologies and infrastructure construction, countries have formulated the technology roadmap for hydrogen production and corresponding infrastructure plans in light of their respective resource endowments, which are in line with hydrogen fuel cell vehicles (FCV) development and scale-up plans.

Hydrogen production: mainstream technologies are fossil fuel reforming hydrogen production, industrial by-product hydrogen (chlor-alkali industry, coal, coke oven gas production), biomass gasification/fermentation hydrogen production, water electrolysis, etc. The technologies and equipments for hydrogen production from coal, natural gas and alkaline water electrolysis, are all ready for commercialization, and currently hydrogen production from fossil fuels represents the overwhelming mainstream. The capacity of China's hydrogen production exceeds 20 million t/a, of which direct hydrogen production from fossil fuels such as coal gasification and natural gas reforming accounts for about 70%, industrial by-product hydrogen production takes around 30%, and water electrolysis and other technologies cover less than 1% [10, 11]. Now, more than 90% of hydrogen in the world is used in petrochemical and ammonia synthesis industries. With growing demand brought by the development of hydrogen fuel cell vehicles, great changes might take place in the technology mix of hydrogen production [12].

Hydrogen storage: The mainstream technologies include compressed gas hydrogen storage, liquefied hydrogen storage, metal hydride hydrogen storage, and adsorption hydrogen storage, etc. Compared with other technologies, compressed gas hydrogen storage and liquefied hydrogen storage are relatively mature, but fall short of industrialization. At present, the technology of gas hydrogen storage with steel hydrogen cylinder below 45 MPa in China has been very mature, and the international R&D and demonstration efforts are all moving towards 70 MPa. Liquid hydrogen storage with high efficiency is the R&D priority of all countries. However, as liquid hydrogen storage and transportation require ultra-low temperature equipment with certain technical barriers, liquid hydrogen for civilian use is mainly applied in Europe, Japan and the United States, whereas China uses liquid hydrogen primarily for aerospace and military purposes.

Hydrogen transportation: road, waterway, and pipeline, etc. are common approaches for transport. Road and waterway transport technologies are subject to the development of hydrogen storage technology and relatively suitable for short-distance and small-scale transport. Gas hydrogen pipeline suits large scale and long distance transport for cost reduction. Pipeline transport is well developed in foreign countries, with Europe and the United States having built 1,500 km and 2,400 km of pipelines for hydrogen transport respectively. In contrast, the length of hydrogen pipeline in China is only 300–400 km, and the longest is the "Baling-Changling" pipeline, which extends about 42 km with a pressure of 4 MPa [13].

Hydrogen fueling: According to statistics of Germany organization (H2station s.org), there were 369 hydrogen fueling stations in operation worldwide by the end of 2018, including 273 of which were open to the public, and the others were in-house stations owned by institutions and enterprises. Germany, Japan and South Korea have all announced plans to build hundreds of refueling stations by 2030, and have each set up joint ventures for this purpose. China has 17 hydrogen refueling stations in operation and 38 stations under construction.

Fuel cell system and the electric reactor technology both meet the demand for vehicle applications. At present, the power system of hydrogen fuel cell vehicles features a cell-fuel cell hybrid system. Aside from cell and fuel cell technologies, atmospheric/variable pressure gas technology, battery management technology, braking energy recovery technology, and battery thermal management technology also serve important roles.

In terms of FCV technology, the overall performance has measured up to commercial scale-up. Fuel cell buses and trucks are under demonstration. Hydrogen fuel cell vehicles will see greater potential in the field of heavy-duty vehicles. *The Technology Roadmap for Energy Efficient and New Energy Vehicles* specifies a timetable for the development of hydrogen fuel cell vehicles. By 2020, 2025 and 2030, China aims to have 5,000, 50,000, and over a million FCVs on the road respectively.

In short, the development of hydrogen fuel cell vehicles entails a robust industrial chain consisting of hydrogen production, storage, transportation, refueling, and the production and management of fuel cell and car body, etc., together with the improvement of infrastructure. At present, fuel cell systems and stack technology could reach automotive level and the performance of FCVs, on the whole, is up for commercialization. Fuel cell buses and trucks are also under demonstration and enjoy a broad prospect for development with mature technology and cost reduction.

The following challenges hinder the development of hydrogen fuel cell vehicle development and merit particular attention:

1. High cost, energy consumption, and carbon emissions in hydrogen production process.
   As CCS technology has not been widely adopted, its effect on carbon emissions from fossil fuel hydrogen production is not immediately apparent compared to direct hydrogen production from fossil fuels. The purity of industrial hydrogen

by-products cannot meet the needs of fuel cells, thus a purified process with separation technology such as pressure swing adsorption (PSA) is needed, which causes a huge amount of energy consumption. Given the mainstream adoption of thermal power in China, clean electric hydrogen remains a long way off.

2. The industrial foundation of hydrogen storage, transportation, and refueling remains weak, with stunted growth in related industries and inadequate technologies.

   The infrastructure of hydrogen refueling stations and hydrogen transport pipelines is seriously underdeveloped in the world. It is estimated the full-fledged hydrogen infrastructure costs at least $2 trillion. At present, worldwide investment in this connection still caters to the pilot stage, and large investment is not yet in sight. The efficiency and safety of the existing hydrogen storage and transport equipment need further improvement. Taking the double-layer vacuum spherical hydrogen storage device as an example, its daily evaporation rate is between 0.2% and 0.5%. Vehicle-mounted hydrogen storage materials also require further development to enhance their capacity weight ratio.

3. Urgent improvement is needed for key fuel cell technologies of China, as well as localization of key components.

While being commercialized, hydrogen fuel cells fall short of meaningful mass production. The design of electric reactor, battery cost, battery power density, conversion efficiency, and service life are all hinder the industrialization of hydrogen for automobiles, and further breakthroughs are essential. Furthermore, localization of key materials and components is another impediment. Key components such as membrane electrode catalyst, proton exchange membrane, air compressor, and hydrogen reflux-pump all rely on import. Besides, the technical capability of fuel cell companies needs further improvement.

7.2.4 Negative CO_2 Emission and Carbon Geoengineering Technologies

CCS and geoengineering are important alternatives for achieving deep decarbonization. Under the deep emission reduction target, CCS can be applied to fossil energy power generation, coal chemical industry, and petrochemical industry to secure deep decarbonization of fossil energy utilization. BECCS is able to capture and store carbon dioxide emissions from biomass power generation and thermal utilization, making for a CO_2 negative emission technology.

1. CCS Technology

CCS technology consists of three steps: capture, transport, and storage. Step one: capture. First adopted in oil refining and chemical industries, etc., carbon capture refers to the process of separating and purifying carbon dioxide to a high concentration state. Three CCS technologies widely accepted are pre-combustion capture, post-combustion capture, and oxygen-enriched combustion. Step two: transport. Mainly through pipeline and low-temperature storage tank, this step involves the transport, in a leak-proof manner, of high concentration CO_2 to the designated venue for storage. The last step: storage/sequestration. By sealing saltwater layer, deep unexploitable coal seam, and waste oil and gas reservoir, this step embeds and stores CO_2 in the deeper stratum of the earth, so that it would be completely isolated from the atmosphere.

Research on CCS cost and demonstration projects have suggested such technology requires huge investment and high operating costs, and call for more breakthroughs in order to be competitive.

Despite being a later mover on CCS, China attaches great importance to its R&D and demonstration, and has committed tremendous amount of funding for its development over the past decade. In terms of the technologies of carbon capture, utilization and sequestration, China is already on par with developed countries. In August 2018, the CCS facility in Jilin province had reached a storage capacity of 600,000 tons of CO_2. Meanwhile, SINOPEC's Qilu CCS project and Yanchang Petroleum's coal chemical CCS facility in northern Shaanxi province are able to capture 400,000 tons and 410,000 tons respectively. In the future, with strengthened efforts in CCS technology demonstration, cost reduction, and steady scale-up, China may stand at the forefront of low-carbon technology.

Despite its huge volume, carbon emissions in China are concentrated and easy to capture. Given the sheer size of China, geographical sites for carbon sequestration are readily available with enormous storage potential. Data show that the carbon sequestration potential of China can reach 3,088 billion tons in theory, and the capacity of sequestration in deep brine layer is 3,066 billion tons, or 99% of the total theoretical capacity. Sun et al. [14] evaluated the theoretical storage of CO_2 in China's sedimentary basins, and concluded that China has a broad prospect for CO_2 sequestration, which is about 184.1 billion tons, over 190 times of total CO_2 emissionsof 2015 in China. Li Xiaochun et al. [15] estimated the CO_2 storage capacity of China's saline aquifer at 1.43505×10^{11} tons by using solubility method.

Several barriers may hold down the progress of CCS in the future:

1. High costs. The high cost of CO_2 capture largely determines the cost of CCS, whose further reduction hinges on the development and maturity of the technology. It's crucial to make full use of exchanges between domestic and foreign R&D institutions, and lay equal emphasis on independent R&D and international cooperation for continuous CCS technology development [16].
2. Lack of relevant policies and legal systems. Nowadays, there are few policies and laws available for CCS, prompting government efforts to ramp up the formulation and implementation of these policies and laws.
3. Leakage risk. CO_2 leakage is a key concern of safety. Human health and life would be under immediate threat in case CO_2 mass fraction exceeds 8%. Even though the possibility of leakage is very small, precautions in all respects should be made [16].
4. Insufficient public awareness. Currently, the notion of CCS remains a novelty to most Chinese people. It is important to promote CCS best practices at home and

abroad through education and publicity. It will alleviate public concerns about CCS leakage, building up popular support, and paving the way for its future development [16].

5. Lack of a sound evaluation system. Many components of CCS technology need to be evaluated. For example, the site selection for CO_2 geological sequestration should be reevaluated in a holistic manner based on the existing assessment system. Only with proper assessment can the best storage site be selected.

2. BECCS Technology

BECCS refers to the technology that combines biomass and CO_2 capture and storage to achieve negative GHG emissions. The difference between BECCS and fossil fuel CCS is that the latter could only contribute to zero emissions goal, while the former could further achieve negative emissions [17]. The technological progress of BECCS involves biomass utilization and CCS. Large-scale biomass utilization technologies, such as biomass power generation, central heating, cellulosic ethanol production, and F-T synthesis, can be installed with CO_2 capture devices to make for potential BECCS.

Statistics show that to date, there are 27 BECCS demonstration projects around the world, and many of which have been called off or shelved. Largely concentrated in the United States and Europe, these projects are based on existing plants of ethanol, cement, pulp and paper, biomass mixed combustion, and biomass pure power [17, 18]. The BECCS Project (IL-ICCS) currently implemented in Decatur, Illinois of the US is the largest project so far. The project, launched in April 2017, captures one million tons of CO_2 each year in the process of converting corn into ethanol. After compression and dehydration, captured CO_2 is injected into a sandstone formation about 2.1 km deep at Mount Simon for permanent storage [19]. Currently, China has not yet built BECCS demonstration project.

Many advanced biomass technologies for large-scale utilization, such as cellulosic ethanol, F-T synthetic biofuels, and biomass gasification combined cycle (BIGCC), are still under R&D and demonstration, and uncertainties abound for their future development. Many CCS projects are also in the stage of demonstration, with multiple challenges ahead in terms of large-scale technology implementation.

BECCS is faced with four different uncertainties, including biomass availability, technology maturity, economic viability of scale-up, and uncertainties of social and ecological impact, which will greatly weaken the contribution of the technology.

The cost of BECCS also primarily depends on the cost of biomass and CCS. BECCS is still under technology demonstration and has not been massively deployed globally. So its cost analysis is mainly conducted based on certain assumptions.

Studies have shown that the cost of CO_2 capture may not support CCS for small biomass energy devices, thus R&D and demonstration are required in the future. First, the carbon chain is very long considering the cost of biomass and CCS; second, biomass and CCS technologies come in great varieties, and great variance exists regarding biomass cost and CCS technologies; third, growing energy crops will result in a shortage of land needed for food production, pushing up food prices. The combination of biomass power generation and CCS has great potential for large-scaleapplication.

Based on biomass availability and the uncertainty of BECCS, estimates [20] on 2050 emission reduction potential have been conducted on the low, median and high level as well as upper limits. The result suggested that the ceiling of CO_2 removed by straw in 2050 in China is 830 million t/a; by forestry residue, 957 million t/a; and by energy plants, 963 million t/a, which add up to a ceiling of 2.75 billion t/a CO_2 removal by BECCS. It's expected that on the way toward the 2 °C and 1.5 °C targets, the potential of emissions redcution by BECCS in China would be 0 ~ 27.51 billion tCO_2 per year, of which the medium-value and high-value are 650 million ~ 1 billion tCO_2.

The following efforts must be made to support the future development of BECCS in China:

First, it is important to build up the scientific understanding of BECCS under the targets of 2 °C and 1.5 °C. Scaled-up implementation of BECCS-related negative emission technologies can reduce costs and help to achieve the temperature targets. However, further research is needed to strengthen the scientific understanding, and take appropriate measures to reduce potential risks in the development of BECCS.

Second, promoting BECCS research and demonstration to enhance scientific understanding and public acceptance. China should strengthen the technical reserve by boosting BECCS demonstration. So far, China has launched commercial demonstration of biomass and CCS, and the future priority is to integrate the two for negative emission. Research and demonstration in such areas as BIGCC+CCS should be to enable future emission reduction.

Third, BECCS should be incorporated into the framework of China's climate change strategy. Although BECCS still faces high uncertainty for its development and application, it's a safe bet that massive deployment of BECCS would be required to achieve the 2 °C and 1.5 °C targets. That is, BECCS should be a possible option for climate change mitigation. Meanwhile, the potential risks of BECCS should be well awared.

3. Geoengineering Technology

Geoengineering is generally defined as "the planned, large-scale human intervention in the climate system in response to global climate change". A growing number of international literatures are using "climate engineering" or "climate intervention" to replace geoengineering to distinguish it from large-scale human activities for other purposes.

In general, geoengineering techniques and methods include carbon dioxide removal (CDR) and solar radiation management (SRM). CDR removes or converts carbon dioxide in the atmosphere, and reduces the greenhouse gas concentration in the atmosphere through biological, physical or chemical methods, such as afforestation and forest ecosystem restoration, biological energy with carbon capture and sequestration (BECCS), biochar for increasing soil carbon content, enhanced weathering or ocean basification, direct air capture and storage (DAC), marine fertilizers, etc. SRM, on the other hand, does not directly reduce the content of carbon dioxide in the atmosphere, but reduces the solar radiation reaching the ground to alleviate the earth's warming through methods such as stratospheric aerosol injection (SAI), increasing cloud albedo modification over land or surface oceans, and increasing the surface albedo modification on land or ocean surfaces.

In the latest *Special Report on Global Warming of 1.5 °C*, IPCC makes a distinction between CDR and CCS. According to the report, CDR refers specifically to the removal of carbon dioxide directly from the atmosphere or the reduction of atmospheric carbon dioxide by artificially increasing ocean or terrestrial carbon sinks. The application of carbon capture, storage and utilization (CCS/CCUS) in the energy and industrial sectors is classified as emission reduction technology, rather than CDR in geoengineering (IPCC, 2018). Thus, in contrast to CCS, CDR stresses the removal of carbon dioxide directly from the air or through enhanced biological or geochemical methods. BECCS belongs to a special CDR technology, which combines biomass and CO_2 capture and storage to bring about negative GHG emissions.

Currently, no example can be found in large-scale geoengineering in the world. Technologies in this regard vary in mechanism, characteristic, and level of maturity, and are under varied stages of development. The most controversial SAI technology in SRM is under theoretical and computer simulation research, and is very sensitive to outdoor environment. The backlash from NGOs has forced such experiments to be postponed or halted. The progress of SAI R&D has been very slow, primarily because of its high risk and hugely uncertain global ramifications from artificially altering the climate system through its massive adoption. Therefore, there has been a groundswell of support for ramping up the international geoengineering governance. Other technologies of SRM are sporadically tested on a small scale, hence limited impact.

In contrast, the research and development of CDR related technologies have boomed, and some technologies have seen commercial demonstration. About 6 operational projects and over 12 construction projects on BECCS can be found worldwide. Professor David Keith from Harvard University actively supports and promotes the commercial application of DAC. As a partner, he founded Carbon Engineering to try out the application of DAC. The Canada-based Carbon Engineering has been running a pilot CO_2 extraction plant for direct air carbon capture since 2015. The company uses a solution of hydroxides to capture carbon dioxide, which must then be heated to high temperatures to release the carbon dioxide and store it and reuse the hydroxide. The process employs available technology and is considered to be relatively low-cost. Based on its design and economic assumptions, it costs between $94 and $232 to capture a ton of carbon dioxide from the atmosphere. The Swiss company Climeworks has opened a commercial facility using amines in small modular reactors. Some experts believe that commercial competition has unfolded in the space of DAC, which can be rapidly and extensively deployed once the cost barrier is shattered.

According to a paper published in Nature Communications in July 2019 by Italian, British and other academics, in spite of the two different technologies, capturing CO_2 directly from the atmosphere and burying it underground is a viable option. Using DAC means global emissions are likely to remain chronically high until 2050, with heavy use of negative emissions only later in the century. Annual negative emissions in the 2080s would be about 30 billion tons (Gt/year), close to this year's global emissions of around 40 billion tons/year. That would mean building approximately 30,000 large DACCS factories. In comparison, there are less than 10,000 coal-fired power stations in the world today. The technology will require as much as a quarter of the world's energy supply by 2100, up to 300EJ (10^{18} J) a year, equivalent to the current annual energy demand of China, US, EU and Japan, or the global supply of coal and natural gas in 2018 [21].

Marine fertilization is also a controversial CDR technology. For ecological and ethical reasons, the Convention on Biological Diversity (CBD) explicitly stipulated in 2010 that geoengineering activities affecting biodiversity and climate are prohibited except for small-scale scientific research.

Geoengineering sparks tremendous controversies worldwide, especially SRM, for its high risk and great uncertainty. CDR is an unavoidable pathway of emission reduction in the near and medium term under the 1.5 °C target, so it is imperative to accelerate research and development and demonstration. Although CDR has entered the commercial demonstration stage, large-scale application may also threaten land use, water resources, and food security, and international governance needs to be strengthened.

China's medium- and long-term low-carbon strategy must attach great importance to geoengineering. First, geoengineering should receive its due attention. Appreciating the importance of geoengineering is, by no means, easing or weakening mitigation and adaptation efforts. Secondly, the research of geoengineering should be strengthened to inform scientific decision-making; the complementation and integration of natural science and social science should be encouraged; and the comprehensive research and talent training relative to science, technology, policy, ethics, law and other aspects should be reinforced. Thirdly, the key technologies of geoengineering CDR and SRM should be distinguished; and the geoengineering technology development strategy should be carefully crafted from a strategic perspective of vision and foresight. Fourth, China should actively participate in the international governance of geoengineering.

References

1. Peng Zheng, Cui Xue, Wang Heng et al (2017) Study on the absorption capacity of micro-grid PV considering energy storage and demand side response. Power Syst Prot Control 21, 7, 45(22):63–69
2. Liao Qiangqiang, Lu Yudong, Wang Dong et al (2014) Technical and economic analysis of backup battery energy storage system on the generation side. Electr Power Constr 35(1):118–121
3. Ye Jilei, Tao Qiong, Xue Jinhua et al (2016) Progress analysis of energy storage economy in wind power grid-connected integrated application. Chem Ind Prog 35(Suppl. 2):137–143
4. Xiao liye (2004) Current situation and development trend of superconducting power technology. Power Grid Technol 28(9):33–37

5. Wang Chao, Su Wei et al (2015) Supercapacitors and their application in the field of new energy. Guangdong Electr Power 28(12):46–52
6. Wang Yu (2014) A brief analysis of carbon dioxide emission reduction and low-carbon industry development approaches in China's petrochemical Industry. Resour Conserv Environ Prot 1:3
7. Wen Qian et al (2014) Carbon emission reduction in China's chemical industry: responsibility and contribution. Chem Ind 4:1–11
8. Chen Hongkun et al (2012) Estimation and analysis of carbon emission in China's oil refining industry. Environ Prot Oil Gas Fields 22(6):1–3
9. Zhang Yuanyuan et al (2016) Comparison of carbon dioxide emissions in typical modern coal chemical processes. Prog Chem Ind 35(12):4060–4064
10. China Automotive Technology Research Center Co., LTD (2018) China hydrogen automotive industry development report (2018). Social Sciences Academic Press, Beijing
11. Liu Jian, Zhong Fufu (2019) China hydrogen energy development and prospect. China Energy 41(02):32–36
12. Hydrogen generation market by generation, application (petroleum refinery, ammonia production, methanol production, transportation, power generation), technology (steam reforming, water electrolysis, & others), storage, and region—global forecast to 2023. Market and Market 2018:9–10.
13. China Institute of Standardization, National Technical Committee of Hydrogen Energy Standardization. Blue Book of infrastructure development of China's hydrogen energy industry (2016). 2016:22.
14. Sun L et al (2018) Assessment of CO_2 storage potential and carbon capture, utilization and storage prospect in China. J Energy Inst 91(6):970–977
15. Li Xiaochun et al (2006) Selection of priority areas for deep saline aquifer CO_2 storage in China. J Rock Mech Eng (05): 963–968
16. Li Sha (2015) Analysis on the development status and problems of CCS technology in China. Shanxi Chem Ind 35(04):32–34 + 39.
17. Karlsson H, Byström L (2011) Global status of BECCS projects 2010. Global CCS Institute & Biorecro AB.
18. Kemper J (2015) Biomass and carbon dioxide capture and storage: a review. Int J Greenhouse Gas Control 40:401–430
19. International Energy Agency (2017) Technology roadmap: delivering sustainable bioenergy. IEA, Paris
20. Chang Shiyan, Zheng Dinggan, Fu Meng (2019) Biomass energy combined with carbon capture and storage (BECCS) under the temperature control target of 2 °C /1.5 °C. Global Energy Internet [2096-5125] (03):277–287.
21. Realmonte G, Drouet L, Gambhir A et al (2019) An inter-model assessment of the role of direct air capture in deep mitigation pathways. Nat Commun 10:3277. https://doi.org/10.1038/s41467-019-10842-5

Chapter 8
Investment and Cost Analysis for Implementing a Low Emission Strategy

The topic of finance has received wide attention among the many issues of global climate governance [1]. The size, source and management mode of climate finance is one of the major elements in climate governance. In the planning of medium- and long-term low emission strategy, it's necessary to quantify the financing needs to facilitate the cost—benefit analysis. In light of existing researches, this chapter estimates the investment needs and energy cost by scenario.

8.1 Energy Investment Needs

Referring to the methodologies provided by the *World Energy Investment* [2] and other related studies, energy investment can be defined in terms of supply-side and demand-side in the implementation of the low emission strategy.

8.1.1 Investment on the Energy Supply Side

The investment on energy supply is calculated based on China Energy Infrastructure Investment Model, which, building on mid- and long-term energy supply and demand data, describes the size, investment and funding sources of energy infrastructure required for supporting energy demand and high quality development, taking into account the pathways, and optimization of energy infrastructure transition (see Fig. 8.1).

The long-term deep decarbonization entails energy transition and corresponding structural changes in energy infrastructure system. The investment priorities shall transition from coal-fired power stations, coal transport corridors, coal ports, and

© China Environment Publishing Group Co., Ltd. 2022
Institute of Climate Change and Sustainable Development of Tsinghua University et al.,
China's Long-Term Low-Carbon Development Strategies and Pathways,
https://doi.org/10.1007/978-981-16-2524-4_8

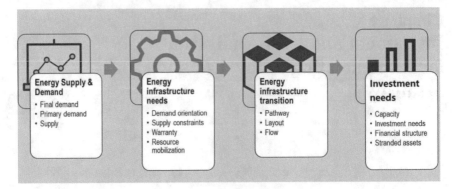

Fig. 8.1 Framework of the China energy infrastructure investment analysis model

coal storage bases into composite energy infrastructure system that features a coordinated complementation of coal, oil, gas and non-fossil fuels, and eventually into a low carbon energy-dominated infrastructure system, including renewable power stations, high efficiency transmission and distribution networks, hydrogen production facilities, hydrogen storage facilities, and so on. Investment in energy infrastructure should balance the long-term and short-term interests. It ought to pursue diversification in energy supply, and encourage joint participation and shared benefits of multiple parties to build a more inclusive and resilient energy infrastructure system.

Energy infrastructure investment mainly covers coal, oil, natural gas and electricity to ensure the demand, supply, distribution, and reserves of these resources (see Fig. 8.2).

The investment needs of energy infrastructure is primarily measured by new investment and investment for retrofitting existing infrastructure. It represents the increase of investment driven by the transformation in energy infrastructure (see Table 8.1).

New investment in infrastructure mainly covers power, oil and gas. First, the new investment demand in each period is calculated separately according to the cost changes of varied power generation facilities and their scale in different stages. Second, based on the empirical relationship between the investment in power generation facilities and the investment in power transmission and distribution facilities, the investment required for conventional grid facilities is estimated for each period, on top of that, the additional cost of renewable energy access, electrochemical energy storage, pumped storage and cross-region transmission are taken as the additional new investment demand for connecting power generation and grid facilities. The additional investment in renewable energy access also corresponds to the need for improving distribution network to cater to increasing proportion of renewables. The investment in trans-regional power transmission facilities mainly focuses on the ultra-high voltage power grid for the trans-regional power flow. Third, based on the analysis of the demand, supply, profit and loss, and flow of other energies, e.g.,

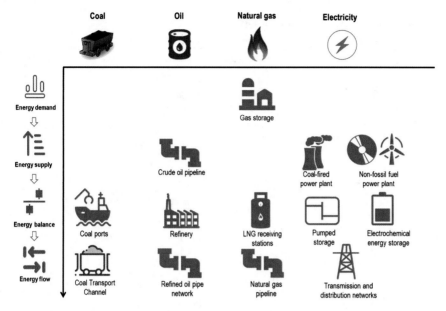

Fig. 8.2 Key investment items

natural gas, the new investment required for natural gas pipeline, gas storage depot, LNG receiving station, and so on, is calculated.

Investment for retrofitting existing infrastructure mostly caters to the goal of low-carbon transition and meeting new demands at low cost, including the flexibility improvement of existing coal power facilities to accommodate high proportion of renewable energy, the installation of carbon capture devices on fossil power generators to attain the carbon emission goal, mainly targeting existing coal and gas power facilities, and the expansion of existing LNG receiving stations to meet new gas demand and ensure gas storage capacity. Investment in LNG storage is included in the construction of LNG receiving stations. The new gas storage demand in different periods will be invested according to the following cost sequence: underground gas storage → expansion of existing LNG receiving station storage tanks → construction of new LNG receiving station storage tanks → construction of new small LNG storage tanks.

All scenarios point to the fact that low-carbon transition will drive more energy infrastructure investment. As shown in the figure, the cumulative energy infrastructure investment required for the policy scenario, reinforced policy scenario, 2°C scenario and 1.5°C scenario from 2020 to 2050 is expected to surge from 54 trillion Chinese Yuan Renminbi (CNY) to 138 trillion CNY at the constant price of 2015 (see Fig. 8.3). The investment demand in energy infrastructure for the reinforced policy scenario is 1.5 times that of the policy scenario, while that for the 2°C scenario and the 1.5°C scenario is 1.8 times and 2.6 times that of the policy scenario respectively.

Table 8.1 Types of energy infrastructure investment

Types of investment	Category	Types of infrastructure
New investment	Electricity	Coal-fired power plant
		Gas-fired power plant
		Nuclear power plant
		Hydropower plant
		Pumped storage
		Onshore wind farm
		Offshore wind farm
		Solar PV power station
		Concentrating solar power station
		Biomass power plant
		Electrochemical energy storage
		Power transmission and distribution network
	Oil	Oil-chemical integration refinery
		Fuel-oriented refinery
	Natural gas	LNG receiving station
		Natural gas pipeline
		Underground gas storage
		Small LNG storage tank
	CCS	Gas-fired power plant with CCS
		Biomass power plant with CCS
	Heat	Heat-supply pipeline
	Other	Hydrogen and other renewable energies
Investment for improving existing infrastructure	Power	Flexibility improvement of existing coal power plant
	Gas	Expansion of existing LNG receiving station
	CCS	CCS installation on existing coal power plant
		CCS installation on existing gas power plant

8.1.2 Investment on the Energy Demand Side

Energy demand-side investment is analyzed by sectors, namely industry, transport, and building sector, respectively.

(1) Industrial sector

Wholistically speaking, the potential of industrial energy saving and carbon reduction could stem from efficiency improvement, energy mix optimization, electrification, raw material and fuel substitution, and improvement of product quality. While a

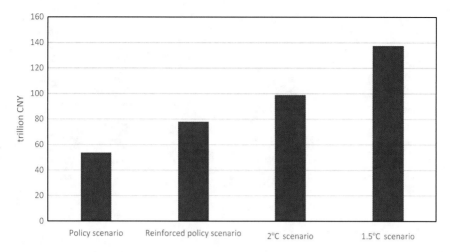

Fig. 8.3 Cumulative energy infrastructure investment from 2020 to 2050 under different scenarios

certain proportion of the investment may be allocated specifically for the purpose of energy conservation and carbon reduction, the majority of the work lies in the investment for the construction of new projects, equipment replacement, and technological innovation, etc. Thus, it is difficult to accurately pinpoint the amount required for energy conservation and carbon reduction. In addition, great discrepancies exist in the potential and pathways of energy saving and carbon reduction between energy intensive and non-energy intensive industries. Furthermore, due to the vast differences in industrial output, technologies, and processes, the levels of energy saving, carbon reduction and investment demand may vary greatly by industry. The following study applies different methods to estimate the investment required for energy intensive and non-energy intensive industries.

For steel, cement and other industries with relatively clearly-defined products and processes, taking into account the project lifetime of about 30 years, the capacity replacement method is adopted to estimate the cumulative investment required from 2020 to 2050. The potentials of energy conservation and carbon reduction are grouped into three categories: "advanced process application", "energy efficiency improvement" and "low-carbon energy substitution". The potential of energy conservation and carbon reduction and investment needs are also assessed separately.

For industries with complex product cycle and production processes such as chemicals, petrochemicals, non-ferrous metals and others, the demand for increased investment is assessed individually for different scenarios based on available cases of energy conservation and carbon reduction, and the economic analysis of cutting-edge technologies, processes and equipment, through tracking changes in investment per unit of energy conservation.

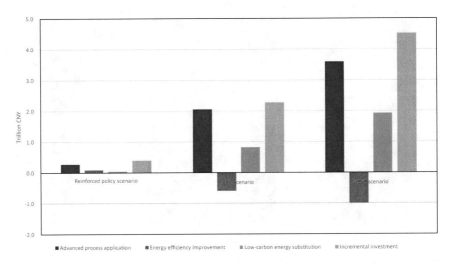

Fig. 8.4   Incremental investment by category

For non-energy intensive industries and manufacturing industries, the need for incremental investment is assessed individually in different scenarios based on the available cases of electricity substitution, energy efficiency improvement of boilers and kilns, and energy-saving enhancement of motor system, through tracking changes in investment per unit of energy conservation.

The incremental investment needed for the industrial sector in the reinforced policy scenario, 2°C scenario, and 1.5°C scenario stands at 393.7 billion CNY, 2270.8 billion CNY and 4517.6 billion CNY respectively (see Fig. 8.4).[1] Applying advanced process and using low-carbon energy are the main sources of incremental investment, which contributes the most to energy saving and emission reduction. The negative incremental investment in energy conservation and efficiency is partly due to the limited potential to improve energy efficiency, which makes large investment a challenge. Meanwhile, the reduced demand dramatically downsizes the capacity of energy intensive products, resulting in a decrease in total investment despite the increasing need for investment per unit of capacity, which also partly accounts for a negative incremental investment.

Four energy-intensive industries of iron and steel, cement, chemicals and petro-chemicals are the main sources of investment in energy conservation and emission reduction. Under the reinforced policy scenario, the four industries account for 24.1%, 7.6%, 19.1% and 22.1% of incremental investment respectively. Under the 2°C scenario, the share is 39.2%, 4.5%, 19.4% and 24.2%, and for the 1.5°C scenario, 29.8%, 5.3%, 20.4% and 22.6% respectively (see Fig. 8.5).

---

[1] The incremental investment in the reinforced policy scenario refers to the increment compared to the policy scenario; the incremental investment in the 2°C scenario is the increment compared to the reinforced policy scenario; and the incremental investment in the 1.5°C scenario is the increment compared to the 2°C scenario.

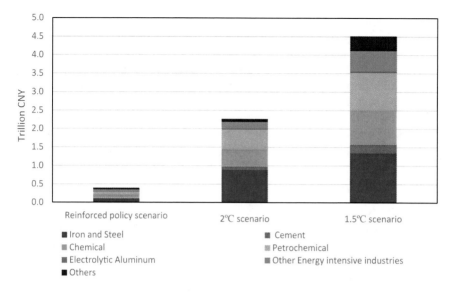

Fig. 8.5 Incremental investment by industry

Figure 8.6 shows the incremental investment under different scenarios in the steel, cement, chemical and petrochemical industries. The main incremental investment in the steel industry goes to advanced process applications, in particular hydrogen steel

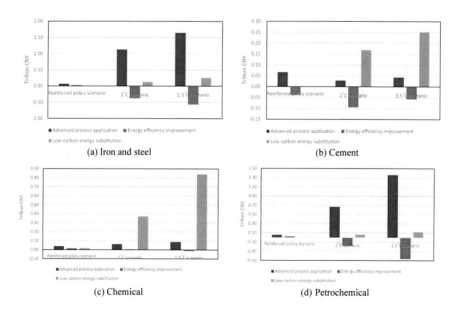

Fig. 8.6 Incremental investment in major energy-intensive industries

making. Most incremental investment in the cement industry flows to low-carbon energy substitution such as carbon capture, storage and utilization. Incremental investment in the chemical industry is mostly spent on the substitution of low-carbon energy, especially the substitution of "green hydrogen" for "grey hydrogen". The bulk of incremental investment in the petrochemical industry is made on advanced process applications such as the production of olefins from light hydrocarbons.

It should be noted that considerable gaps exist in the production capacity of energy-intensive products among varied scenarios. Generally speaking, more ambitious targets in energy conservation and emission reduction would make for smaller production capacity of energy-intensive products. Despite a substantial rise in investment per unit of capacity due to investment in energy conservation and emission reduction - the investment per ton of steel capacity stands at 4,106 CNY, 6,844 CNY and 10,750 CNY respectively, and the investment per ton of cement capacity is 840 CNY, 1,230 CNY and 1,590 CNY respectively in the reinforced policy scenario, 2°C scenario and 1.5°C scenario - negative incremental investment occurs in some industries owning to the reduction in total capacity.

(2) Transport sector

Analysis of the transport sector takes into account the increased cost of EVs relative to traditional vehicles and the construction of charging piles. The cost increase of EVs compared to traditional vehicles is estimated based on the number of new EVs on the road and the ratio of varied types of vehicle. The medium and long term investment of charging piles is projected based on the vehicle-to-pile ratio and the proportion of diverse types of charging piles. The vehicle-to-pile ratio has improved from 7.8:1 in 2015 to 3.1:1 in 2019 [3, 4]. In the long run, the vehicle-to-pile ratio will be around 1:1.

On new investment in EVs, it is estimated that the average production cost of EV in 2020 is 126,000 CNY - a difference of 62,000 CNY compared to gasoline vehicles [5]. Considering the decreasing production costs of specific car models, it is projected that by 2050, the average production cost of EVs will be lowered to 83,000 CNY, and the difference with gasoline vehicles will be reduced to 16,000 CNY. Therefore, the total investment in the four scenarios from 2020 to 2050 is approximately 9.0, 11.9, 15.0 and 18.6 trillion CNY respectively.

The new investment of charging piles is estimated based on the ratio of public DC, public AC, and private charging piles and their cost changes. The average cost in 2020 and 2025 is 4,400 CNY and 3,300 CNY per unit, assuming that it remains unchanged afterward. In 2019, a total of 1.22 million charging piles were operational nationwide. It is projected that the total investment between 2020 and 2050 is about 1.5, 2.1, 2.5 and 3.0 trillion CNY under the four scenarios in this study.

(3) Building sector

The emission reduction potential of the building sector mainly comes from maintaining green and energy-saving behaviors, improving the performance of building envelope, enhancing the efficiency of equipment system, optimizing energy mix, and tapping into renewable energy. Currently, there are various methods to estimate the investment need for low-carbon buildings, with different boundaries and definitions. This study, centering around the building hulk and the northern heating system, makes the projections for improving the performance of building envelope and the northern heating system.

The investment for performance improvement of building envelope includes the increased cost of new buildings due to higher design standards and the cost of existing building envelope renovation, of which the energy-saving renovation of existing buildings in rural areas accommodates the usage of "part of the space" of rural residents and gives priority to economic renovation [6].

Analysis on the improvement of heating systems in northern China accommodates the difference between urban and rural areas. For the urban areas, the focus is the varied types of CHP and low-grade waste heat from industrial production process. The use of waste heat requires the construction of heating pipelines for long distance transport. And in areas where waste heat is out of reach, heat pumps with high efficiency are preferred. Rural areas shall slash the use of bulk coal and inefficient biomass, and encourage the use of new and efficient biomass boilers and heat pumps.

The cost data of this study is obtained by referring to previous engineering survey and related cases [6, 7, 8, 9].

According to the estimates, 1.1 trillion CNY more investment is required in the reinforced policy scenario than the policy scenario for the two main undertakings proposed in this study; and additional 0.5 trillion CNY is needed in the 2°C scenario compared to the reinforced policy scenario; and the 1.5°C scenario requires 0.06 trillion CNY less than the 2°C scenario.

As for the improvement of envelope performance, more aggressive targets for energy conservation and emission reduction lead to an enormous downsizing of new buildings, hence the reduced cost in this regard. Yet the cost of energy conservation renovation for existing buildings is on the rise. By 2050, the additional cost for the four scenario is estimated to be 0.65 trillion CNY, 0.15 trillion CNY and -0.18 trillion CNY respectively.

Regarding heating upgrade in the north, the investment required by all scenarios in the northern urban areas is on a steady rise due to the waste heat utilization, heat pump installation and elimination of all boilers. For rural areas, the four scenarios mostly feature a rising trajectory, except for the 1.5°C scenario that shows a drop in overall investment due to the fact that fewer rural residents choose to embrace the earlier electrification after they have already switched from bulk coal and inefficient biomass to gas-fired wall-mounted furnace. The scenario difference of investment in heating in the north is 0.48 trillion CNY, 0.37 trillion CNY, and 0.12 trillion CNY respectively in 2050.

Specific scenario differences of investment to 2050 are shown in Table 8.2.

8.1.3 Total Investment Needs

Considering investment on both energy supply and demand side, the cumulative energy investment from 2020 to 2050 stands at 71 trillion CNY, 100 trillion CNY, 127 trillion CNY, and 174 trillion CNY respectively under different low-emission

Table 8.2 Investment differences between scenarios (in billion CNY)

	Reinforced → Policy	2°C → Reinforced	1.5°C → 2°C
Performance improvement of building envelop	649.0	152.3	-180.9
New buildings	104.4	-111.9	-309.3
Renovation of existing buildings	544.7	264.2	128.5
Northern heating renovation	482.7	369.1	120.7
Urban areas	477.6	325.4	131.0
Rural areas	5.1	43.7	-10.3
Total	1131.8	521.4	-60.2

scenarios (Table 8.3, Fig. 8.7). Compared with the policy scenario, the cumulative investment in the reinforced scenario, 2°C scenario and 1.5°C scenario rises by 29 trillion CNY, 57 trillion CNY, and 104 trillion CNY respectively. And the average

Table 8.3 Total investment from 2020 to 2050 (in trillion CNY)

	Energy supply	Industry	Building	Transport	Total
Policy scenario	53.71	0.00	6.29	10.51	70.51
Reinforced policy scenario	77.89	0.39	7.42	13.99	99.69
2°C scenario	99.07	2.66	7.94	17.57	127.24
1.5°C scenario	137.66	7.18	7.88	21.66	174.38

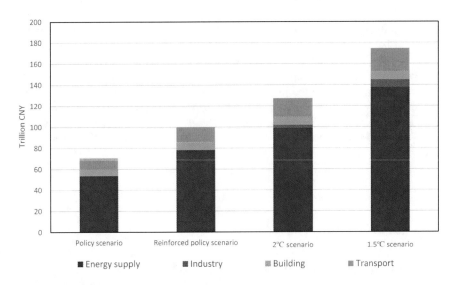

Fig. 8.7 Total energy investment by scenario

annual investment from 2020 to 2050 is 2.4 trillion CNY, 3.3 trillion CNY, 4.2 trillion CNY, and 5.8 trillion CNY respectively.

8.2 Energy Cost Analysis

This research makes further estimates on the comprehensive power supply cost and society-wide energy cost.

8.2.1 Cost of Electricity Supply

Electricity supply is one of the major costs of production and living. Generally speaking, electricity price is the sum of the cost and profit of electricity supply. The electricity supply cost in this study mainly refers to the weighted average cost per kilowatt-hour of each generation technology. The long-term trend of power supply cost varies greatly under different scenarios. The policy scenario features an overall decline in the cost; the reinforced policy scenario is characterized by a stable and slight increase in the cost over a fairly long period of time until a slump after 2033; both the 2°C and 1.5°C scenarios witness a pattern of an increase before a drop, with the cost hitting the peak in 2028 and 2033, respectively which is 1.4 and 1.42 times of that in 2018 respectively (see Fig. 8.8). In the long run, the power supply cost

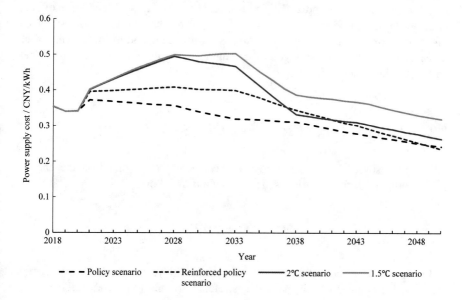

Fig. 8.8 Trend of changes in power supply cost

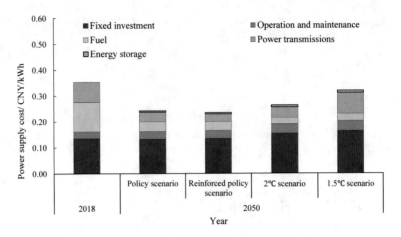

Fig. 8.9 Cost composition of power supply in 2050

witnesses a downward trend, and the cost in 2050 under the policy, reinforced policy, 2°C, and 1.5°C scenarios is 69%, 66%, 75%, and 90% of that in 2018 respectively (see Fig. 8.8). Looking at the cost composition of electricity supply in 2050, higher costs occur in the 2°C and 1.5°C scenarios compared to the policy and reinforced policy scenarios, this is mainly attributed to the higher fixed investment costs, operation and maintenance, and power transmission costs (see Fig. 8.9).

Future cost reduction in power supply is primarily driven by the reduced fuel cost. The cost of fuel in 2018 accounts for about 29% of the total cost of electricity supply. In the reinforced policy and 2°C scenarios, fuel cost falls to 8.4% and 8.5% of cost of electricity supply respectively as coal-fired power plant capacity downsize (see Fig. 8.9). Despite minor increases in the cost of operating and maintenance, power transmission, and energy storage in the long-term, the dramatic reductions in fuel cost offset these increased costs.

8.2.2 Cost of Energy Use in the Entire Society

This study estimates the energy cost of the whole society based on the energy consumption of different varieties. The formula is as follows:

$$\text{EC}_j = \sum_{i}^{5} E_{ij} \times EP_{ij}$$

EC_j is the energy cost of the whole society in the jth year; i stands for five energy varieties, including coal, oil, natural gas, electricity and others; E_{ij} refers to the consumption of different energy varieties in different years; EP_{ij} refers to the price

Table 8.4 Energy cost of the whole society (in trillion CNY)

Scenario	2020	2030	2050
Policy scenario	9.1	11.7	11.3
Reinforced policy scenario	9.1	12.5	10.7
2°C scenario	9.1	13.3	12.2
1.5°C scenario	9.1	13.1	13.5

of different energy varieties in different years. The base year data of energy price is from existing literature [10], and its variation trend over time is assumed based on the trend of cost variation of different energy categories.

As shown in Table 8.4, it is estimated that the energy cost of the whole society under the four scenarios sees continued increase from 2020 to 2030; the policy scenario, reinforced scenario and 2°C scenario report a drop from 2030 to 2050; and a steady climb is observed from 2020 to 2050 in the 1.5°C scenario. Under the four scenarios, the total energy cost of the whole society in 2050 is approximately CNY 11–14 trillion.

8.3 Summary

Overall, there are notable discrepancies in the amount of investment under the various scenarios. Between 2020 and 2050, cumulative energy investment in policy, reinforced policy, 2°C and 1.5°C scenario amounts to 71 trillion, 100 trillion, 127 trillion, and 174 trillion CNY respectively, with the average annual investment of 2.4 trillion, 3.3 trillion, 4.2 trillion, and 5.8 trillion CNY respectively. Compared with the 2°C scenario, the cumulative investment in the 1.5°C scenario increases by an additional 47 trillion, which averages to 1.6 trillion CNY additional investment on an annual basis.

In terms of average power supply cost, both 2°C and 1.5°C scenarios witness an increase before decreasing from around 2030. The costs in 2050 under the 2°C scenario and 1.5°C scenario are markedly higher than that under the policy and the reinforced policy scenario. The energy cost of the whole society sees continued growth from 2020 to 2030, and might experience a fall from 2030 to 2050, reaching roughly 11–14 trillion CNY by 2050.

References

1. Chai QM, Fu S, Wen XY et al (2019) Financial needs in implementing China's nationally determined contribution to address climate change by 2030. China Popul Resour Environ 29(4):1–9
2. IEA. World Energy Investment. 2016, Organization for Economic Co-operation and Development (OECD), International Energy Agency (IEA), 2016.

3. China Electric Vehicle Charging Infrastructure Promotion Alliance. Charging pile operation data in December 2019. (2020–01–14). [2020–04–24]. http://www.china-nengyuan.com/news/151232.html.
4. China Electric Vehicle Charging Infrastructure Promotion Alliance. Operation of national EV charging Infrastructure in 2019. 2020.
5. Ouyang D.H. Study on EV user's consumer decision factors and total ownership cost. Tsinghua University. 2020.
6. Research Center for Building Energy Efficiency of Tsinghua University. Annual development research report of China's building energy efficiency 2020. China Building Industry Press, Beijing.
7. Building Energy Efficiency Research Center of Tsinghua University (2019) Annual report on China building energy efficiency 2019. China Building Industry Press, Beijing
8. Building Energy Efficiency Research Center of Tsinghua University (2016) Annual report on China building energy efficiency 2016. China Building Industry Press, Beijing
9. Building Energy Efficiency Research Center of Tsinghua University (2016) Technical analysis for China's building energy conservation. China Building Industry Press, Beijing
10. Wang QY (2019) Energy data. 2019. (2020–04–13). [2020–08–18]. https://www.efchina.org/Reports-zh/report-lceg-20200413-zh.

Chapter 9
Long-Term Low-Carbon Transition Pathways

The previous chapters provide systematic analysis on the pathways of emissions reduction and the subsequent economic impact under different scenarios in end-use energy consumption sectors, power and energy systems, and non-CO_2 greenhouse gases. In view of the urgency of addressing climate change, it's crucial for China to explore the development pathway of low-carbon transition. In addition, the emission reduction targets and actions should be reinforced to implement the nationally determined contribution commitment by 2030. While, research should be conducted to determine the long-term low-carbon emission targets and approaches suitable for China by 2050. Building on the analysis of previous chapters, we propose to study and propose with recommendations for implementing and updating NDC targets before 2030 under the reinforced policy scenarios. After 2030, further actions should be taken to strengthen the intensity of deep decarbonization, in line with the Paris Agreement target of keeping global temperature rise below 2 °C while pursuing efforts to limit the increase further to 1.5 °C, study and explore the transition pathway to deep decarbonization by 2050.

9.1 Implementing and Reinforcing NDC Targets and Actions for 2030

9.1.1 Energy Consumption and CO₂ Emissions

Since China's economy entered the "new normal" stage, the high quality economic growth has spurred the transition from quantity-driven extensive growth to quality-focused intensive development, industrial structure accelerated in transforming and upgrading, while growth drivers also shifted. Attributed by the actions taken in the 12th Five-Year Plan and roadmap planned in the 13th Five-Year Plan, the share of

© China Environment Publishing Group Co., Ltd. 2022
Institute of Climate Change and Sustainable Development of Tsinghua University et al.,
China's Long-Term Low-Carbon Development Strategies and Pathways,
https://doi.org/10.1007/978-981-16-2524-4_9

the primary, secondary and tertiary industries in China's GDP improved from 9.3, 46.5 and 44.2% respectively in 2010 to 7.1%, 39.0% and 53.9% respectively in 2019, served to curb the fast growth of energy consumption [1]. In the long run, economic growth is expected to fall back to a medium- to low-growth level. As per the previous analysis and forecast by mainstream domestic research institutions, growth will slide from the current 6% during the 13th Five-Year Plan period to around 4.5–5.0% by 2030.

Among China's current NDC targets, the most important ones involving CO_2 emissions include:(1) CO_2 emissions intensity per unit of GDP to decrease by 60–65% by 2030 from 2005 level; (2) non-fossil energy to account for roughly 20% of the primary energy consumption mix by 2030; (3) CO_2 emissions to peak around 2030 and ideally earlier. These NDC targets were evaluated and proposed before 2015 based on the fulfillment of the goal set out during the Copenhagen Climate Conference in 2009, goals entailed a 40–45% drop in emission intensity by 2020 from 2005 level, and non-fossil energy to account for 15% the primary energy consumption mix. These are also achievable targets in line with China's economic and social reality at that time and in the foreseeable future. *China Statistical Abstract 2020* published China's total primary energy consumption by the end of 2019 is at 4.86 billion tce. The improved energy mix features a further drop of coal in the primary energy consumption mix to 57.7%, non-fossil energy and natural gas increase to 1.0 and 0.3 percent to 15.3% and 8.1% respectively, while oil making up 18.9%, on a par with that in 2018 [2]. By this measure, China has achieved its goals set out in Copenhagen of reducing CO_2 intensity per unit of GDP by 40–45% and for non-fossil energy to account for 15% of primary energy consumption by the end of 2019 ahead of schedule, which laid a solid groundwork to implement and further strengthen the NDC targets.

In 2020, the Covid-19 pandemic plunged global economy into a severe contraction, presenting grave challenges to social stability. At present, it's hard to gauge the exact length and impact from the virus, but it has aggravated global geopolitical issues, sparking a new wave of oil price volatility, and with an unprecedented disruption in global energy market. To reduce the shock of the pandemic on the economy, the world will focus more on economic stimulus to secure social stability in the short-term. And a growing number of scholars have called for "greener" and more "sustainable" economic recovery. With the impact of Covid-19 in mind, this research also adjusted the forecast of China's economic growth amid the recent low-carbon transformation.

According to the *Strategy for Energy Production and Consumption Revolution (2016–2030)* [3] jointly released by National Development and Reform Commission and National Energy Administration, power generation from non-fossil energy will reach approximately 50% of total electricity use by 2030. Study in the report also finds that over 50% of primary energy will be used for electricity production by 2030, which means that non-fossil energy is on track to comprise of roughly 25% of primary energy consumption by 2030 with increasing electrification in the future [4]. With China's attention shifting to low-carbon transformation of the energy mix, non-fossil energy has gained some weight in primary energy consumption, but it still falls short of developed countries. In the short term, newly added non-fossil fuels can't adequately meet the needs of total additional energy demand with decarbonization efforts. Natural gas, as a "relatively low carbon" "clean" energy option, will grow in importance, serving as a "transition" toward a non-fossil centered energy mix.

By 2030, coal, oil, natural gas, and non-fossil energy as a percentage of China's primary energy consumption will be improved to 45%, 17%, 13%, and 25% respectively from 57.7%, 18.9%, 8.1% and 15.3% respectively in 2019. Among them, the share of non-fossil fuels will rise by another five percent compared with the current NDC target of 20%, and coal-use can be further reduced through enlarging the size of natural gas in primary energy consumption. By enhancing the energy mix, CO_2 intensity per unit of energy consumption will drop from 2.085 $kgCO_2$/ kgce to 1.75 $kgCO_2$ / kgce, down 16%, paving the way for further reduction in CO_2 intensity per unit of GDP.

The Covid-19 puts a brake on the economic growth in 2020, triggering a slowdown of the annual GDP growth to 5.7% during the 13th Five-Year Plan period. Based on the current progress of recovery, the annual GDP growth is expected to be above 5% in the 14th Five-Year Plan period (see Table 9.1). This new period features the implementation of a new development philosophy which will steer new sources of economic growth and new infrastructure investment towards the digital economy and high-end technology sectors, with the saturation or reduction in the demand for energy intensive raw materials and products such as steel and cement and a reduced elasticity in energy consumption. Energy consumption per unit of GDP will experience a decline equivalent to that in the 13th Five-Year Plan period, or no less

Table 9.1 Energy consumption under the enhanced NDC target and its CO_2 emissions

		2005	2010	2015	2020	2025	2030
Annual GDP growth rate (%)			11.3	7.9	5.9	5.3	4.8
5-year decline in energy consumption per unit of GDP (%)			19.1	18.5	14.3	14.0	14.0
Energy consumption (billion tce)		2.61	3.61	4.34	4.94	5.50	5.98
Energy consumption structure	Coal (%)	72.4	69.2	63.7	57.0	51.0	45.0
	Oil (%)	17.8	17.4	18.3	18.5	18.0	17.0
	Natural gas (%)	2.4	4	5.9	8.5	11.0	13.0
	Non-fossil fuel (%)	7.4	9.4	12.1	16.0	20.0	25.0
CO_2 emissions per unit of energy consumption ($kgCO_2$ / kgce)		2.32	2.25	2.16	2.03	1.90	1.75
CO_2 emissions ($GtCO_2$)		6.06	8.13	9.37	10.03	10.45	10.46
5-year decline in CO_2 emissions per unit of GDP (%)			21.5	21.2	19.7	19.5	20.6
Decline from the 2005 level (%)					50.3	60.0	68.2

than 14%. Total energy consumption could reach 5.5 billion tce and 5.98 billion tce by 2025 and 2030 respectively. Given the economic growth prospect in this study, with the prediction of total energy consumption and energy mix, CO_2 emissions in 2030 will reach a total of 10.46 billion tCO_2, with CO_2 intensity per unit of GDP falling by 68.2% from 2005 level, surpassing the upper limit of the 60–65% reduction in China's NDC target. CO_2 emissions will reach a plateau around 2025 and hit the peak before 2030. Table 9.1 illustrates energy consumption and CO_2 emission levels before 2030 with intensive energy conservation and carbon reduction policies and measures, which will fully exceed the 2030 NDC target.

9.1.2 Policies and Measures for Enhanced NDC Targets and Actions

As required by the Paris Agreement, countries should report and update their NDC targets by the end of 2020. This includes not only the emission reduction targets of each party, but also the goals and actions of enforcing the strategic measures and policies on this front, which encapsulates the actions to fully implement the national strategy on climate change. Given the achievements and strides China has made in this regard, the following key considerations can be made to strengthen the 2030 NDC targets and efforts:

1. Strive to exceed the 2030 NDC targets. With non-fossil energy getting momentum in recent years, its share in primary energy consumption is expected to climb from 20 to 25% in 2030. A low-carbon energy system will facilitate a further reduction of China's CO_2 intensity of GDP. Through the current efforts, progress and active preparations in the 14th and 15th Five-Year Plan periods, China is on track to cut its CO_2 intensity of GDP by more than 65% around 2030 from the level in 2005, and peak CO_2 emissions before 2030.

2. Put forth guiding goals for curbing total energy consumption and CO_2 emissions. A cap on total energy use has been introduced and executed in the 13th Five-Year Plan period, and a similar cap on CO_2 emissions should also be imposed during the 14th Five-Year Plan period to be aligned with the policies and measures to ensure a peak around 2030, which should gradually integrate or replace the cap on total energy consumption. The cap on emissions should not put a ceiling on energy consumption of local governments and enterprises. Rather, it should encourage them to tap into new and renewable sources of energy in a bid to propel a revolution in energy production and consumption. The cap could also be aligned with the ongoing efforts to establish national carbon market to gradually move from the right to energy use system toward the enterprise carbon allowance system, demonstrating China's reinforced efforts in climate change.

In view of the great uncertainty in future economic development and energy consumption, especially the impact of Covid-19, GDP growth in the 14th Five-Year Plan period still hangs in the balance. Therefore, a relatively looser cap should be adopted to allow some leeway. The total energy consumption should be capped at 5.5 billion tce, total emissions at 10.5 billion tons at the end of 14th Five-Year Plan period, and a CO_2 emission peak before 2030 capped below 11 billion tons. A dual control mechanism with intensity reduction as the mainstay and caps as the supplement should be created, which is also consistent with the characteristics and reality of China's current economic development.

3. Developed areas in the eastern coast of China should peak CO_2 emissions earlier. The 13th Five-Year Plan has indicated "support to optimized development areas to secure an earlier CO_2 emission peak". After 2025, China as a whole can enter a peak period of CO_2 emissions. Developed coastal areas in the east are well-positioned to lead in the country, with some cities potentially reaching peak before 2025. Provinces and cities in the southeast region rich in renewable energy can also strive to take the lead for CO_2 emission peak. It's important to develop situational plans for different regions based on local realities, encourage the developed areas in the eastern coast to set out peak targets in order to catalyze the midwest region to transform development patterns. Encourage provinces and cities to create and enforce related targets and policy measures, thereby driving low-carbon economic and social transformation via the CO_2 emissions peak targets.

4. Push for CO_2 emission peak before 2025 in energy-intensive heavy chemical industries. Despite a reduced weight in GDP, the industrial sector still represents an mix of energy-intensive industry that consumes more than 60% of the total energy for end-uses in China, making it the most promising sector for energy saving. With adjustment, transformation and upgrading of the industrial structure, the output of heavy and chemical products will be saturated and with potential to decline during the 14th Five-Year Plan period, prompting an earlier peak than transportation and construction, etc. China can make active efforts to set the goal of peaking CO_2 emissions of the heavy and chemical industry sector in the 14th Five-Year Plan, striving for an earlier peak in the entire industrial sector through the leadership of the heavy and chemical industry sector, thus setting the stage for a nationwide peak by 2030.

5. Curtail non-CO_2 greenhouse gases emissions. Non-CO_2 greenhouse gas emissions are mostly short-lived emissions. Despite the diminishing greenhouse impact in their entire lifespan, their complicated sources make for high cost of technologies for emission reduction in the long run, which represent one of the hardest obstacles to overcome for the net-zero emission target in the future. China's current implementation and commitment of NDC mainly focus on energy-related CO_2 emissions, with plenty of room for improvement in the capacity for managing non-CO_2 sources. Thus, the 14th Five-Year Plan period should start to strengthen the management and control of CO_2 and other greenhouse gas emissions from industrial production, agriculture, forestry, waste management, etc., especially for the production and use of fluoride gas, and gradually establish a monitoring, reporting, and verification system for all sources of greenhouse gas emissions, build capacity for the further implementation of mitigation measures and actions, and adapt to the implementation of the Paris Agreement.

6. Deepen the administrative and market-based reform of energy and CO_2 emission management system. Now, the burgeoning energy technologies and industrial innovations have sped up the transformation of the energy industry, but some institutional and policy barriers are yet to be overcome. It's essential to advance reform of the energy management and pricing systems, boost the development of distributed renewable energy. Meanwhile, carbon market should be strengthened through prompt efforts in expanding the unified national carbon market to the eight designated energy-intensive sectors, and properly managing carbon allowance in an effort to promote low-carbon transition of the energy and economy with market-based tools during the 14th Five-Year Plan period.

9.2 Analysis of Long-Term Emission Reduction Pathway Driven by the 2 °C Target

Based on previous analysis, more space is needed for carbon emissions and more time is need for low carbon transformation in light of the current economic and social development situation, it is difficult to switch to the 2 °C emission reduction pathway in the near future. To enable long-term 2 °C pathway without sacrificing near-term economic growth, a feasible option is to step up emission reduction efforts based on the reinforced policy scenario before 2030, laying solid groundwork for the transition toward the 2 °C. In addition, achieve carbon peak before 2030, with additional emission reduction efforts and use the goal of staying within 2 °C by 2050 to force the leap toward the 2 °C target. This section builds the feasible long-term low-carbon transition scenario based on enhancing NDC targets before 2030 elaborated in Sect. 9.1, and facilitating greater emission reduction in the entire economy and society toward the 2 °C scenario after 2030.

9.2.1 Comprehensive Analysis of the 2 °C Target Path Scenario

The Paris Agreement sets its long-term temperature goal to hold global average temperature increase "well-below to 2 °C above preindustrial levels and pursuing efforts to limit the temperature increase to 1.5 °C above pre-industrial levels". Under the guidance of "the two-step strategy of new era", China should also undertake its mitigation efforts and contributions in combating climate change in accordance with its development toward a great, modern socialist country. In this context, China's domestic long-term strategy of low-carbon development should be in line with the long-term goal of the Paris Agreement by limiting the temperature rise to well below 2 °C [5]. Based on the analysis of the reinforced policy scenario and the 2 °C scenario in the previous sections, the emission reductions pathway can be gradually

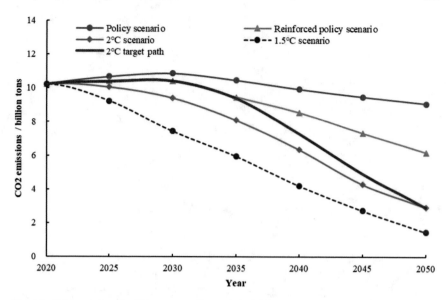

Fig. 9.1 CO_2 emissions in different scenarios and the 2 °C target path

achieved in two stages as China embraces high-quality economic transition: before 2030 (Stage one), before 2030, we should strive for emission reduction actions that are more ambitious than the enhanced NDC targets based on the reinforced policy scenario in order to peak carbon emissions earlier, whch makes it highly likely to transit toward the 2 °C scenario at the same time; after 2030 (stage two), the priority should be given to the low-carbon energy system and improved energy efficiency to speed up the transition toward the 2 °C target, thus achieving the goal of emission reduction effectively under the 2 °C scenario by 2050. Accordingly, this section provides the 2 °C target oriented transition scenario pathway (2 °C target path shown in Fig. 9.1). The data of economic growth, energy mix assumptions, and energy-related CO_2 emissions in major years are illustrated in Table 9.2.

Before 2030, the enhanced NDC plan proposed in Sect. 9.1 should be adopted to support the expected economic and social growth with a lower carbon-intensity energy mix. In the meantime, steps should be taken to boost technological efficiency and cut back emissions, further reduce energy consumption elasticity while optimizing industrial structure, actively pursue more ambitious emission reduction before 2030 than the enhanced NDC targets, laying the foundation for the transition to 2 °C pathway after 2030. Energy-related CO_2 emissions can be brought down to 10.46 billion tons by 2030, which signifies the start of a downward trend after peak before 2030. From 2030 to 2050, a lower carbon-intensity energy mix can be achieved with the share of non-fossil energy in primary energy consumption increasing from 25% in 2030 to 73.2% in 2050, and the share of coal, oil, and natural gas decreasing from 45%, 17%, and 13% in 2030 to 9.1%, 7.7%, and 10.0% in 2050 respectively (see Fig. 9.2). The new energy system with non-fossil fuels as the mainstay would by

Table 9.2 Economic growth, energy consumption and CO_2 emissions in the 2 °C target path

Item (unit)		2005	2010	2015	2020	2025	2030	2035	2040	2045	2050
Annual GDP growth rate (%)		-	11.3	7.9	5.9	5.3	4.8	4.4	4.0	3.6	3.2
GDP index (2005 = 1)		1	1.71	2.50	3.33	4.31	5.45	6.75	8.22	9.81	11.48
Energy consumption (billion tce)		2.61	3.61	4.34	4.94	5.50	5.98	6.06	5.89	5.58	5.20
Energy consumption structure	Coal (%)	72.4	69.2	64	57	51	45	36	28	19	9.1
	Oil (%)	17.8	17.4	18.1	18.5	18	17	17	13	9	7.7
	Natural gas (%)	2.4	4	5.9	8.5	11	13	15	14	12	10
	Non-fossil fuel (%)	7.4	9.4	12	16	20	25	32	45	60	73.2
CO_2 emissions per unit of energy consumption (kgCO₂/kgce)		2.32	2.25	2.16	2.03	1.90	1.75	1.55	1.24	0.88	0.56
Annual decline in CO_2 emissions per unit of energy consumption (%)		-	0.58	0.83	1.27	1.30	1.59	2.46	4.38	6.50	8.62
CO_2 emissions (Gt CO_2)		6.06	8.13	9.38	10.03	10.45	10.46	9.38	7.29	4.93	2.92
Annual decline in energy consumption per unit of GDP (%)		-	4.15	3.82	3.09	2.97	2.97	3.96	4.39	4.53	4.44
Annual decline of CO_2 emissions per unit of GDP (%)		-	4.72	4.62	4.33	4.23	4.57	6.32	8.58	10.74	12.74
Decline from the 2005 level (%)		-	21.5	38.0	50.3	60.0	68.3	77.1	85.4	91.7	95.8

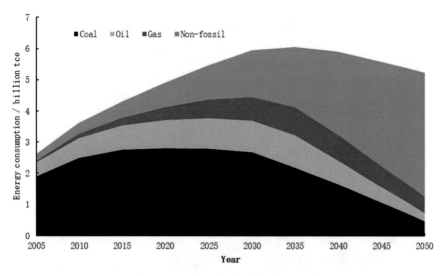

Fig. 9.2 Primary energy consumption and its mix in the 2 °C target path

then provide major buttress to the economic and social development, with increasing decoupling of economic growth and energy consumption. By 2050, energy-related CO_2 emissions can be reduced to 2.92 billion tons. Considering CO_2 emissions from industrial process, carbon sink, and CCS, the net CO_2 emissions would be 2.18 billion tons with a per capita CO_2 emission of about 1.5 tons, which is lower than the world's per capita level under the 2 °C target path, paving the way for the prompt achievement of net-zero emission after the middle of this century. Table 9.3 shows the sectoral breakdown of energy consumption and CO_2 emissions under 2 °C target path.

Between 2030 and 2050, direct CO_2 emissions from the industrial and electricity sector remain the largest contributor to energy-related CO_2 emissions. In 2050, the direct CO_2 emissions from the electricity sector (the carbon sink from CCS not

Table 9.3 Energy consumption and CO_2 emissions by sector in the 2 °C target path

End users	2020		2030		2050	
	Energy consumption (billion tce)	Emissions ($GtCO_2$)	Energy consumption (billion tce)	Emissions ($GtCO_2$)	Energy consumption (billion tce)	Emissions ($GtCO_2$)
Industrial	1.61	3.77	1.88	4.15	0.69	1.19
Building	0.55	1.00	0.51	0.88	0.26	0.31
Transport	0.49	0.99	0.54	1.09	0.30	0.55
Power	2.17	4.06	2.83	3.95	3.92	0.83
Other sectors	0.12	0.21	0.22	0.38	0.03	0.04
Total	4.94	10.03	5.98	10.46	5.20	2.92

Table 9.4 Total CO_2 emissions and their composition in the 2 °C target path (unit: $GtCO_2e$)

	2020	2030	2050
CO_2 emissions from energy consumption	10.03	10.46	2.92
CO_2 emissions from industrial production	1.32	0.94	0.47
CCS/BECCS	0.00	0.00	-0.51
Forest carbon sinks	-0.58	-0.61	-0.70
Net emissions	10.77	10.79	2.18

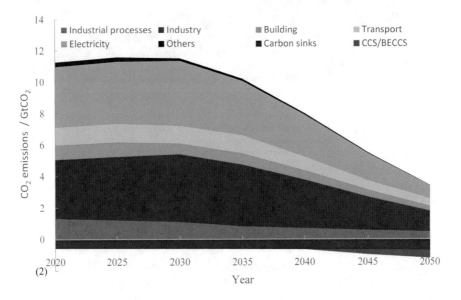

Fig. 9.3 CO_2 emissions by sector in the 2 °C target path

included), driven by a lower carbon energy mix, are down by roughly 80% from the 2020 level, providing low-carbon power to the end users. CO_2 directly emitted from energy consumption stands at 2.92 billion tons, and emissions from industrial processes are about 470 million tons, the size of forest carbon sinks and CCS will reach 700 million tons and 510 million tons respectively in 2050. The total net CO_2 emissions in 2050 reach 2.18 billion tons, about a 80% decrease from the peak year. The total CO_2 emissions under 2 °C target path are shown in Table 9.4 and Fig. 9.3.

9.2.2   Pathway Analysis of the 2 °C Target Path Scenario

By comparing the CO_2 emission trajectories of the 2 °C target path and the four emission reduction scenarios mentioned earlier (see Fig. 9.1), the results show that

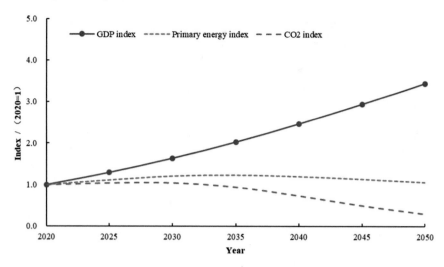

Fig. 9.4 Indexes of GDP, primary energy consumption, and CO_2 emissions (2020 = 1)

the 2 °C target path is somewhere between the reinforced policy scenario and the 2 °C scenario, in line with China's current development stage and characteristics. That is, before 2030, to ensure a smooth transition of the economy and society still require certain space of carbon emissions, but the low-carbonization process of the energy structure is expected to be further strengthened; after 2030, the entire economy and society will experience a rapid transition, and meet the 2 °C scenario by 2050.

In the 2° C target path scenario, by 2050, the total GDP is 3.5 times larger than that of 2020. After peaking around 2035, the total energy consumption gradually falls, but new and renewable energy continue to develop rapidly. CO_2 intensity per unit of energy consumption spirals down. As CO_2 emissions and GDP growth are decoupled before 2030, the decrease of CO_2 emissions per unit of GDP after 2030 will reach to 10% annually after 2040 (see Figs. 9.4–9.6). The change in energy mix provides increasingly greater contributions as time goes on.

The year-by-year trend of energy conservation and energy substitution in the reduction of CO_2 intensity in GDP under the 2 °C target path are showed in Fig. 9.5. Without considering the second-order infinite quantity [6], the annual decline of CO_2 intensity of GDP can be approximately expressed as the sum of the annual drop of energy intensity of GDP (or referred to as the energy-saving effect) and the annual decrease in CO_2 intensity of energy consumption (or referred to as the energy substitution effect). Data in the figure suggests that after 2030, with the swift transition to the 2 °C target, CO_2 intensity of GDP sees faster annual decline from 4.5% between 2025 and 2030 to 12.7% between 2045 and 2050. With the rapid advancement of low-carbon energy mix, non-fossil energy gradually replaces coal to become the mainstay in primary energy composition, making greater contribution to carbon reduction. With the spread of energy-saving technologies, the future energy intensity of GDP is likely to maintain an annual fall of 4–5%. This signifies that after

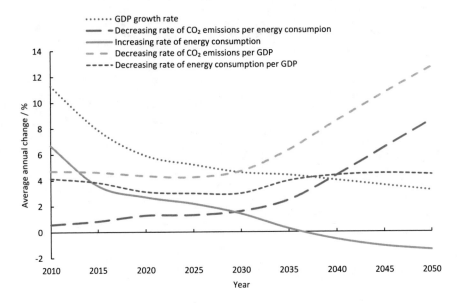

Fig. 9.5 Analysis on the decline of GDP growth and CO_2 emissions per unit of GDP

2040, energy substitution is set to be a greater contributor to the reduction of CO_2 intensity of GDP than energy conservation.

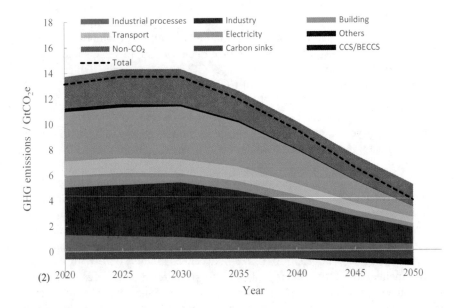

Fig. 9.6 Greenhouse gas emissions under the 2 °C target path

Two curves representing the expected GDP growth and the deceasing rate of CO_2 emissions per capita GDP meets at point A, the year when the CO_2 emissions peak occurs. Point B is the focal point of the annual growth rate of energy consumption and the annual rate of decline of CO_2 intensity of energy consumption. It suggests that the annual CO_2 intensity in energy consumption is declining at a greater rate than the annual growth rate of energy consumption, which also signals the CO_2 peak from an energy point of view. Point C represents the year with zero annual growth of energy consumption when economic growth is officially decoupled from energy consumption, that is, economic growth can be achieved without using additional energy—a new stage of endogenous growth. Data in the chart shows that economic growth is on track to decouple from energy consumption in the period from 2035 to 2040 under the 2 °C target path. Energy savings from technological progress and profound economic restructuring can meet the energy demand of economic growth. Overall, it shows a continuous downward trend of total energy consumption amid economic growth.

Compared with the policy scenario, the 2 °C target path is characterized by dramatic emission reduction in end users and the power sector, shown in Table 9.5.

Compared with the policy scenario, the 2 °C target path enables a 76.7% reduction in emissions by 2050, with the biggest contributors being the industrial sector and power sector (34.7% and 34.2% respectively) and CCS contributing 7.1% to as well.

9.2.3 Result of the 2 °C Target Path Through Different Statistical Methods

In measuring renewable energy consumption, this report converts the power generated from renewables as hydro, wind, and solar into primary energy based on coal equivalent calculation method, which is China's traditional statistical methodology. Some other countries and international agencies (e.g.: the IEA) do so based on the method of calorific value calculation. In the case of around 40% efficiency of traditional fossil energy generation, there is a roughly 60% discrepancy between the two methods might be found in the amount of primary energy. When renewables does not take up a major share in primary energy consumption, no major discrepancy is detected. But with the surge and increasing uptake of renewables in the future, significant variances will occur on the assessment of total energy consumption, energy conservation and substitution, prompting the need for further diagnosis and comparison. Table 9.6 compares the two methods of converting hydro, wind, solar and other renewables into primary energy [7].

Both statistical methods have considerable bearing on total energy demand and assessment on the contribution of energy conservation and energy substitution to CO_2 emission reduction. Converting renewable power into primary energy by coal equivalent calculation equivalent method underestimates the total energy demand, when comparing to the traditional method of coal consumption from power generation. Meanwhile, with the increase in share of renewables in primary energy, the method

Table 9.5 CO_2 emission reduction by sector under 2 °C target path and policy scenario (Unit: billion tons)

	2020	2030				2050			
		Policy scenario	2 °C target path	Emission reduction	Contribution (%)	Policy scenario	2 °C target path	Emission reduction	Contribution (%)
Industrial	3.77	4.54	4.15	0.39	45.3	3.69	1.19	2.50	34.7
Building	1.00	0.97	0.88	0.09	10.5	0.83	0.31	0.52	7.2
Transport	0.99	1.16	1.09	0.07	8.1	1.11	0.55	0.56	7.8
Others	0.21	0.40	0.38	0.02	2.3	0.16	0.04	0.12	1.7
Power	4.06	4.01	3.95	0.06	7.0	3.29	0.83	2.46	34.2
Industrial process	1.32	1.17	0.94	0.23	26.7	0.92	0.47	0.45	6.3
Forest carbon sinks	−0.58	−0.61	−0.61	0	0.0	−0.62	−0.7	0.08	1.1
CCS/BECCS	0	0	0	0	0.0	0	−0.51	0.51	7.1
Total CO_2 emissions	10.77	11.64	10.78	0.86		9.38	2.18	7.20	

Table 9.6 Primary energy consumption with different statistical methods

	2005		2020		2030		2050	
	Coal equivalent calculation	Calorific value calculation	Coal equivalent calculation	Calorific value calculation	Coal equivalent calculation	Calorific value calculation	Coal equivalent calculation	Calorific value calculation
Hydro, wind, solar power generation (trillion kWh)	0.5		2.1		3.6		9.6	
Conversion to primary energy (billion tce)	0.16	0.07	0.62	0.25	1.07	0.44	2.88	1.18
Non-fossil fuel (billion tce)	0.19	0.10	0.79	0.42	1.48	0.86	3.81	2.11
Total energy demand (billion tce)	2.61	2.52	4.94	4.57	5.98	5.35	5.20	3.50
Coal (%)	72.4	75.1	57.0	61.6	45.0	50.3	9.1	13.5
Oil (%)	17.8	18.5	18.5	20.0	17.0	19.0	7.7	11.4
Gas (%)	2.4	2.5	8.5	9.2	13.0	14.5	10.0	14.9
Non-fossil fuel (%)	7.4	3.9	16.0	9.3	25.0	16.1	73.2	60.2
CO_2 emissions (Gt CO_2)	6.06		10.03		10.46		2.93	

(continued)

Table 9.6 (continued)

	2005		2020		2030		2050	
	Coal equivalent calculation	Calorific value calculation	Coal equivalent calculation	Calorific value calculation	Coal equivalent calculation	Calorific value calculation	Coal equivalent calculation	Calorific value calculation
CO_2 emissions per unit of energy consumption ($kgCO_2$/kgce)	2.32	2.41	2.03	2.19	1.75	1.96	0.56	0.83
Decline of CO_2 emissions per unit of energy consumption(%)			0.89	0.62	1.44	1.12	5.52	4.18
Average annual decline of energy consumption per unit of GDP (%)			3.69	3.96	2.97	3.29	4.30	5.63
Decline of CO_2 intensity per unit of GDP (%)			4.55	4.55	4.40	4.40	9.60	9.60

Table 9.7 Comparison of metrics in 2050 with 2030 under two statistical methods

	Coal equivalent calculation	Calorific value calculation
Decline in total energy consumption (%)	13.0	34.6
Decline in energy consumption per unit of GDP (%)	58.7	68.9
Decline in CO_2 consumption per unit of energy consumption (%)	67.9	57.3
Decline in CO_2 consumption per unit of GDP (%)	86.7	86.7
Decline in total CO_2 emissions (%)	72.0	72.0

overestimates the decline in energy consumption per unit of GDP and underestimates the decline of CO_2 emissions per unit of energy consumption, thus transferring the emission reduction contribution from energy substitution to energy conservation and inflating the contribution of energy conservation to CO_2 emission reduction.

From 2030 to 2050, the generating capacity of hydro, wind, and solar will surge from 3.6 trillion kWh to 9.6 trillion kWh, with their share of power generation soaring to 73% from 38.9%. The total energy demand in 2050 measured by the two statistical methods is 5.2 billion tce and 3.5 billion tce respectively, with calorific value calculation method 1.7 billion tce less, or 32.7% lower than the method of coal equivalent calculation. This is not due to greater energy saving, but the difference in statistical method. From 2030 to 2050, energy intensity per unit of GDP is down by 58.7% and CO_2 emissions per unit of energy consumption is down by 67.9%, measured by the coal equivalent calculation method; whereas the calorific value calculation method calculate the decline at 68.9% and 57.3% respectively. During the same period, total CO_2 emissions from energy consumption will decrease by 7.53 billion tons, down 72.0%. By the coal equivalent calculation method, energy conservation contributes approximately 10% and energy substitution about 90% to the decline; whereas the calorific value calculation method puts their contribution at 33% and 67%, respectively (Table 9.7). Thus the latter method suggests an illusion of a greater contribution by energy saving. In this sense, when comparing and analyzing with similar international studies and global data, one should be mindful of the statistical methods used, as the results and conclusions produced by different methods mostly are not comparable.

9.2.4 Analysis on Transition of All Greenhouse Gas Emissions in 2 °C Target Path

Table 9.8 and Fig. 9.6 illustrate the structure and total amount of all GHG emissions under the 2 °C target path. Until 2030, total net greenhouse gas emissions remain at plateau at roughly 13.57 billion tCO_2, in which energy consumption-related CO_2

Table 9.8 Makeup of greenhouse gas emissions under the 2 °C target path (unit: GtCO$_2$e)

	2020	2030	2050
CO$_2$ emissions from energy consumption	10.03	10.46	2.92
CO$_2$ emissions from industrial processes	1.32	0.94	0.47
Non-CO$_2$ greenhouse gas emissions	2.44	2.78	1.76
Forest carbon sinks	−0.58	−0.61	−0.70
CCS + BECCS	0.00	0.00	-0.51
Net emissions	13.21	13.57	3.94

emissions stand at 10.46 billion tCO$_2$, emissions from industrial processes at 0.94 billion tCO$_2$, non-CO$_2$ emissions at 2.78 billion tCO$_2$e, and forest carbon sinks at 610 million tCO$_2$e. Energy consumption-related CO$_2$ emission remains the most important source of greenhouse gas emissions. In 2050, with the rapidly decarbonized energy mix, energy consumption-related CO$_2$ emissions, emissions from industrial processes, and non-CO$_2$ emissions are 2.92 billion, 470 million, and 1.76 billion tons respectively, down 71.9%, 57.3%, and 36.7% from the level in 2030. With the prospect of 700 million tons of forest carbon sinks and 510 million tons of CCS/BECCS, net greenhouse gas emissions in 2050 are estimated at 3.94 billion tCO$_2$e, 71% less than in 2030. In 2050, without taking into account agricultural and forestry carbon sinks and CCS, non-CO$_2$ greenhouse gases emissions make up 34% of total greenhouse gas emissions, compared with 16% today. Non-CO$_2$ emission reduction becomes an important source for curbing greenhouse gas emissions, providing the rationale for earlier technological planning.

In the long run, further technological breakthroughs are needed to reduce non-CO$_2$ GHG emissions, and effective new alternatives should be developed, which might rapidly drive down the cost of emissions reduction. Under the 2 °C target path, it is more important to strategically deploy research on non-CO$_2$ emission reduction technologies in advance, and explore more effective market mechanisms such as the carbon market to accelerate non-CO$_2$ emission reduction.

Compared with the policy scenario, the total greenhouse gas emissions without CCS and carbon sinks in the 2 °C target path by 2050 decreases by 61%, and the total greenhouse gas emissions with CCS and carbon sinks decreases by 69%, as shown in Table 9.9.

9.3 Analysis of Net Zero CO$_2$ Emissions Pathway Driven by the 1.5 °C Target

At present, the call for establishing and achieving the target of holding temperature rise below 1.5 °C has been getting stronger. Since the release of the IPCC *Special Report on Global Warming of 1.5 °C*, many countries have set out to layout and plan relevant strategies and goals for net zero emissions by the middle of this century. In

Table 9.9 Greenhouse gas emissions under the 2 °C target path and policy scenario in 2050 (unit: $GtCO_2e$)

	Policy scenario	2 °C target path
CO_2 emissions from energy consumption	9.08	2.92
CO_2 emissions from industrial process	0.92	0.47
Non-CO_2 greenhouse gas emissions	3.17	1.76
Forest carbon sinks	−0.62	−0.70
CCS + BECCS	0.00	-0.51
Net emissions	12.55	3.94

particular, European Union, aims to be the first regional block to have committed to going "climate-neutral" by 2050, and plays a leadership role of the global climate change progress. China, as the biggest emitter of greenhouse gases with aiming the goal of building a great modern socialist country by 2050, should undertake emission reduction obligations which commensurate with the rising national power and international influence. Based on the detailed analysis of the 2 °C target-oriented scenario in the previous section, this part is going to provide a further study and discussion on the 1.5 °C target-oriented scenario, including transition pathway towards net zero CO_2 emission goal by 2050 and also deep decarbonization of other greenhouse gases.

9.3.1 Scenario Analysis on 1.5 °C Target-oriented Net-zero Development Path

To bring about net-zero CO_2 emission of the energy system by 2050, as shown in Fig. 9.7, efforts should be made to achieve the carbon peak as soon as possible by 2030, and promptly transition to the emission pathway of the 1.5 °C scenario. Compared with the 2 °C target path, the 1.5 °C target-oriented net-zero development path (1.5 °C target path) shows more notable decline in direct CO_2 emissions of the energy system after 2030. From 2030 to 2050, CO_2 falls by 9.3% annually, 3.2 percentage point higher than the 6.1% in the 2 °C target path, meaning that the period should be characterized by greater decarbonization of the energy mix, and newly added non-fossil fuel consumption should rise by an extra 0.5 percentage point to reach an annualized 5.5% growth.

To achieve the 1.5 °C target path, the key lies in accelerating the rise of non-fossil fuels in primary energy consumption mix while curbing total energy use. Based on the 2 °C target path analysis in the previous section, greater efforts should be made in decarbonization of the energy mix to enable net-zero emission goal by 2050. The relevant indicators of economic, energy, and CO_2 emission are provided in Table 9.10.

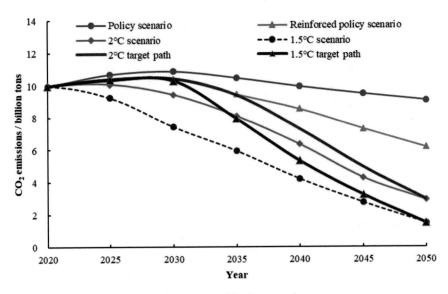

Fig. 9.7 Comparison of the 1.5 °C target path with other scenarios

To strive for net zero CO_2 emission by 2050, total energy consumption should peak around 2030, ideally under 5.87 billion tce, and energy consumption should also show a downward trend after 2030. In the meantime, decarbonization of the energy mix must accelerate, with coal, oil, natural gas, and non-fossil energy accounting for 45%, 17%, 13%, and 25% of primary energy consumption respectively in 2030, and 5.4%, 3.0%, 5.5%, and 86.1% respectively in 2050, with non-fossil energy making greater contributions. From 2030 to 2050, the annual energy consumption per unit of GDP can decrease at around 4%-5%, and the decline in carbon intensity of energy consumption brought by the decarbonization of energy mix can be improved from 4.4% in 2030 to 13.7% in 2050, playing a crucial role in achieving the net-zero emission in 2050. The total amount and composition of China's primary energy consumption under the 1.5 °C target path are shown in Fig. 9.8. After 2030, China should push for a quick reduction in its consumption of fossil energy, especially coal, and the installation of coal and other fossil energy units would risk massive early retirement and huge stranded costs.

Figure 9.9 and Table 9.10 illustrate the CO_2 emissions of major sectors under the 1.5 °C target path. After 2030, more notable carbon reduction is observed in all sectors. The industrial and power sector, in particular, need to strengthen their efforts in emission reduction. By 2050, with the contribution of 780 million tons of forest carbon sink and 880 million tons of CCS (BECCS), the economy-wide CO_2 emissions are reduced to about 60 million tons, basically achieving a net zero emission target.

Table 9.10 Indicators of economic, energy, and carbon dioxide emission under the 1.5 °C target path

Item (unit)		2005	2010	2015	2020	2025	2030	2035	2040	2045	2050
Annual GDP growth rate (%)			11.3	7.9	5.9	5.3	4.8	4.4	4.0	3.6	3.2
GDP index		1	1.71	2.50	3.33	4.31	5.45	6.75	8.22	9.81	11.48
Energy consumption (billion tce)		2.61	3.61	4.36	4.94	5.50	5.98	5.65	5.43	5.22	5.00
Energy consumption structure	Coal (%)	72.4	69.2	64	57.0	51.2	45.0	35.5	24.0	14.1	5.4
	Oil (%)	17.8	17.4	18.1	18.5	17.8	17.0	13.6	9.6	6.1	3.0
	Gas (%)	2.4	4	5.9	8.5	10.7	13.0	11.2	9.0	7.2	5.5
	Non-fossil fuel (%)	7.4	9.4	12	16.0	20.3	25.0	39.7	57.4	72.7	86.1
CO_2 emissions per unit of energy consumption (kgCO_2/kgce)		2.32	2.25	2.16	2.03	1.90	1.75	1.40	0.98	0.62	0.29
Annual decline of CO_2 intensity per unit of energy consumption (%)			0.58	0.83	1.24	1.37	1.59	4.37	6.92	8.85	13.76
CO_2 emissions (billion tCO_2)		6.06	8.13	9.38	10.03	10.45	10.46	7.93	5.33	3.21	1.47
Annual decline of energy consumption per unit of GDP (%)			4.15	3.82	3.09	2.97	2.97	4.93	4.60	4.26	3.92
Decline of CO_2 emissions per unit of GDP (%)			4.72	4.62	4.33	4.23	4.57	9.08	11.20	12.77	17.10
Decline from 2005 level (%)			21.5	38.0	50.3	60.0	68.3	80.6	89.3	94.6	97.9

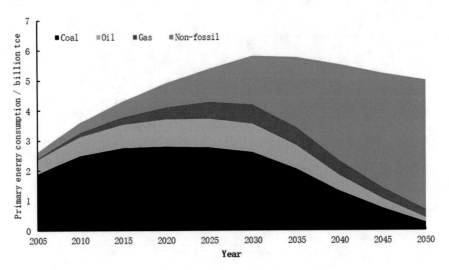

Fig. 9.8 Primary energy consumption and composition under the 1.5 °C target path

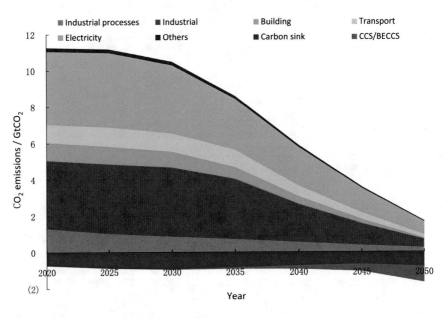

Fig. 9.9 CO_2 emission by sector in the 1.5 °C target path

9.3.2 Comprehensive Analysis on GHG Emissions in the 1.5 °C Target Path

Under the 1.5 °C target path, China's total greenhouse gas emissions and their composition in the future are shown in Table 9.11 and Fig. 9.10. In comparison with the 2 °C target path, total greenhouse gas emissions could drop to 1.33 billion CO_2e by 2050, of which energy-related CO_2 emissions make up approximately 1.47 billion tons and industrial process CO_2 emissions make up about 250 million tons. Forestry carbon sinks, CCS (BECCS), by and large, can offset CO_2 emissions from energy consumption and industrial processes, thereby achieving net zero CO_2 emission, other greenhouse gases will see an over 50% reduction from the peak, laying a solid foundation for net-zero emission of all greenhouse gases after 2050.

Table 9.11 All greenhouse gas emissions and their composition under the 1.5 °C target path (unit: $GtCO_2e$)

	2020	2030	2050
CO_2 emissions from energy consumption	10.03	10.31	1.47
CO_2 emissions from industrial process	1.32	0.88	0.25
Non-CO_2 greenhouse gas emissions	2.44	2.65	1.27
Forestry carbon sinks	−0.72	−0.91	−0.78
CCS + BECCS	0.0	−0.03	−0.88
Total emissions	13.07	12.90	1.33

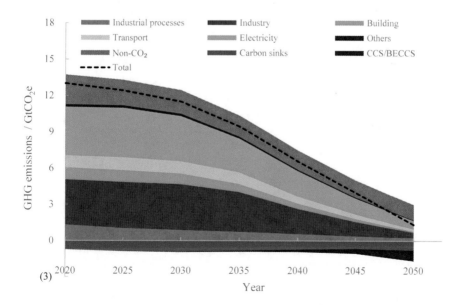

Fig. 9.10 Greenhouse gas emissions by sector under the 1.5 °C target path

9.3.3 Analysis on the Transition from the 2 °C Target Path to the 1.5 °C Target Path

By 2050, the net CO_2 emissions under the 2 °C target path will be 2.18 billion tons, and the net greenhouse gas emissions will be 3.94 billion tCO_2e. As for the 1.5 °C target path, in addition to achieving net-zero CO_2 emissions all greenhouse gases emissions would also need to be reduced to 1.33 billion tCO_2e. Figure 9.11 depicts the efforts needed for further mitigation in various sectors under 1.5 °C target path. Data suggests that industrial sectors should see significant reduction in greenhouse gas emissions, building and transport should also strive for extreme reductions, and the remainder can be offset by further deployment of CCS/BECCS technologies. Under the 1.5 °C target path, non-CO_2 emissions are expected to drop further with technological breakthroughs, but the remaining emissions still run as high as 1.33 billion tCO_2e, which would become the main challenge of net-zero emission of all greenhouse gases. In the 1.5 °C target path, strengthening the R&D of groundbreaking non-CO_2 emission reduction technologies in advance is crucial for China to achieve neutrality in all greenhouse gases. The initial cost of non-CO_2 emission reduction is relatively low, and emission reduction can be quickly obtained. However, the cost of deep decarbonization is expected to be on a steep rise, requiring introduction of groundbreaking technologies. At the same time, breakthroughs and development of CO_2 direct removal (CDR) technology is necessary to secure net-zero emission of all greenhouse gases soon after 2050.

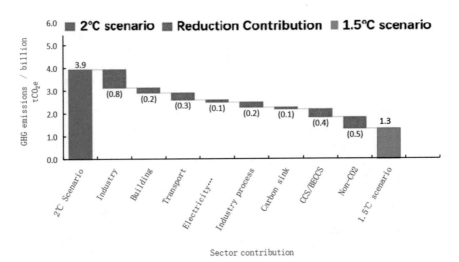

Fig. 9.11 Changes of sectoral contribution of emission reduction with shifting from the 2 °C target path to the 1.5 °C target path

9.4 Conclusions and Suggestions

(1) Difficulties and Challenges for Deep Decarbonization

Achieving "net-zero emission" by the middle of this century has become the goal of many developed countries, and the global efforts to cap the temperature rise within 1.5 °C have also gained more attention. At the end of 2018, the European Commission announced the target of "carbon neutrality" by 2050, and introduced "European Green Deal" policies package at the end of 2019 [8]. The UK also revised its Climate Change Act in 2019, incorporating "carbon neutrality" by 2050 into the Act, building the legal basis for deeper decarbonization of the UK [9]. The active advancement of the 1.5 °C-oriented net zero emission target in developed countries will not only accelerate their own low-carbon transformation of economy and society, but also bring challenges and pressure to developing countries and less developed nations for their efforts on this front.

The above analysis suggests that, China's long-term low-carbon transition pathway should compose of the short-term economic and social development with the long-term global temperature control targets set out in the Paris Agreement. It not only calls for greater reduction in greenhouse gas emissions in the 14th and 15th Five-Year Plans, but also further R&D of disruptive technologies including CCS/BECCS, large-scale energy storage, hydrogen energy, and smart grid to accelerate deeper decarbonization after 2030.

If the 1.5 °C target path proposed in this report is to be achieved by 2050, strengthened efforts should be made on top of the 2 °C target path. Compared with the 2 °C target path, the consumption of coal, oil, and natural gas should fall by another 42.9%, 62.5%, and 47.0%, whereas non-fossil energy should grow by another 13.2%, and the share of non-fossil in primary energy consumption should rise to 86.1%. To enable deeper emission reduction, aside from applying necessary CCS in the power sector and industrial processes, BECCS technology should be applied in the power sector, and hydrogen production from renewable energy and hydrogen energy as secondary energy should be used in fuel cell freight and steelmaking processes. However, these new technologies are hampered by the barriers of technological development, as well as uncertainties in the market penetration, which also brings major challenges to deep decarbonization.

(2) Work Plans for Long-term Low-carbon Transformation

China's long-term low-carbon transformation pathway should not only be consistent with the stage and characteristics of its socio-economic development, but also linked with the process of socialist modernization. It needs to accommodate both domestic and international interests and deploy futuristic competitive strategy. Before 2030, NDC targets need to be enhanced to balance social-economic development efforts in and energy conservation and carbon reduction, and cap both the intensity and total amount of CO_2 emissions to push for high-quality transformation. In addition, great efforts should be taken to step up the R&D of technologies and capacity building to lay a more comprehensive and solid groundwork for deeper decarbonization after 2030. Post- 2030, China should aim for the net-zero CO_2 emissions, and deep decarbonization for non-CO_2 greenhouse gases. Manage all metrics of China's socio-economic development, and strive to fulfill the responsibilities of a major power commensurate with its comprehensive national strength by 2050. Key recommendations for China's long-term low-carbon transformation.

(1) Strengthen the strategy-oriented development. The transformation of future socio-economic development should be strategy-oriented, with the low-carbon strategy embedded in every aspect of socio-economic development. Low-carbon strategy should be aligned in a timely and phased manner with the national long-term development strategy, and assigned to every five-year planning as well as sub-national strategies, thereby connecting the long-term low-carbon strategy and the short-term socio-economic development to form a new paradigm where the two promote and coordinate each other.

(2) Prioritize technological development. Technological development and revolutions will direct influence on the progress and results of global low-carbon transformation. Therefore, attention must be placed on the R&D of technologies, with adequately preparation for comprehensive low-carbon transformation. Particular focus should be placed on the R&D and demonstration of forward-looking technologies such as large-scale renewable energy grid connection, hydrogen energy, energy storage, smart grid, CCS/BECCS/DACCS, etc., to secure the bottleneck breakthroughs and technology industrialization, further enhance the international competitiveness of China's advanced technologies, and showcase China's contributions to the global low-carbon transformation.

(3) Boost capacity building. With the improvement of China's comprehensive national power, climate change actions should gradually expand to the entire economy and all greenhouse gases in order to demonstrate China's contribution as a responsible power. This requires an all-round strengthening of China's capacity to tackle climate change, including capacity building for greenhouse gas measurement, verification and reporting, enhanced government's capacity to climate change response and more active engagement and practices of the private sector.

(4) Enhance institutional framework. Strengthen institutional building, promote climate change legislation, provide stronger legal ground for climate change efforts, and boost synergies in addressing climate change and improving environmental quality. Build stronger carbon trading market, and better harness market forces to cut carbon emissions. Facilitate the coordination and integration of climate policies with those related to socio-economic development and environmental protection, and leverage the synergy of policy systems.

(5) Promote comprehensive and active all stakeholders participation. Step up education, publicity and training to further enhance the awareness, understanding, and support of the general public in addressing climate change, and encourage low-carbon consuming behavior and lifestyle. Create an enabling

environment for low-carbon participation and provide broader space for various stakeholders to take low-carbon actions actively.

References

1. National Bureau of Statistics (2019) China Statistical Yearbook 2019. China Statistics Press, Beijing
2. National Bureau of Statistics (2020) China Statistical Yearbook 2020. China Statistics Press, Beijing
3. National Development and Reform Commission, National Energy Administration. Revolution in Energy Production and Consumption (2016–2030) [2020-4-5]. https://www.ndrc.gov.cn/fggz/zcssfz/zcgh/201704/t20170425_1145761.html
4. Hailin W, Jiankun He (2019) China's pre-2020 CO_2 emission reduction potential and its influence. Front Energy 13(3):571–578
5. Jiankun He (2018) Post-Paris Agreement Global Climate Governance and China's Leadership. Environ Manag China 10(1):9–14
6. Jiankun He (2011) Economic analysis and effect evaluation of China's CO_2 emission reduction target. Sci Res 1:9–17
7. Yi J, Lanbin L, Xiu Y (2006) Discussion on different energy accounting methods in energy statistics. China Energy 28(6):5–8
8. European Commission. A European green deal [2020-03-29]. https://ec.europa.eu/info/strategy/priorities-2019-2024/european-green-deal_en#documents
9. Committee on Climate Change. Net Zero the UK's contribution to stopping global warming. [2020–04–10]. https://www.theccc.org.uk/publication/net-zero-the-uks-contribution-to-stopping-global-warming/

Chapter 10
Strategic Linchpins and Policy Safeguards

10.1 Policies and Measures for China's Low-Carbon Development

10.1.1 Put in Place and Perpetuate a Complete and Comprehensive Policy Framework

China has created a policy system for low-carbon development that covers a wide range of sectors. Since the end of last century, the central government has deemed climate change policy as part of its sustainable development policy, in particular as part of its energy reform and economic development policy. Despite a milestone in itself, this approach lacks clarity on the importance of climate change, which results in the absence of clear policies for climate change mitigation and adaptation [1]. Since the 12th Five-Year Plan, more independent policies have emerged in the field of climate change, with setting clear and legally binding targets for curbing carbon intensity for the very first time. A variety of low-carbon pilot projects, carbon trading pilot projects, and MRV capacity building have also been launched. During the 11th Five-Year Plan period, the Chinese government established the first energy conservation policy system with binding targets as the centerpiece and performance assessment as the safeguard, contributing to lower greenhouse gas emissions as well as reduced energy intensity. With some revisions, most of the energy saving policies were kept in the 12th and 13th Five-Year Plans. For instance, energy saving mandates were expanded from the industrial sector to transport and buildings; the provincial breakdown of energy saving targets were more science-based; targets were extended from capping energy intensity at the national level to capping total energy consumption and promoting industry-wide energy saving; and more focus was placed on the role of market forces [2, 3]. During the 12th Five-Year Plan period, China set a target for CO_2 intensity for the first time, i.e. a reduction of 17% per 10,000 yuan of GDP, with continued enforcement of the accountability system. The 13th Five-Year Plan

© China Environment Publishing Group Co., Ltd. 2022
Institute of Climate Change and Sustainable Development of Tsinghua University et al.,
China's Long-Term Low-Carbon Development Strategies and Pathways,
https://doi.org/10.1007/978-981-16-2524-4_10

went a step further to set a national target of reducing CO_2 emission intensity by 18%.

Now, a complete and comprehensive climate policy framework has been created in China, comprising of not only administrative mandatory policies and best practices with Chinese characteristics, but also economic incentives (pricing, trading cap, fiscal subsidy, etc.), regulations (laws, regulations, and standards), and policies for R&D on low-carbon technologies. Meanwhile, there are other policies affecting climate, such as electric power market reform, reform of taxes and fees, etc. At present, the formulation of China's *"Law on Tackling Climate Change"* is underway in a steady manner. The 10th Meeting of the Standing Committee of the 11th National People's Congress adopted *the Resolution on Actively Tackling Climate Change of the Standing Committee of National People's Congress* on August 27, 2009, vowing to make the strengthening of climate change legislation part and parcel of the efforts to build and enhance the socialist legal system with Chinese characteristics, and listing it on the legislative work agenda. Subsequently, legislation on enhancing climate change management was incorporated in the 2010/15 legislative plan in the National Development and Reform Commission. In 2011, the Leading Working Group for Climate Change Laws and Legislation was set up, comprised of 17 government agencies, including the Environmental Protection and Resources Conservation Committee of the National People's Congress, the Legislative Affairs Commission of the National People's Congress, the Legislative Affairs Office of the State Council, etc. The National Development and Reform Commission served as the lead for legislative research, survey and drafting, stakeholder opinions soliciting. The Climate Change Law (Draft) consists of areas including General Provisions, Principles, Supervision and Management, Mitigation, Adaptation, Publicity and Education, etc. [4]. At the moment, no consensus has been reached on the draft law, with ongoing debates over the legislative framework, design of the core system and the coordination with relevant laws.

At the local level, Shanxi, Qinghai, Shijiazhuang and Nanchang have successively carried out individual local legislation to address climate change and low-carbon development. In 2010, Qinghai province promulgated the *Measures of Qinghai Province to Tackle Climate Change,* defining the province's climate change adaptation and mitigation legal system. Shanxi Province released its measures in 2011 for tackling climate change based on its own reality, including identifying key areas of mitigation and adaptation, and lying out specific steps for greenhouse gas concentration monitoring, greenhouse gases list compilation list, verifying regional and corporate greenhouse gas emission output, and minimizing climate change's impact. Shijiazhuang introduced its city-wide regulation on low-carbon development in 2016, which provided stringent provisions on the management of the operation, distribution and use of coal and oil, encouraged the development and utilization of new and renewable energy, and devised incentive policies for building, transportation, lighting, indoor temperature regulation, consumption reduction of daily necessities and other daily residential consumption.

10.1.2   Goal-Oriented Planning, Administrative Mandates, and Market Measures

As oppose to western countries' rule-based governance system centered around legislation, law enforcement, and judicature, China's climate change policy is generally characterized by goal-based governance built upon the planning system, stressing goal-oriented planning, administrative mandates, followed up by market measures [5]. Though the special legislation on climate change is still being drafted, China has incorporated a relatively comprehensive low-carbon planning regime with the "Five-Year Plans" at the core, with detailed steps at the local, and sectoral and territorial level [6]. The Five-Year Plans are one of the fundamental tools of national macro-control. China views climate change management as a long-term strategic task for social and economic sustainable development. Since the 12th Five-Year Plan, China has incorporated carbon intensity as a binding target in the national economic and social development plan, and launched national, regional and sectoral plans, programs and strategies to cope with climate change. Revolving around the national binding target of emission reduction per unit of GDP, efforts have been made to propel the low-carbon transition in energy and economy. The *National Plan for Tackling Climate Change (2014–2020)* was released in 2014. The 12th and 13th Five-Year Plan both had work plans for curbing greenhouse gases, with low-carbon development plans for various industries. These have provided a major boost to China's climate change efforts.

Administrative means are the most distinctive policy tools of China. In spite of China's increasingly full-fledged climate policies, the administrative means centering around "goals" are still predominant [7, 8]. Among the different administrative measures, performance assessment policies of emission reduction, energy conservation and the phase-out of outdated production capacity for local governments and major emitters stand out as typical policy instruments that have adapted to China's current system and institutional mechanism, ensuring the attainment of China's climate change goals. On the other hand, administrative means alone have their deficiencies and adverse consequences. First, administrative means are mostly expensive and economically inefficient. Secondly, this top-down stress escalation may not be properly translated into the spontaneous desire of local governments and enterprises to preserve energy and cut emissions. Thirdly, as the targets of energy conservation and carbon reduction are assigned to multiple levels of government, those at or below the county level bear the responsibilities incommensurate with their administrative authority. Caving in to mighty political pressure from above, grassroot governments may resort to extreme measures such as power cuts to reach their goals, resulting in adverse consequences [9].

Driven by international mechanisms, market-oriented policy system for climate change have flourished from its infancy, marking its milestone with the creation of the carbon market. Holistically speaking, China's market mechanism for climate change has experienced three stages of development [10]. The first stage, spanning from 1992 to 2008, was the exploratory phase featured by mostly project-level, process-based and single-sided international transactions. As a developing country, China can sell its "Certified Emissions Reductions (CERs)" to developed nations helping them to meet their emission reduction targets through the Clean Development Mechanism (CDM). The second stage, from 2009 to 2015, is marked by efforts to foster the carbon market through innovations by bottom-up, allowance-based local pilots. In 2011, the Chinese government explicitly proposed in the 12th Five-Year Plan to "gradually set up a carbon trading market," recognizing the importance of carbon market in energy conservation and emission reduction for the first time at the national level. Later, to effectively implement the goals set out in the 12th Five-Year Plan, the National Development and Reform Commission issued the *Notice on the Pilot Work of Carbon Trading* at the end of 2011, approving the pilots in Beijing, Tianjin, Shanghai, Chongqing, Hubei, Guangdong and Shenzhen. From 2013 to 2014, the above seven provinces, municipalities, and cities successively promulgated the *Carbon Emission Trading Management Measures*, officially launching the seven pilots. The National Development and Reform Commission has provided each provincial and municipal commission much leeway for building the seven pilots that were distinctively designed in terms of the coverage, allowance allocation and trading rules, gaining rich and diverse practical experience for creating a nationwide carbon market. With the adoption of the Paris Agreement in 2015 and the inclusion of the emission peak target in the policy agenda, China is now embarking on the third stage of market development to combat climate change. In December 2017, the *National Carbon Trading Market Development Plan (Power Generation Sector)* was introduced, specifying that the power generation sector would be the springboard to kick off the national trading system. The first 1,700 power generation companies included in the national carbon market emit over 3 billion tons of CO_2e annually, accounting for about 1/3 of the country's total emissions. By the end of 2018, nearly 800 million tons of carbon emissions had been traded in China. In March 2019, the Ministry of Ecology and Environment released the *Provisional Regulations on the Management of Carbon Trading (Draft for Comments)*—a milestone in the institutional development of the national carbon market, providing the policy basis and legislative guarantee for building a nationwide carbon market.

10.1.3 Strengthen the Coordinated Governance of Environment and Climate Change

In recent years, it has become evident and trendy that coordinated governance between environment and climate change have become increasingly prominent. In the mid and late period of the 12th Five-Year Plan, air quality issues, in particular, PM2.5 and ozone drew tremendous concern both domestically and internationally, prompting swift actions by the Chinese government to introduce a spate of ambitious policies for combating air pollution, including the *Action Plan on Air Pollution Prevention* and *Control, Three Year Action Plan for the Blue Sky Campaign and*

Comprehensive Work Plan for Energy Conservation and Emission Reduction of the 13th Five-Year Plan. These policies, through decarbonizing the energy mix, developing clean energy, enhancing energy efficiency, and gradually decrease the use of fossil fuels, have managed to secure the coordinated governance of air pollution and climate change [11]. For example, the central government has enacted policies for curbing and reducing coal consumption, and started with the most polluted areas for stringent restriction on coal use, which were gradually expanded to the entire nation. Thanks to an array of interventions to curtail coal consumption, more than 950 million tons of coals were saved from the closure of small thermal power plants in China between 2005 and 2020. Approximately 15.18 million tons of SO_2 emissions were prevented from the reduced coal use. With over two billion tons of CO_2 emissions forestalled [12]. In a new round of organizational restructuring in March 2018, the climate change and emission reduction responsibilities of the National Development and Reform Commission were integrated into the newly established Ministry of Ecology and Environment, further benefit the coordinated governance of environment and climate change.

10.2 Strategic Linchpins and Policy Safeguards for China's Long-Term Low-Carbon Development

Combating climate change involves every facet of the economy, society and environment. Climate change strategy must also be aligned with the overall national development strategy and integrated into the economic and social development strategies of all industries and sectors, to enable coordinated governance and win–win cooperation among economy, society, environment and climate change. China's 2050 long-term low-carbon development strategy should, first and foremost, support the country's development, and assist in building a modern socialist country that is prosperous, strong, democratic, culturally advanced, harmonious and beautiful. Meanwhile, the strategy should seek to strengthen ecological civilization, green low-carbon circular development, harmonious coexistence between human and nature and sustainable development. It should fulfill the target of deep decarbonization aligned with the global efforts to limit temperature rise well below 2 °C and ideally with 1.5 °C (see Fig. 10.1). From the perspective of national macro strategy, China's climate change strategy and policy support need to highlight the following aspects:

Fig. 10.1   Strategic framework for long-term low-carbon emissions transition

10.2.1   Defining Climate Change Strategy in National Priority and Embed the Pathway of Deep Decarbonization in the Overall Development Goals and Strategies

The 19th CPC National Congress incorporated progress of ecological civilization in the foundation for building socialism in the new era, making a two-stage strategic roadmap. First stage, from 2020 to 2035, aims to basically achieve socialist modernization with fundamental improvement of the ecological environment and the achievement of a beautiful China in general. Second stage, lasting from 2035 to mid-century, seeks to build China into a great modern socialist country that is prosperous, strong, democratic, culturally advanced, harmonious, and beautiful with all-round ecological progress and fulfillment of a beautiful China. Climate change efforts need to be aligned with the two-stage goal, the formulation of long and medium-term and short-term climate plans needs to be integrated into the comprehensive strategies for the economical social environment. During the first stage in 2020–2035, it is important for the national economic and social development strategy to harness the synergy of improving environmental quality and reducing CO_2 emissions, attain China's NDC goals while lowering PM2.5 concentrations in major cities and regions to below 35 $\mu g/m^3$, thus achieving the coordinated governance and a win–win scenario of the economy, energy, environment and climate change. Therefore, it's essential to specify the earlier peak of CO_2 emissions as a major strategic target at this stage, and put forth the peak emission targets and measures to spur the energy revolution and economic transformation. Ideally, 2035 should witness notable reduction in CO_2 emissions from the peak year, with a continuous cut of absolute emissions and the absolute decoupling of sustainable economic and social development from CO_2 emissions, which are to be aligned with a more ambitious emission reduction targets in the second stage [13, 14].

After 2035, as China achieves basic modernization, with ecological environment fundamentally improved, urgency is placed on climate change response and greenhouse gas reduction, which is a more daunting challenge. To fulfill the 2 °C target, the annual decline of global greenhouse gas emissions must exceed 7%, far higher than that of the developed nations before 2030. With this in mind, during the second stage from 2035 to 2050, China's climate change strategy should go beyond the domestic need for resource utilization and environmental protection, it should accommodate the pathway of emission reduction needed for preserving the ecological safety of the earth, honor the historic mission by making contributions to humanity. Using the benchmark of keeping temperature rise below 2 °C with continuous pursuit for 1.5 °C, formulate goals, strategies and pathways for a steep reduction in absolute emissions across all greenhouse gases in the entire economy, and bring to fruition near-zero CO_2 emission and deep emission reduction from other greenhouse gases by the middle of the century. The process of deep decarbonization of the energy system will also effectively roll back the emissions of conventional pollutants, playing an instrumental role in securing a $PM_{2.5}$ concentration below 15 $\mu g/m^3$ in major regions and cities in China by 2050. With the continuous development of China's comprehensive national strength and international influence, China should proactively assume responsibilities and obligations that echo its position in the greater world, placing climate change goals at the center of the strategy of building a socialist modern society, steer the world onto the path of climate-friendly low-carbon economic development, taking on commitment to global ecological civilization, the shared interests of all humanity. Demonstrate world-leading influence and leadership in combating climate change, by historic contribution to the preservation of the earth's ecological safety, and the survival and development of all humankind.

The key to implementing the low-carbon development strategy today lies in changing the mindset of policy-makers at all levels. It is crucial to adhere to the development philosophy of the new era, with innovation at the core of development, shift growth drivers, and transform the way of development. It's imperative to enable harmonious coexistence between human and nature and embark on a path of high-quality green and low-carbon circular development. Policy-makers at all levels must shift their mindset in development concepts, from sole pursuit of the quantity and speed of GDP growth to a more comprehensive effort on the quality and efficiency of economic growth, formulate and enforce binding targets and policy "red lines" for resource conservation, environmental protection and CO_2 emission reduction. Governments at all levels should balance between economic growth, employment, and life quality with resource conservation, environmental protection and carbon emissions, carefully balancing GDP growth with its related losses of resources and environment, to avoid a "dilemma" and enjoy a "win–win". Efforts should be made to take climate change as the centerpiece of building ecological civilization, placing sustainable development strategy featuring green low-carbon transformation high on the agenda of economic development, change the mindset and assessment criteria of policy-maker, and strengthen the accountability of energy conservation, carbon reduction, and ecological protection.

Decision-makers at every position should fully recognize the importance and strategic positioning of global cooperation in addressing climate change. It is an indisputable scientific fact that human activities have caused global climate change, and tackling climate change is a common cause for global cooperation in protecting the ecological safety of the earth and the survival and development of humankind. President Xi Jinping repeatedly stressed that "climate change affects the well-being of everyone and the future of humanity." China's strategy for promoting low-carbon transformation and combating climate change is not only an inherent demand for sustainable development domestically, but also a major step forward to proactively assume international responsibilities, and actively contribute to the progress of humankind. The shared interests of all should be the center in the strategic goals and roadmap for low-carbon transformation of China. It's imperative to proceed from and to take the leadership of the global trend for low-carbon development in combating climate change, (develop competitive advantages for low-carbon technologies and execution capabilities for it to become a milestone in building of a strong modern socialist country.)

10.2.2 Adhere to New Development Concept, Spur Industrial Restructuring and Upgrading, and Facilitate Green, Low-Carbon, Circular Economical Transformation

Since the Chinese economy entered a period of "new normal", the central government has put forth five development notions, namely, "innovative development, coordinated development, green development, open development and shared development," using innovative development as the growth drivers, and green development as the strategic target. The 19th National Congress also enshrined strengthened ecological civilization and the building beautiful China as a fundamental strategy of China's socialist modernization, indicating the need to establish an industrial system that is of green, low-carbon and circular development, build a clean, low-carbon, safe and efficient energy system, and promote comprehensive conservation and utilization of resources. This will help accelerate the strategic restructuring of the economy, catalyze industry transformation and upgrade, enhance development efficiency and quality. It is also a strategic action to reduce CO_2 emissions.

Since the 11th Five-Year Plan period, China has seen remarkable achievements in energy conservation and carbon reduction, with energy intensity and CO_2 intensity per unit of GDP on rapid decline. Given the unique characteristics of economic structure and energy mix at the current stage of industrialization, the energy and CO_2 intensity per unit of GDP are still at a relatively high level. The technical efficiency of energy conversion and utilization and energy consumption per unit of product in China are not far behind that of developed countries, with many areas at world leading level. Yet China's energy consumption per unit of GDP in 2017 was still 1.6 times that of the world's average, and two to four times that of developed countries.

This was mainly due to an excessively high proportion of the secondary industries, especially the heavy chemical sector. Energy consumption from the industrial sector accounts for 2/3 of the total national end-use, as opposed to no more than 1/3 in developed countries in general. Moreover, China's manufacturing products are at the middle and lower end of the value chain with high energy consumption and low added value. The structural factors of the above industries and products are the main contributors to high energy consumption of GDP [15].

At present, the new wave of industrial revolution marked by digitalization and smart technologies is flourishing worldwide with cross influence of two major trends of economic green transformation and digital transformation. The growing construction of data centers during digital transformation will spur up the demand for electricity, and will also lead to tremendous improvement in the efficiency of economic growth and resource utilization, cutting energy consumption intensity per unit increase of GDP as a whole. Therefore, it's vital to grasp and speed up the new economic development marked by digitalization, optimize allocation of resources, and accelerate the spread and application of such digital technologies, including artificial intelligence, 5G, cloud computing, and Internet of Things, during the low-carbon transformation of energy and economy.

From 2005 to 2018, about 1/3 of the decline in GDP energy intensity was attributed to the technological energy saving from the improved efficiency of energy technologies, whereas around 2/3 was from the structural energy saving from the changes and upgrading of industrial structure and product portfolio. In the future, the strategic restructuring shall focus on accelerated high technology industry and modern service industry, reducing the share of the industrial sector, especially the heavy and chemical sector with high energy consumption, while moving up manufactured goods to the higher end of the global value chain. This is an important strategy for low-carbon economic transformation.

It is necessary to establish and perpetuate an energy-saving and low-carbon industrial system by adopting new industrialization trends, vigorously developing circular economy, optimizing the industrial structure, revising the catalogue of industrial restructuring, restoring the expansion of energy intensive and high emission industries, speeding up the phase-out of backward production capacity, vigorously fostering the service and strategically emerging industries, devising targets and action plans for emission control in key industries, and studying and formulating standards for greenhouse gas emissions in key industries. Measures should be made to enhance energy efficiency through energy conservation, restrain emissions from key industries such as electricity, iron and steel, nonferrous metals, building materials, and chemicals, and etc. Strengthen the management of carbon emissions in new projects, actively control greenhouse gas emissions from industrial production, promote the circular transformation of industrial parks and build a green, low-carbon and circular industrial system.

Currently, China's exports mainly comprise of manufacturing goods, and the export of services is relatively low. However, these manufacturing products are mostly medium and low-end with high energy consumption and low added-value, causing persistently high implied energy and CO_2 levels of export in goods, which accounted for more than 20% of China's total emissions from 2005 to 2014, a percentage only declined until recent years. The implied CO_2 emissions from export in goods made up 15.3% of total domestic emissions in 2018, while implied emissions from imports stood at 542 million tons, and implied emissions from net exports contributed 10.8% to total domestic emissions. Thus, it is imperative to actively place efforts in limiting the extensive expansionary economic growth pattern driven by continuous rise in investment, capacity building of the heavy and chemical industry and expansion in export of manufacturing goods. The export structure should be optimized to curb and reduce the export of energy-intensive products, promote export of higher-end products in the value chain, reduce implied carbon transfer to foreign countries, encourage and boost export of high technology and high-end service industries, and work to strike a balance of imports and exports in terms of implicit carbon, a crucial step to lower domestic CO_2 emissions, and should be deemed a major component of China's low-carbon development strategy. With the ongoing economic globalization, China's efforts in developing export-oriented economy, improving the international competitiveness of manufacturing products and technologies, and moving to the higher end of international industrial value chain also constitute the country's solution to reducing CO_2 emissions and securing low-carbon transformation [16, 17].

It is of great significance to develop a circular economy with improved efficiency of resource utilization to cut energy consumption and CO_2 emissions. Resource utilization efficiency is an important bridge toward green and low-carbon circular development. It emphasizes reduction, reuse and recycling of resources to obtain more higher added-value and sustainable products and services with less resource consumption and emissions. This also marks a key component of green and low-carbon transformation and sustainable development.

By current universal standard, China consumes vast amount of material resources. In 2017, China made up 1/3 of the world's resource extraction. In terms of per capita intensity, China's per capita material resource extraction in 2017 was 23.6 tons, ranking the third in the world, about twice the world's average per capita of 12.2 tons. The resource output of China's economic system is less than 1/10 of that of Germany and Japan. One reason is structural. China is under rapid urbanization and massive infrastructure construction that consume large quantities of material resources, different from post-industrialized developed nations. Second, China is relatively behind in the technology and management of primary resource reduction, efficient processing and conversion, waste recycling, and etc. due to technical impediments. China's overall resource consumption will remain relatively high for a long time in the future. Growing urbanization will keep increasing the stock of material resources in the society, with explosive growth in urban wastes, prompting accelerated efforts in the recycling and reuse of used and scrapped cars, home appliances, photovoltaic components, and electric vehicle batteries. "Urban mines" will partly replace natural minerals to become an important source of metals and non-metallic material resources, with recycling making greater contributions to the guarantee of strategic resources. Under this scenario, it is necessary to make the development of circular economy and the improvement of resource utilization efficiency a crucial component of the long-term low-carbon development strategy, and strive to double the output of major resources by 2035 and measure up to the world's advanced level of resource utilization efficiency by 2050. Similar to the proposal of the EU Green Deal to decouple economic growth and resource utilization to embark on the path of sustainable development featuring a green and low-carbon cycle [18].

10.2.3   Further Improvement in Energy Conservation and Efficiency for End Use and Facilitate the Substitution of Electricity and Hydrogen for Direct Use of Fossil Energy

The industrial sector is the biggest end user of energy in China, accounting for about 65% of the total end-use of energy nationwide in 2018, of which 70% comes from energy-intensive raw material sectors. China's output of steel, cement, household appliances and other products makes up about half of the world's total. Aside from accelerating industrial restructuring, transformation and upgrading, focus must be placed on the phase-out of backward production capacity, swifter industrial technology upgrade and improve energy efficiency. In particular, for steel, cement, coal and petrochemical industries where emission reduction remains a challenge, revolutionary and deep decarbonization technologies such as hydrogen steel making and hydrogen chemical technologies should be developed. Electric and hydrogen energy should take up greater share in end-use energy composition to bring about the deep decarbonization in the industrial sector. As a manufacturing powerhouse, China will not completely follow the path of developed countries like the Americas and Europe to relocate the manufacturing industry to less-developed regions in its future modernization endeavors. By the middle of this century, China will maintain its strategic positioning of a major and comprehensive manufacturing nation, with the added value of industrial sector to percentage of GDP no less than the current average level of developed countries. Despite the trend of urbanization, the projected rate of growth and proportion of energy consumption and CO_2 emissions in the construction and transportation sectors will overtake that of the industrial sector, but the share of energy end consumption in the industrial sector will remain higher than that in the construction and transportation sectors. Given the great potential of industrial sector for energy conservation and carbon reduction, it should enable an earlier peak of its CO_2 emissions prior to 2030 by ramping up industrial transformation and upgrading, improving energy efficiency and promoting electric energy replacement [19].

In 2017, the end energy consumption of the construction sector accounted for 16% of the total end use in China. To fully unleash the potential of energy conservation in the construction sector, improve related policies on building energy efficiency, and push for the 65% energy efficiency standard for all new buildings by 2020, enhance the mix of energy end-use of buildings, so that renewables could contribute 25% to all energy consumption of buildings by 2020. Accelerated steps should be taken to revamp existing buildings and heating metering for energy conservation, government should lead the efforts with imposed cap on energy consumption of public buildings. More importantly, the total number of public and civil buildings should be limited. In 2015, the construction industry consumed around 864Mtce of energy, approximately 20% of the national total energy consumption. In the future, the total building area should be kept at about 75 billion square meters with roughly 1100Mtce in energy consumption. Heating in northern China should better leverage the combined use of waste heat and pressure in the power and industrial sector, strengthen distributed renewable energy and electric energy replacement, boost the clean and efficient utilization of biomass energy in rural areas, and strive for emission peak around 2030, support the nationwide peak target, and bring about near-zero emission in the sector by the middle of this century [20].

A comprehensive green and low-carbon transport system should be established to improve the transport structure. Develop railways, urban rail transit, waterways and other modes of transport that are low energy consumption and enhance transport demand management. Fuel consumption limits should be raised across the board for vehicles and vessels with a timetable for phasing out gasoline vehicles. Green mobility should be encouraged with promotions for more trips on public transport. The composition of vehicle fuels should be improved by tapping into biofuels, especially for electric vehicles and hydrogen fuel-cell vehicles. Oil and gas should be replaced with electricity and hydrogen on a large scale. Smart management of transportation must be strengthened to develop combined transport of land and sea with enhanced efficiency. In 2018, 509 Mtce of energy was consumed in the transport sector, equivalent to 10.7% of the country's total energy consumption, with freight taking up over 60%. The future will see increasing proportion of energy consumption in passenger transport with reduced consumption in freight. By adopting proper measures, China is on track to achieve the peak around 2035 in the transport sector, and near-zero emission by the middle of this century [21].

With the end-use sector, promoting the substitution of electricity and hydrogen energy for direct combustion of fossil fuels and utilization of raw materials is a vital step in the low-carbon transition strategy. With the support of near-zero emission energy system dominated by new and renewable energy, replacing fossil energy with zero-carbon electricity and hydrogen generated from electricity is a crucial initiative for the end users to achieve near-zero emission in the future. The share of power in end-use energy consumption in China was about 24% in 2019, projected to rise to over 30% in 2030, and around 55% in 2050 under the 2 °C scenario. Non-fossil energy as a percentage of primary energy will also climb from 15.3% in 2019 to 25% in 2030 and over 70% in 2050. With the increase of non-fossil power in total electricity consumption, more electric power substitution in the end use sectors will be instrumental to lower CO_2 emissions. On the other hand, hydrogen as a clean, zero-carbon secondary energy, can be used as a raw material in the process of industrial production to replace fossil fuels such as coal, oil, coke and gas. It can also be used as a means of energy storage to replace and reduce fossil fuel combustion and direct use of the end users, and should therefore be included in China's long-term low-carbon development strategy as an important strategic resource [22].

10.2.4 Speed up the Decarbonization of Energy Mix and Ensure Clean, Safe, and Economical Energy Supply

Since the 11th Five-Year Plan, China has paced up the development of new and renewable energy and promoted the low-carbonization energy mix. The share of coal in the total energy mix has dropped from 72.4% in 2005 to 57.7% in 2019, non-fossil energy has risen from 7.4% to 15.3%, and CO_2 intensity per unit of energy consumption has fallen by 13.3% in 2019 from the level in 2015. However, the coal-based energy mix is yet to be fundamentally transformed, with the CO_2 intensity per unit of energy consumption more than 20% higher than the world average and 35% higher than that of the European Union. Accelerating the development of new and renewable energy and reducing CO_2 intensity per unit of energy consumption are important measures to cut CO_2 emissions on the premise of ensuring safe and efficient energy supply. By the middle of this century, China is to build a near-zero-emission energy system with new and renewable energy as the mainstay, actively develop technologies like Carbon Capture and Storage (CCS) and Bioenergy with Carbon Capture and Storage (BECCS). These technologies can be applied for massive industrialization after 2030 and enable more than 500 million tonnes of CO_2 capture and storage a year by 2050, paving the pathway for earlier carbon neutrality.

By 2030, non-fossil power should comprise about 50% of the total power generated, with non-fossil fuels taking up about 25% of the primary energy mix before rising up to 90% and over 70% respectively by 2050 under the 2 °C scenario, and over 90% and over 85% respectively under the 1.5 °C scenario. This should be part and parcel of China's long-term low-carbon transformation strategy. China now ranks first in the world in terms of installed capacity and new production capacity of hydro, wind and solar power, which will all exceed 400 million kilowatts by 2030. Taking safety into consideration, nuclear power should be developed on a large scale with an installed capacity exceeding 100 million kilowatts by 2030, and 200 to 300 million kilowatts by 2050. With its high load factor, nuclear power holds the key to the safe and stable operation of grids with a high uptake of renewable energy. With the development of technological innovation and new energy industry, the cost of wind and solar electric power has fallen rapidly to a level on par with that of traditional fossil power in most cases. Energy storage, electric vehicles and other technology costs are also seeing rapid decline, ready to embrace massive industrialization and experience faster growth driven by technological innovation and policy support. Natural gas, as a cleaner, more efficient and lower-carbon source of fossil energy than coal and oil, enjoys its unique advantages and importance in the low-carbon transition of the energy system, providing grounds for the exploitation and utilization of more conventional and unconventional natural gas to replace and reduce the consumption of coal and oil [23, 24].

At present, China is still highly dependent on oil and natural gas imports—a potential hazard for energy security. With the swift energy transformation, additional future energy demand will be met mainly by newly developed energy sources in order to peak emissions by 2030. The consumption of coal, oil and natural gas will successively experience steady decline, and the oil and gas imports will see a decline. To safeguard the security of national energy supply, focus should be placed on the macro trend of energy transformation to accommodate short and long-term development. The replacement of fossil fuels with renewable energy is the ultimate solution to the security of energy supply. Excessive dependence on coal-to-oil and coal-to-gas is not desirable. Attempts to cut back imports of oil and gas by promoting coal consumption run counter to the world trend of energy transformation, with no prospects for technological competitiveness internationally. The rapid industrialization and lower cost of new and renewable energy is increasingly disrupting the traditional energy market, with growing interest from the finance and investment community for the zero-carbon energy system. Traditional energy infrastructure and capacity expansion might risk early retirement and elimination in the future, incurring huge stranded costs. By the middle of this century, China should put in place a sustainable energy system based on new and renewable energy to ensure the turnaround of environmental quality, achieve the security of energy supply, end the country's dependence on oil and gas imports, thereby creating a clean, safe, economically appropriate modern system of sustainable energy supply to support the goal of building a strong socialist modern country and a "beautiful China" [25].

10.2.5   Push Forward the Reduction of Non-CO_2 GHG Emissions, and Take Measures to Reduce Economy-Wide GHG Emissions

The Paris Agreement calls for absolute emission reduction of all greenhouse gases in the entire economy of developed nations and encourages developing nations to do the same step by step. China's NDC targets and action plans under the Paris Agreement mainly involve CO_2 emissions related to energy activities, which embody the principle of common but differentiated responsibilities. With the rise of China's comprehensive national strength and future progress in global response to climate change, China should steadily press ahead on emission reduction from all greenhouse gases across the economic sectors to rein in and reduce non-CO_2 greenhouse gas emissions, placing it as a crucial component of the long-term low-carbon development strategy, set clear objectives and strategies, and develop holistic plans for coordinated progress.

In 2014, China's emissions of non-CO_2 greenhouse gases such as methane, nitrous oxide and fluorine-containing gases accounted for approximately 16% of the country's total greenhouse gas emissions. These emissions involve a wide range of industries from scattered sources such as methane leakage in oil and gas production, gas emissions from coal mines, replacement of refrigerants for air conditioners and waste disposal. Since the initial cost of emission reduction is fairly low with immediate positive economic benefits, mitigation measures should be vigorously promoted. But if deep emission reduction or near-zero emission is to be achieved, its marginal cost of emission reduction will mostly be on a nonlinear rise. With the prospect of substantial CO_2 reduction in the energy system, non-CO_2 and other greenhouse gas emissions will make up a larger share and become the biggest challenge for emission reduction, thus taking a more vital role in the future deep decarbonization scenario. To achieve near-zero and net-zero emissions of all greenhouse gases by the middle of this century, China needs to spur innovation and industrial development of non-CO_2 emission reduction technologies. During the 14th Five-Year Plan period, solutions should be crafted and implemented to cut non-CO_2 greenhouse gas emissions to gradually set up a policy framework and management mechanism for reducing all greenhouse gases across economic sectors.

Methane is a short-lived greenhouse gas whose warming potential is 84 times that of CO_2 in 20 years (22 times in 100 years). Methane emission reduction can contribute greatly to controlling immediate temperature rise and is gaining more attention in this regard. Relatively mature emission reduction technologies and measures are available for methane leakage in coal, oil and gas production with low cost and high potential economic benefits. It is necessary to vigorously disseminate these technologies, propose emission reduction targets and plans, formulate industry standards, and incorporate them in national and regional development plans to take the lead in achieving significant results in non-CO_2 emission reduction.

Methane emissions are the most important source of non-CO_2 greenhouse gas emissions in China, comprising 56% of all non-CO_2 greenhouse gas emissions in 2014. It is also the most important area for emission reduction with the biggest potential and effective technology for implementation. Under the macro trend of energy transition, activities in coal and oil and gas production will be reduced, resulting in lower methane emissions. There is also a need for better management and technological progress to facilitate methane recycling and reuse, strengthened management of rice paddies, improved animal feeding and solid waste disposal for the effective control and reduction of methane emissions.

It's vital to step up technological innovation, develop alternative technologies and processes, enhance end-use recycling, reuse, disposal and decomposition, and improve farming methods to effectively curb and reduce greenhouse gas emissions such as nitrous oxide and fluorine-containing gases. By employing a mixture of measures, China can manage to simultaneously peak non-CO_2 greenhouse gases and CO_2 emissions. By 2050, non-CO_2 greenhouse gas emissions, ideally, should fall by more than 50% from the peak year.

The initial cost of reducing non-CO_2 greenhouse gas emissions is fairly low, with some technologies bearing positive economic benefits, but the marginal cost shows a steep rise. Technically and economically feasible tools and measures are not yet available for deep emission reduction of non-CO_2 greenhouse gases, which, nonetheless, are hopeful for a simultaneous peak with CO_2 by 2030 on condition of redoubled efforts. But by 2050, there will still be more than one billion tons of CO_2e in non-CO_2 greenhouse gas emissions that cannot be effectively reduced. To achieve prompt carbon neutrality of all greenhouse gases after the mid-century benchmark, non-CO_2 reduction will be a gigantic obstacle, which calls for the development of negative emission technologies (such as BECCS) and other CO_2 removal technologies (CDR) in the energy sector to offset the remaining emissions. Therefore, the vigorous development of deep emission reduction technologies and tools for non-CO_2 greenhouse gases is vital to the future goal of carbon neutrality.

10.2.6 Attach Importance to Urban Energy Conservation to Drive the Change of Consumption Patterns and the Construction of a Low-Carbon Society

As China wraps up massive infrastructure construction from industrialization and urbanization, the proportion of energy consumption and CO_2 emissions needed for the production of basic investment goods and raw materials decreases, and the share of energy and CO_2 emissions supporting the production of social and public consumer goods and direct household consumption increases. In this context, promoting policies of energy saving and low-carbon consumption in urbanization will serve an increasingly important role in realizing long-term emission strategy and pathway.

Urbanization is not only a powerful driver for spurring domestic demand, fueling economic growth and improving people's well-being, but also a major contributor to increased energy consumption and CO_2 emissions. China should seek balanced and coordinated development among regions in the process of industrialization and urbanization, devise holistic plans for urban and rural areas, and ensure social fairness and harmony. At the same time, it is necessary to explore the model of social production and consumption centering around ecological civilization and pave the path of a new model of urbanization [26].

President Xi Jinping stressed the need to "give high priority to energy conservation during urbanization, foster diligent and thrifty consumption, and speed up the formation of an energy-conserving society." Energy consumption is bound to rise with the construction of urbanization infrastructure and the migration of rural residents. Given the circumstance, endeavors should be made during urbanization with focus on low-carbon urban planning, infrastructure, lifestyle and consumption, and promote public awareness, and consumption habits. The overall transition strategy should revolve around city size and layout, transportation system and way of travel, building structure and energy-saving standards, energy supply and consumption pattern. Apart from more stringent standards for building energy efficiency, enhanced energy efficiency of home appliances and improved automotive fuel economy, among other energy saving measures, more focus should be placed on the general layout and planning of low-carbon cities, the transition from luxury spending and material pursuits of oversized luxury housing, large-engine vehicles, etc. The change of consumption awareness and habits can effectively curb the final energy demand of the whole society, and spark the transition of economic and social development pattern. Caution must be exercised to avoid copying developed nations' traditional development model of carbon-intensive infrastructure in urban construction and luxury spending, and avoid the dilemma in which one can only seek low energy consumption and low-carbon emission on specific facilities and individual technologies because high energy intensity and high carbon are commonplace in cities in the first place. Campaigns should advocate ecological civilization and a low-carbon society, explore new models of low-carbon eco-cities and green lifestyle, so as to embark on a new journey of ecological, low-carbon urbanization.

While stressing the shift of development concept and innovation of development model as we build ecological civilization and materialize low-carbon development transformation, priority should also be given to encouraging and reinforcing the transformation and innovation of consumption awareness in order to lead a new trend of consumption. With basic material needs met, clean air, clean drinking water, and a livable environment are deemed more important than personal material comforts. A high quality life is the shared experience and public interest of all, which will boost the accumulation and sharing of public wealth, and fairness and win–win outcomes for all countries and social strata. Therefore, it is crucial to steer the whole society's interest from the path of industrial civilization featuring the one-sided pursuit of economic output and productivity onto the course of ecological civilization characterized by the harmony between human and nature, economy and environment, human and society and sustainable development, and shift from the preoccupation with material wealth maximization to a more healthy, moderate and simple way of consumption with more focus on the ethical and cultural advancement to avoid extravagant consumption and waste, thus prompting the transformation of economic and social development through innovation in people's mindset [27].

The transformation of the way of consumption requires direction from public opinion and media outlets. The dissemination of information should be reinforced to popularize the knowledge and raise public awareness for climate change, and encourage the public to take voluntary actions together through grass-root organizations and associations within various initiatives and joint actions, so that residents can willingly change their consumption behaviors and lifestyles. Moderate and frugal low-carbon consumption pattern and behaviors should be viewed as social morality and the communication and education of low-carbon consumption concepts and behaviors should be strengthened. Public institutions must lead by example by carrying out energy-saving and low-carbon campaigns in institutions, campuses, hospitals, gymnasiums and military barracks to promote moderate consumption and the use of energy-saving and low-carbon products, curb extravagance and waste. The recycling and classification of waste products should also be improved.

Urban development policies and mechanisms should entail incentives for low-carbon consumption to guide, regulate and encourage reflected individual consumption behaviors, and create an enabling environment for low-carbon consumption in accommodation, transportation, food consumption, among others. The shift in the awareness and pattern of public consumption will also trigger the transformation of production pattern. The public preference for low-carbon household appliances and daily consumer goods will incentivize energy saving and carbon reduction in the production process. Through carbon labeling and low-carbon certification of manufacturing products, consumers and producers can join hands in prompting the low-carbon transformation of the entire society.

Efforts to build smart cities have become the latest urban trend. The application of smart technologies in managing cities and running infrastructure is conducive to efficiency improvement, resource allocation enhancement and energy and material consumption reduction. It promotes the smart distributed energy system, optimizes the match of power source, grid and load, boosts the development of the urban energy network, and energy block chain, and ultimately drives and secures low-carbon urban development. Thus, smart technologies and low carbonization programs should be integrated as two strategic pillars and path for implementation during China's new urbanization.

10.2.7 Promote Coordinated and Balanced Regional Development and a Differentiated Low-Carbon Transition in Line with Local Conditions

Like many developing countries in the world, China faces serious regional disparities. The eastern coastal provinces and municipalities are more economically developed and classified as priority development zones, whereas the central and western regions lag behind, with some being key development zones and some others falling under the category of key ecologically functional zones or restricted development zones. Different regions vary in their energy and resource endowments, hence varied positioning of industries. For instance, Shanxi and northwest China are rich in coal resources, so the bulk of their income comes from high-energy-consuming raw material and coal chemical industry with high energy consumption and high carbon intensity per unit of GDP. With abundant hydropower resources, southwest China is the national hydropower base that is well positioned to foster energy-intensive heavy chemical industry, big data processing and other energy-intensive IT industries. In comparison, the eastern coastal regions lack energy and resources but outperforms in terms of economy and technology. Thus these regions should speed up economic transformation and develop high-tech and modern service industries. On the other hand, China's renewable energy resources are unevenly distributed, with the northwest rich in wind and solar and southwest rich in hydro. Besides meeting the regions' own needs, these renewables can be transmitted to the mid-east regions of China to complement the distributed energy power there, thus achieving an ambitious uptake of renewable electricity in the country. Therefore, during the national low-carbon energy and economic transformation, disparities regarding the positioning, economic structure, energy and resource endowment of each region, province and municipality must be taken into account in order to devise holistic plans to ensure the coordinated, balanced and inclusive development among all regions and advance nationwide low-carbon development.

Pursue innovations in the low-carbon development model by promoting trials of low-carbon provinces, autonomous regions and cities, carrying out pilot low-carbon cities (towns), industrial parks, communities, businesses and transportation, exploring diverse low-carbon development models with distinctive characteristics, and studying effective solutions to carbon emissions in different regions and cities so that such cities will feature rational spatial distribution, intensive utilization of resources, low-carbon production, high efficiency and green and livable environment. Research should focus on trying out and promoting a low-carbon certification and honor roll system in order for products with low-carbon footprint to be certified and labeled.

Implement customized regional strategies and policies to address climate change. Differentiated targets, tasks and approaches for climate change mitigation and adaptation should be determined in light of the resource endowments and industrial structure of the east, central, northwest, southwest and northeast China as well as the federal government determined positioning of the different functional priority zones. Cities under optimized development in eastern China must put strict limit on the consumption of coal and other fossil fuels, with the aim to peak CO_2 emissions first in China. Cities under prioritized development must step up control of carbon intensity, with the old industrial bases and resource-intensive cities in northeast China ensuring faster transition to green and low-carbon development. Southwest China, a region rich in renewable energy resources, could prioritize pursuing 100% renewable energy in the cities, and some provinces and cities can also aim for an earlier peak. Major agricultural products production regions for should build more small to medium-sized cities and towns, encourage proper concentration of population, actively promote a controlled level of agricultural mass production and industrialization. In key ecologically functional zones, ecological red lines should be strictly enforced, and stringent industrial development catalogue formulated to restrict new high-carbon projects. Industries that do not meet their main functions should be withdrawn from the system. Distinctive low-carbon industries should be fostered based on local conditions. Differentiated and inclusive low-carbon transformation should be implemented with a particular focus on the marginal impoverished rural population to improve their work and living conditions and secure the supply of high quality clean energy. Distributed renewable energy programs should be tailored to the local reality and its certified emission reductions (CCER) should be included in the national or provincial/municipal carbon market as an offset mechanism to spur sustainable development in less developed regions.

At present, it is essential to facilitate an orderly peak in CO_2 emissions among regions. Provinces and municipalities in the priority development zones in the east should reach the peak during the 14th Five-Year Plan period. The central, southwest and northeast should hit the peak around 2030, and those in the northwest around 2035, ensuring a roughly nationwide peak prior to 2030.

Regional strategic priorities and measures for near-zero CO_2 emissions by 2050 will vary. Drawing upon the experience of low-carbon pilot cities, a number of zero carbon cities, development zones and communities should be built based on local realities. In the southwest region that possess abundant renewable resources, some cities can experiment 100% renewables ahead of their peers. The eastern and central regions should gradually build and scale up near-zero-emission cities underpinned by renewable energy.

10.2.8   *Harmonize Measures for Economic Development, Environmental Protection, Climate Change, and Biodiversity Improvement*

The Paris Agreement stresses the inherent linkage between addressing climate change and the sustainable development. China's climate change strategy also highlights the coordinated management and solutions for overall economic and social development, ecological and environmental protection and greenhouse gas emission reduction to drive the transformation toward green, low-carbon and circular economic development. Advancement in ecological progress is needed for the harmonious coexistence and sustainable development between human and nature. Therefore, increasing focus has been placed on the coordinated and win–win strategies for tackling climate change, boosting economic growth, improving people's livelihood, enhancing environmental quality and protecting biodiversity, of which nature-based solutions (NbS) have drawn increasing attentions worldwide.

Through climate change management and accelerated transition to low-carbon energy and economy, a near-zero-emission energy system primarily based on new and renewable sources of energy should be up and running by the middle of this century, curbing and reducing the emission of conventional pollutants at the source, helping to meet environmental quality standards. For instance, $PM_{2.5}$ concentration in major regions will be limited to below 15 $\mu g/m^3$, and most regions below 10 $\mu g/m^3$ by 2050 [28].

NbS refers to the practice of following the rules of natural ecological system through the protection, remediation, improvement and strengthened management of the ecosystem to enhance its functionality, thus effectively and adaptively easing the strain of climate change on society, upgrading climate resilience, preserving biodiversity while producing benefits for human well-being and biodiversity. Through enhancing carbon storage capacity of forests, grasslands, wetlands and agricultural land, emission sources would be reduced and carbon sinks increased, generating new economic growth and jobs, improving water, soil, air quality and food safety, among other synergies. The UN Climate Summit held in September 2019 identified NbS as one of the nine priority areas for coping with climate change, where China and New Zealand jointly hosted the thematic forum on "Nature-based Solutions", arousing strong interests from across the world. It also affirmed that NbS is part and parcel of the implementation of the Paris Agreement, stressing the importance of NbS in boosting harmonious coexistence between human and nature, and ecological progress. It is economically efficient and sustainable, hence an effective approach to a holistic human-centered response to climate change. Since then, NbS has increasingly become a major strategic and policy option for countries' climate change efforts.

NbS involves agriculture, forestry, grassland, wetland, and etc. It preserves the ecology, improves the environment, prevents and controls desertification, enhances biological diversity, upgrades the quality of life and health, and increases the ecological and economic value of natural systems with greater economic output, among others. It is a strategic move to bring about the harmonious development of human and nature, thus a win–win outcome.

The broader ideology of eco-civilization, envisaged by President Xi Jinping, contains a range of expressions, including "the harmonious coexistence between man and nature", "green is the new gold", "mountains, waters, forests, fields, lakes and grassland are a community of life", "good ecological environment represents the greatest well-being of the people", "preserving ecological environment through the most stringent system and the rigorous rule of law", etc. These ideologies provide guiding principles for the implementation of NbS, a notion that echoes ecological civilization featuring "ecology first" and "harmonious development of human and nature", respects nature and relies on the functional enhancement of natural ecosystems to cope with the global ecological crisis—a new paradigm and major avenue for coordinated and sustainable development of the economy, society and environment. NbS should be guided by and embedded into the framework of ecological civilization in order to showcase its rigor and dynamism, both theoretically and practically speaking. Bearing in mind the harmonious coexistence between human and nature, one should by no means detach the concept from ecological civilization in improving the functionality of natural ecosystem to solve the earth's ecological crisis including climate change and biodiversity, and refrain from stressing or expanding the measures and potential of NbS in an isolated and single-sided manner.

To fulfill the promise of NbS, it is essential to uphold ecological civilization, holistically advance the institutional development of ecological civilization and the coordinated governance of economy, resources, environment and climate change. It is also important to combine adaptation and mitigation of climate change with policy measures in resources-conservation and ecological and environmental protection. For example, formulating balance sheet of natural resources, identifying ecological red line, providing comprehensive governance of mountains, waters, forests, fields, lakes, grassland and other ecological systems, carrying out projects to remediate ecology, reducing and sequestering carbon, preventing and controlling soil erosion and desertification, and preserving biodiversity conservation. NbS should be integrated with efforts on agricultural modernization and "beautiful countryside", as well as improvement of crop cultivation techniques and methods, reduction of fertilizer use, development of ecological agriculture, and cutting of methane emissions from rice fields and livestock. Efforts should be made to utilize and improve marginal land, promote afforestation or planting of energy crops, boost forest management and grassland restoration, convert farmland to forest and grassland, strengthen the modernization of rural areas, enhance the comprehensive and safe use of rural waste, and step up the commercial use of biomass resources, etc. Overall, NbS for climate change should be incorporated in the overall national strategy for economic and social development as well as the strategy and measures of relevant government authorities so as to maximize the synergy of economic development, social progress, environmental protection, desertification control, biodiversity conservation, greenhouse gas reduction and carbon sink increase, thus catalyzing ecological progress and sustainable development in China.

Major steps should be taken to increase carbon sink, such as afforestation and greening, voluntary tree-planting, continued natural forest protection, grain for green, sandstorm sources control in Beijing and Tianjin, shelterbelts planting, comprehensive treatment of stony deserts, water and soil conservation, strengthening of forest tending and management, etc. Work should also be done to ramp up prevention and control of forest disasters, strengthen protection of forest resources, reduce emissions from deforestation, reinforce protection and restoration of wetland to improve carbon storage, continue to return grazing land to grassland, ensure the balance between pastures and livestock, curb grassland degradation, restore grassland vegetation, step up the prevention of grassland disasters and farmland conservation, and improve soil carbon storage capacity.

The goal of limiting global temperature rises to 2 °C or even 1.5 °C means that carbon neutrality should be achieved by the second half or middle of this century. NbS is of vital importance for achieving carbon neutrality amid deep global emission reductions. During global low-carbon transformation, deep emission reduction can be materialized in energy, industry, transportation, construction and other sectors through the development and application of various emission reduction technologies, but to enable net zero emission, the marginal cost of emission reduction will experience a nonlinear spike. The initial emission reduction of non-CO_2 greenhouse gases such as methane can be achieved at a fairy low or even negative cost, but no feasible technology is available to support near zero emission. Therefore, the implementation of NbS can harness the potential of agriculture, forestry, land use and land use change for reducing carbon sources and increasing carbon sinks, sequestering more carbon to offset the remaining greenhouse gas emissions from sectors where further reduction is a challenge, and secure carbon neutrality as a whole. China's additional annual carbon sink in agriculture and forestry could be kept at approximately 700 million tons by 2050, holding the key to offsetting the remaining emissions from hard-to-crack sectors.

The implementation of NbS hinges on the innovation of science and technology, calling for scientific research to understand and grasp the law of natural ecosystems, protect, restore, and enhance the service of natural ecosystems, and leverage the force of nature to tackle global ecological crisis, especially climate change, as opposed to the sole dependence on technology-based solutions.

Indeed, amid the urgent low-carbon energy and economic transformation under the global actions against climate change, parallel efforts should focus on continued innovation and development of emission reduction technologies to be used in conjunction with NbS. Most of the technical measures for saving energy, improving energy efficiency and strengthening energy substitution also produces the synergy of resources conservation, environmental protection and sustainable development, which represent the main options for low-carbon transformation. The technological solutions must go hand in hand with NbS to complement and reinforce each other with greater focus on the synergy and joint actions for addressing climate change and sustainable development. On the other hand, implementing NbS also requires technology-based innovation and industrialization. For example, technologies for the comprehensive and safe utilization of agricultural and rural waste, commercial biomass fuel, comprehensive treatment of saline and alkaline land, energy crop planting and management and modern ecological agriculture would require more R&D and realization of advanced technologies for close coordination and joint progress [29].

10.2.9   Improve the Institutional Arrangement for Addressing Climate Change and Form a Policy System, Investment and Financing Mechanism, and Market System

First, push for climate change legislation to provide the legal support for the implementation of climate change strategies, mechanisms and policy systems and the attainment of long-term emission reduction targets. The institutional building for low-carbon development is one of the centerpieces of that for ecological civilization. Given the characteristics of China's current policy framework, it's crucial to sharpen the legal and market tools while maintaining the advantageous administrative measures. it is of fundamental importance to put in legislation efforts as soon as possible in order to lay the basic legal groundwork for nation-wide climate change actions, involving a wide spectrum of fields and issues. Dedicated laws would provide clear legal directions to actions at the national level. What's more important is to specify medium—and long-term quantified emission reduction targets and strategic measures within the climate change law to galvanize low-carbon energy and economic transformation. Internationally, the UK was the first country to pass the Climate Change Act, which has incorporated a carbon neutrality target by 2050 through a recent revision. Japan and Mexico also introduced their climate legislations. The EU has scheduled legislation in 2021 via Green Deal to support carbon neutrality by 2050. China is in dire need of a dedicated law to ensure continued progress of energy conservation and emission reduction. Carbon market, for example, currently operates under only departmental regulations of related ministries and commissions with low legal recognition and limited terms of reference, thus unable to secure the progress and operation of carbon market.

Develop of green finance and establish green investment and financing mechanisms. The transformation of low-carbon development entails financial support, and there are increasing incentives for investing in low-carbon technologies and industries worldwide. The EU Green Deal states that to meet the 2030 climate and energy targets, an average of an additional € 260bn in investment will be needed annually, equivalent to approximately 1.5% of EU's GDP in 2018. In terms of funding channels, 25% of the EU budget and at least 30% of the European Investment Fund will be spent on combating climate change. The European Investment Bank (EIB) will also ramp up climate-related investment from 25 to 50% by 2025, with a focus on mobilizing private sector for green investment and engaging international investors through the new international sustainable finance platform. In addition, the EU will further push the reform of budget and taxation system to encourage public investment, consumption and taxation in green priority areas through pricing and incentive mechanisms, repeal subsidies that goes against green development (e.g.: subsidies for fossil energy) to make for the shift of tax revenues from labor to pollution.

China also attaches great importance to the development of green finance, creating and implementing a selection of mechanisms and policies for green investment and financing. A guideline on green project investment was formulated to enhance the green credit system, offer green credit asset securitization (ABS) and green bonds, establish mechanisms for green loan incentives and guarantee, improve green review and supervision for investment projects, enhance government green purchase system, and incentivize enterprise and social investment. Along with the development of carbon market, the auctions of emission allocation will be gradually increased, with the future proceeds from the auction or the carbon tax revenue used in an integrated manner to advance the institutional progress of national low-carbon development and key action agenda. At present, the global financial community is devising the guidelines and norms of green investment, rejecting and withdrawing investment in coal, oil and other high-carbon industries and projects. An enabling policy environment for encouraging green investment is being formed in various countries. China's energy and economic low-carbon transformation will entail huge investment—approximately 40 trillion RMB for NDC targets and action plans by 2030 and around 100 trillion RMB for energy and power systems to fulfill the deep decarbonization target by 2050. On the other hand, the gigantic investment in energy and economic transformation also serves as a strong buttress for economic growth, tapping into new sources of low-carbon economic growth and new green jobs to bring about high-quality economic development. Therefore, the government should constantly develop and enhance the green investment and financing mechanism, which should also be embedded in the "Belt & Road" cooperation projects and mechanisms [30].

Presently, it is of pressing importance that the reform of energy system should lay particular emphasis on carbon market development. With the urgency of global actions against climate change, the scarcity of carbon emission and its nature as a production factor are growing increasingly evident. Market levers should be employed to quantify the value of carbon emissions and allowance, prompting emission reduction at the corporate level to enhance the economic output per unit of carbon emissions. The European Union, California of the United States, South Korea and other countries and regions have successively established the carbon markets, with more on the way to advance efforts on this front. During the 12th Five-Year Plan period, China launched a unified national carbon market, apart from the trials of carbon trading in five cities and two provinces, reinforcing market incentives for low-carbon transformation. The uniform national carbon market started with the power sector and would gradually expand to other energy-intensive industries including petrochemicals, steel, building materials, non-ferrous metals, paper making, civil aviation and chemicals, accounting for about half of the country's total CO_2 emissions. Swifter progress is needed for carbon market in China with enhanced systems of corporate emission accounting, monitoring, reporting and verification and emissions. Carbon pricing signal should be deployed to push mitigation measures in businesses, incentivize private investment and lower the cost of the entire society while safeguarding national emission targets. By quantifying and monetizing the value of carbon allowance, carbon market will spawn a string of financial products in the future. The market itself and carbon finance will also evolve into major instruments in the international competition for low-carbon development [31].

Institutional safeguards for green and low-carbon development should be strengthened. The accountability system of governments at all levels for energy conservation and carbon reduction should be reinforced, and targets for energy saving and carbon reduction included in national and local five-year development plans. Meanwhile, innovations must be explored in the macro-control of energy system to establish and improve the energy legal system, reform and enhance the fiscal, taxation and financial policy systems for low-carbon development, facilitate green finance and reform of the pricing mechanism for energy products and the system of resource and environmental taxes and fees. Advancement for reform of the energy market mechanism to put in place a unified and open market structure and system that are just, fair, and conducive for competition. Market monopolies in certain areas should be eliminated, and disorderly market competition should also be rectified and forestalled. Currently, the pricing mechanism for China's fossil energy fails to fully reflect its social costs. To illustrate, the social costs of air and water pollution and public health damage from coal burning are equivalent to over 50% of the current coal price, but these social costs are yet reflected in the price, and a sound taxation and fee system to transfer the income accordingly is also absent. The internalization of the social cost of resource and environmental losses through the reform of the resource and environmental tax system and the carbon market will help save fossil energy, encourage the development of new and renewable energy, and spur the transformation of energy mix. The reform of energy prices, such as time-of-use and tiered pricing, serves to conserve energy and ensure low-income families equitable access to quality energy services, thereby boosting harmonious social development. On the other hand, technical standards for energy conservation, commodity labeling of energy efficiency, and industry access policies must be tightened. The management of corporate energy use rights and that of carbon allowance should be coordinated with the development of the national carbon market. The medium- and long-term national strategies and goals should be followed to reinforce government binding targets, mandatory standards, fiscal, tax, and financial policies, which should be deployed in parallel with market mechanisms to provide stronger institutional guarantees for the energy revolution and low-carbon development.

It is crucial to promote the institutional building and policy safeguards of technological innovation and industrialization of advanced technologies. Technological innovation is the bedrock for low-carbon transformation. China is working in parallel with developed countries on the R&D of a spate of advanced energy technologies, with its own characteristics and advantages. It is important to ramp up the R&D

and industrialization of advanced energy technologies, leverage the market size of China to sharpen the competitive edge of superior technologies of energy enterprises, harness technological innovation for industrial and technological upgrade and energy system reform, gain an upper hand in technological competition amidst the revolution of world energy system, and enhance the competitiveness of in-house technologies, industries and markets. With the progress of low-carbon energy transformation in the future and the increasing urgency of emission reduction, revolutionary and strategically advanced technological breakthroughs must be made to overcome the nonlinear spike of the marginal cost of emission reduction in most cases under the deep decarbonization scenario of holding temperature rise within 2 °C. More efforts are needed in the R&D and proliferation of deep CO_2 emission reduction technologies, such as large-scale energy storage and smart grid technologies in the context of a large uptake of renewable electricity, BECCS technology for negative CO_2 emission, the technology for producing, storing and utilizing hydrogen as a clean secondary zero carbon energy, technology for producing zero carbon chemical, steel, cement, petrochemical and other raw materials. The R&D and demonstration of technologies for deep emission reduction of such non-CO_2 greenhouse gases as methane should also be strengthened so as to enable deep emission reduction of all greenhouse gases. At the same time, financing services and policy incentives should be provided for small and medium-sized start-ups to encourage all forms of technological innovation. China should proactively develop and demonstrate the pioneering technologies and ensure their prompt breakthroughs and industrialization, making them technically mature and economically viable, in order to gain long-term technological competitiveness in leading the global climate governance and low-carbon transformation under the macro-trend of climate change management and low-carbon economic and social transformation [32].

10.2.10 Promote Global Climate Governance and International Cooperation, and Advance the Construction of Global Ecological Civilization and a Community with a Shared Future for Humankind

China has taken an active part in global climate governance, pioneering international cooperation on climate change, and becoming a major participant, contributor and leader in global ecological progress. As President Xi Jinping pointed out, it is the common cause and interest of all humankind to jointly facilitate global ecological progress and tackle climate change. The Xi Jinping Thought on Ecological Civilization and the vision of building a community with a shared future for humankind embody China's wisdom and proposal for building a fair, reasonable and win–win governance system and cooperation process.

China has been working actively to build ecological civilization, put in place a green, low-carbon and circular economic system, create a clean, low-carbon, safe and efficient energy system, and bring about the coordinated governance and win–win scenario of the economy, energy, environment and climate change. The theories and practices in this regard will provide the world, especially developing countries with successful experiences and useful examples in implementing their strategies of low-carbon development.

China adheres to the goals and principles set forth in the UNFCCC and the Paris Agreement, and has promoted the progress and implementation of the Paris Agreement in a comprehensive, balanced and effective manner. It observes the principle of fairness and common but differentiated responsibilities and respective capabilities, and accommodates the varied national conditions to ensure the balanced progress of the Paris Agreement in mitigation, adaptation, funding, technology, capacity building and transparency, advance win–win cooperation across the world and, in the meantime, build the synergy of climate change management and sustainable development in all countries, especially developing nations. On the other hand, China is also actively engaged in non-UNFCCC cooperation agenda, including initiatives that set standard development of green and low-carbon technologies and products or industry codes of conduct, and joint actions in various international arenas to incentivize low-carbon transformation [33].

China has actively advanced international cooperation on climate change, especially South-South cooperation, promoting "Green Belt & Road" and highlighting climate change policies in "Belt & Road" projects. Use the South-South Climate Cooperation Fund with a Chinese commitment of 20 billion yuan for more green financing options and transfer of green and low-carbon technologies. Stringent environmental protection and energy efficiency standards should be advocated and enforced in "Belt & Road" cooperation projects to avoid high carbon infrastructure and capacity expansion, and joint efforts should be made to explore the strengthening of bilateral or multilateral cooperation on carbon pricing and carbon market mechanisms under the market mechanism framework of Article 6 of the Paris Agreement, so as to jointly promote the implementation of the Paris Agreement and its global cooperation. Meanwhile, China should step up cooperation with developed countries or their local governments or cities, and boost cooperation among businesses, civil society and scientific research institutions, so that climate change governance will set an example for building a community with a shared future for humankind [34].

References

1. Ye Qi, Li Ma (2007) Towards a more active climate change policy and management. Chin Popul, Resour Environ 17(2):8–12
2. Lo K, Wang MY (2013) Energy conservation in China's Twelfth Five-Year Plan period: Continuation or paradigm shift? Renew Sustain Energy Rev 18:499–507
3. Zhao X, Wu L (2016) Interpreting the evolution of the energy-saving target allocation system in China (2006–13): A view of policy learning. World Dev 82:83–94

4. Jiwen C (2015) Proposed draft of public participation provisions in the Climate Change Law of the People's Republic of China. Law J 2:11–18
5. Zhao X, Young OR, Qi Y et al (2020) Back to the future: can Chinese doubling down and American muddling through fulfill 21st century needs for environmental governance? Environ Policy & Gov 2:59–70
6. Young OR, Guttman D, Qi Y et al (2015) Institutionalized governance processes: comparing environmental problem solving in China and the United States. Glob Environ Chang 31:163–173
7. Xiaofan Z, Huimin Li, Xin Ma (2020) Assessment of China's administrative measures and actions to address climate change since the 12th five-year plan period. China Popul, Resour Environ 30(4):9–15
8. Zhao Xiaofan, Li Huimin. Target management in 40 years of energy saving policy//Qi Ye, Zhang Xiliang (2018) China low-carbon development report. Social Sciences Academic Press. Beijing, 24–50
9. Li H, Zhao X, Yu Y et al (2016) China's numerical management system for reducing national energy intensity. Energy Policy 94:64–76
10. Junfeng Li, Qimin C, Cuimei Ma et al (2016) China's climate change policy and Market Outlook. China Energy 38(1):5–11
11. UN Environment (2019) Synergizing action on the environment and climate: good practice in China and around the globe. United Nations Environment Programme, Nairobi, Kenya
12. Yang X, Teng F (2018) Air quality benefit of China's mitigation target to peak its emission by 2030. Climate Policy 18(1):99–110
13. Jinping Xi (2017) Securing a decisive victory in building a moderately prosperous society in all respects and winning the great victory of socialism with Chinese Characteristics for a new era. People's Publishing House, Beijing
14. Jiankun He (2017) Global climate governance and China's low-carbon development countermeasures. J China Univ Geosci (Social Sciences edition) 17(5):1–9
15. Yande D, Yanbing K, Xiaoping X et al (2017) 2050 China's energy and carbon emission scenario & energy transformation and low-carbon development roadmap. China Environment Press, Beijing
16. Jiankun He (2017) Implementing the strategy of energy revolution to promote green and low-carbon development. Scientific Chinese 28:37–39
17. Alun Gu (2015) Changes in China's export trade cost after the introduction of carbon price. China Popul, Resour Environ 25(001):40–45
18. China Academy of Engineering—Tsinghua University "China's Sustainable Development of Mineral Resources Strategy Research" project Team (2016) Research on Mineral Resources Strategy for Sustainable Development in China. Science Press, Beijing
19. Xiangwan Du (2016) Research on China's energy strategy. Science Press, Beijing
20. Building Energy Conservation Research Center of Tsinghua University (2017) Annual development research report of China building energy conservation 2017. China Building Industry Press, Beijing
21. Zhihuan Fu et al (2019) Strategic research of China's transport sector. People's Publishing House, Beijing
22. Zhenhua X et al (2017) General report on China's macro-strategy for low-carbon development. People's Publishing House, Beijing
23. Jiankun He (2017) Low-carbon transformation under the new normal economy. Environ Econ Res 2(001):1–6
24. Xiangwan Du (2014) Energy revolution: For a sustainable future. Chinese Population. Resources and Environment 24(7):1–4
25. Development Research Center of the State Council, Shell International Co., LTD (2019) China's energy revolution in the context of global energy transformation. China Development Press, Beijing
26. Yi Q et al (2018) Research on ecological civilization, new urbanization and green consumption. Science Press, Beijing

27. Yi Q, Jiankun He (2018) Theory and practice of ecological civilization. Tsinghua University Press, Beijing
28. Institute of Climate Change and Sustainable Development, Tsinghua University (2019) Collaborative governance of environment and climate—Successful cases in china and other countries. Tsinghua University, Beijing
29. Jinnan W, Hongqiang J, Jun He et al (2017) Strategy and task of building ecological civilization of socialism with Chinese characteristics in the new era. China Environ Manag 9(6):9–12
30. Jiankun He (2018) The situation of global climate governance after the Paris agreement and China's leading role. Environ Manag China 10(1):9–14
31. Xiliang Z, Sheng Z (2017) Research on global carbon market. People's Publishing House, Beijing
32. Qili H, Qingtang Y, Tao H (2015) Research on some issues of energy production revolution. Chinese J Eng Sci 17(9):105–110
33. Jiankun He (2016) New mechanism of global climate governance and the low-carbon transformation of China's economy. J Wuhan Univ (Philosophy and Social Sciences edition) 69(4):5–12
34. Kefa C et al (2017) Support and safeguards for promoting the revolution of energy production and consumption. Science Press, Beijing

Chapter 11
Global Climate Governance and International Cooperation

11.1 The History of Global Climate Governance

Since the 1980s, global climate change has drawn increasing attention from the international community. In 1988, the Intergovernmental Panel on Climate Change (IPCC) was set up by the United Nations to organize thousands of scientists around the world to assess the scientific facts, causes, impacts, options for adaptation and mitigation and cost effects of climate change. Five assessment reports have been issued so far, and the sixth one is under preparation. The IPCC assessment report is a scientific report based on facts and research findings, and also the most important scientific basis for international climate negotiations and global governance decisions.

As global climate change endangers the survival and development of mankind, global cooperation in combating climate change is growing into a consensus of the international community. At the United Nations Conference on Environment and Development in 1992, the United Nations Framework Convention on Climate Change was adopted, kicking off the global process of tackling climate change. The United Nations Framework Convention on Climate Change (hereinafter referred to as the Convention) envisaged the goals of the human society to address climate change: curb global greenhouse gas emissions from human activities; stabilize atmospheric concentrations of greenhouse gases; prevent a spike in surface temperatures; ensure that the earth's ecosystems can adapt naturally to climate change; and protect grain production from threats. At the same time, the Convention has also put forward the basic principles to deal with climate change, among which two principles are the most important. The first is "Common but Differentiated Responsibilities", primarily in view of the high emissions since the industrial revolution in the developed countries, who were deemed the major contributor to the current climate change and should bear the historical responsibility to take the lead in emission reduction and provide developing countries with funding, technology and capacity building support for them to adapt to and mitigate climate change. The second principle—"Sustainable

© China Environment Publishing Group Co., Ltd. 2022
Institute of Climate Change and Sustainable Development of Tsinghua University et al.,
China's Long-Term Low-Carbon Development Strategies and Pathways,
https://doi.org/10.1007/978-981-16-2524-4_11

Development", pointed out that the economic development and poverty eradication of developing countries are still the top and overriding priorities at present. Therefore, developing countries need to closely integrate climate change efforts with sustainable development.

In accordance with the goals and principles set forth in the Convention, the Kyoto Protocol was adopted at the Third Conference of the Parties held in Kyoto, Japan, in 1997, stipulating quantified obligations for developed countries to cut greenhouse gas emissions. Developed countries as a group should reduce their greenhouse gas emissions by at least 5% from 1990 levels, during the first five-year "commitment period" from 2008 to 2012. This overall obligation to reduce emissions has been broken down into each developed country (i.e. Annex I countries in the Convention). The signing of the Kyoto Protocol marks the beginning of substantial progress for human society to cope with climate change [1].

The 13th Conference of the Parties was convened in Bali, Indonesia. The conference launched the "Bali Roadmap", charting the course for new negotiating process to determine how developed nations would further quantify obligations of emissions reduction after the end of the first commitment period in 2012, and to discuss how to meaningfully engage developing countries in the global actions to reduce emissions. With the consultations and hard work of all parties in Copenhagen, Cancun, Durban and Doha climate talks, the second commitment period (2013–2020) for developed countries was finally adopted, and so was the Durban Platform for Enhanced Action, which launched a new process to negotiate a post-2020 climate regime for more ambitious emissions cut in all countries at the end of the Paris climate conference in 2015. The second commitment period of the Kyoto Protocol lost its substantive relevance as the United States, Canada, Japan, Australia, Russia and other developed countries did not sign/ratify it, and the Paris Agreement has become the new institutional framework for global climate governance beyond 2020.

Unanimously adopted by the Conference of the Parties, the Paris Agreement is a legally binding document that applies to all countries under the Convention in accordance with the targets and principles of the Convention. It defines the institutional framework of global cooperation and actions on climate change beyond 2020 and embodies the political will and common actions of all countries involved to tackle climate change [2]. Though playing down the "dichotomy" of the developed and developing countries in terms of their responsibilities and obligations, the Paris Agreement reflects the principle of "Common but Differentiated Responsibilities" in a range of elements, including mitigation, adaptation, capital, technology, capacity building, transparency and global stocktake, and demonstrates flexibility for developing countries by providing a balanced arrangement.

The Paris Agreement sets the long-term global goal of "holding the increase in global average temperature to well below 2 °C above pre-industrial levels and pursuing efforts to limit the temperature increase to 1.5 °C above pre-industrial levels" in a bid to substantially reduce the risks and impacts of climate change. To achieve this goal, the Paris Agreement calls for reaching "global peaking of greenhouse gas emissions as soon as possible" and achieving "a balance between anthropogenic emissions by sources and removals by sinks of greenhouse gases in the second half of this century" which means achieving global net zero emissions by the second half of the century. This global long-term goal is built upon Nationally Determined Contributions (NDCs), which requires all Parties to propose their NDC targets and action plans in a "bottom up" fashion under the guidance of "holding the global temperature rise to well below 2 °C and preferably to 1.5 °C". In order to achieve the global long-term goals, the transparency of national mitigation actions and funding support should be enhanced. Meanwhile, a global review or stocktake on the worldwide actions should be conducted every five years to evaluate the collective progress of agreement implementation and long-term goals, prompt further update of NDCs, strengthen actions and funding support, and narrow and bridge the emissions gap with long-term global temperature goals.

The new international institutional framework established by the Paris Agreement signifies a key turning point in the model and philosophy of global governance. On the one hand, it represents an unprecedented consensus and a shared political will across the world to deal with the "urgent and potentially irreversible threat that climate change poses to human societies and the planet". On the other, the Paris Agreement, by designing bottom-up mechanisms to encourage countries to set ambitious goals and action plans, rather than imposing top-down responsibilities and obligations, demonstrates the shift in thinking from the "zero-sum game" to a "win–win outcome". The new institutional framework seeks to spur low carbon economic transformation in all countries, and facilitate the coordination between climate change management and promotion of equitable access to sustainable development and poverty eradication, and achieve the dual benefits of "development" and "carbon reduction". At the same time, it stresses the joint actions on sustainable development and climate change initiatives, promoting win–win cooperation between the countries.

China has made outstanding contributions to the adoption and enforcement of the Paris Agreement. The US-China and France-China joint statements on climate change before the conference had already built consensus on the core and focal issues of the negotiations, which ultimately served as the basis for consensus among all Parties. The withdrawal of the Trump administration from the Paris Agreement has undermined global climate governance, and might shake the confidence and actions of the international community on climate cooperation, and fuel unilateralism and protectionism, and affect the willingness and strength of other developed countries in providing climate financing and technical support. In contrast, the global "bottom-up" collaborative efforts are booming. For instance, twenty US states that represent over 60% of US population and GDP formed the US Climate Alliance and vowed to uphold the Paris Agreement within their borders. Due to China's remarkable achievements in energy conservation and carbon reduction at home and its responsible and constructive attitude in global climate governance, the international community has full expectations for China's leadership.

Clouded by unilateralism of some developed countries, the UN Climate Change Conference held in Madrid (COP25) in December 2018 failed to complete the unfinished tasks at COP24 in Katowice of reaching agreement on the detailed rules for implementing Paris Agreement (that is, market mechanism under Article 6 of the

Paris Agreement). The failure to produce the desired outcome sparked concerns of the international community about the mechanism and capability of global climate governance under Paris Agreement. The global outbreak of COVID-19 in the end of 2019 has shifted the priority of policy agendas in all countries to shoring up economic weakness, improving industrial chains and accelerating economic recovery, which would inevitably put the brake on green and low-carbon transformation. At the same time, the pandemic has further heightened the sense of sovereignty and national consciousness of all countries, as evidenced by the all-round escalation of rivalry between China and the United States and other developed countries, with mounting "deglobalization" and "anti-globalization" sentiments. This means that the comprehensive, balanced and effective implementation of Paris Agreement is facing extremely daunting challenges.

A lookback of global climate governance history reveals a process of constant evolution of China's strategic positioning in global climate governance. When UNFCCC was adopted in 1992, China's carbon emissions were still fairly low, and the strategic positioning was to safeguard its legitimate rights and interests as a developing country. The negotiation was conducted in a way to ensure sufficient emission space before 2030 to support economic and social development. With the continuous rise of its comprehensive national strength, China's positioning and mission for global climate governance have undergone notable evolvement. Especially after the 19th National Congress of the Communist Party of China, the government introduced the new strategic positioning of "steering international cooperation on climate change and becoming a major participant, contributor and torchbearer in the global endeavor for ecological civilization". China's current and near-to-mid-term climate strategies mainly center around the demand for sustainable development of domestic resources and environment and focus on the synergy of carbon emission reduction and environmental protection. But by the middle of the twenty-first century when China evolves into a modern socialist powerhouse with the largest economy in the world, it must assume the responsibility of a major country to contribute to the development of human societies, protect the ecological safety of the Earth, and provide more public goods for the world. This is also the historical mission of the Communist Party of China. Therefore, China's long-term low-carbon development strategy must go beyond the domestic needs for sustainable development and proactively take on the international responsibility of addressing climate change. While China's current low-carbon strategy can well achieve its nationally determined contributions (NDCs), it falls short of the goal of 2 °C and the mission of protecting the ecological safety of the Earth. The 2 °C scenario must be taken as the target in a prompt manner to drive the transformation. Moreover, the current policies should be reinforced in the next decade (14th and 15th Five-Year Plan periods), and the transformation should be sped up after 2030 to move toward the 2 °C scenario. Failure to do so would exponentially increase the difficulty and cost of emission reduction in the future.

11.2 Several Core and Focal Issues in Global Climate Governance

11.2.1 The Principle of Equity and Common but Differentiated Responsibilities

In the process of global cooperation on climate change, each country faces different prominent problems and conflicts in achieving its own sustainable development on account of varied national conditions and stages of development, hence different standpoints and requests are reflected in global climate governance. But climate change is a global challenge. In balancing the overall interests of the world and the local interests of individual countries, the governance system would inevitably involve the sharing of responsibilities and obligations, which, more often than not, sparks conflict and tensions between countries. Therefore, the establishment of a fair and effective international institutional framework becomes the cornerstone for the world's multilateral mechanism on climate change. With the progress of climate governance, the notion of equitable access to sustainable development has become a request and principle shared by all developing countries in the negotiations of the international climate change system, and an integral part in building the system of international sustainable development.

Climate equity has always been at the heart of climate negotiations and the bedrock for the international governance system of fairness, justice and win–win cooperation. The Paris Agreement is the institutional framework applicable to all countries. It is guided by the principles of the Convention, including equity, common but differentiated responsibilities (CBDR) and respective capabilities. Though watering down the "dichotomy" between developed and developing countries in terms of their responsibilities and obligations laid out in Kyoto protocol, it brings to the fore the principle of CBDR in the elements of mitigation, adaptation, capital, technology, capacity building and transparency, etc., with flexibility for developing countries. On mitigation, for instance, the Paris Agreement stipulates that developed countries should continue to take the lead by taking economy-wide absolute emission reduction targets, and that developing countries should continue to ramp up their mitigation efforts, and are encouraged to move over time toward economy-wide absolute emission reduction in the light of different national circumstances. On funding, for another example, the Paris Agreement sets that "developed country Parties shall provide financial resources to assist developing country Parties with respect to both mitigation and adaptation in continuation of their existing obligations under the Convention", and requests developed countries to raise at least $100 billion every year before 2025 to support developing countries, and encourages other Parties "to provide or continue to provide such support voluntarily" [3].

The CBDR principle established in the Convention stems from the notion of "corrective justice" and reflects "distributive justice". The principle comprehensively delineated the different responsibilities between developed and developing countries relative to global warming, stating that developed countries bear the historical responsibility of global warming, and that the need of developing countries for greenhouse gas emissions must be accommodated to secure their transformation toward sustainable development. Under this circumstance, the Convention requests developed countries to take the lead in emission reduction, and provide developing countries with funding, technology and capacity building for them to adapt to and mitigate climate change. In order to underline the importance of balancing responsibilities and capabilities for climate equity, the principle of "respective capabilities" has always been accompanied by the principle of CBDR in the Convention and the subsequent negotiations as well as the outcome of Rio+20. Especially after the Cancun Conference when the notion of climate justice featuring "equitable access to sustainable development (EASD)" cropped up, there was a perspective of development that said the global climate governance justice system that focused on the past, present and future had been formed, that is, the CBDR principle which primarily traces the history is mainly used to identify historical responsibilities; the reality-based principle of "respective capabilities" mainly caters to the differences in development stages; and the future-oriented principle of "Equitable Access to Sustainable Development" mainly accommodates the long-term right to development [4].

The Convention is a fundamental legal document for global climate governance and the cornerstone for the establishment of an international climate governance system. Its basic principles have been observed and reaffirmed under the Paris Agreement. The creation of the global climate system must showcase the concept of "fairness", requiring equal consultation and broad participation of all countries and comprehensive and balanced solicitation and reflection of requests from countries under different national conditions and development stages. It should also adhere to the concept of fairness, historical responsibilities and respective capabilities in terms of code of conduct, and embody the CBDR principle. The Paris Agreement also emphasizes the "intrinsic relationship that climate change actions, responses and impacts have with equitable access to sustainable development and eradication of poverty". The equitable access to sustainable development is the shared request and principle for developing countries in global climate governance talks, and the central element in the global climate governance system [5].

In the process of industrialization and modernization, developed countries recklessly consumed the planet's mineral resources and discharged wastes in a way that severely exceeded the Earth's capacity, and thus bears the historical responsibility for climate change in the past century. In contrast, on the way to industrialization and modernization, developing countries no longer enjoy the same resources and environment endowment, and could not follow the path of high resource consumption and eco-environmental damage for the sake of development as their developed counterpart did. Their modernization endeavors will be severely hampered by the capacity of the Earth's environment and carbon emission limit, prompting a path of resources-saving, environmentally-friendly, and low carbon development to bring about green and low carbon transition for the energy system and economic development. Now and in the future, only with ambitious emission reduction by developed countries can it reserve sufficient room for carbon emissions by developing countries to support their economic transformation and sustainable development. The pressing low carbon transformation of developing countries also requires financial and technical support from developed nations to ensure equitable access to modern high-quality energy services for the people in the developing world, eliminate poverty, improve people's livelihood, promote fairness and sustainability and, in the meantime, achieve low carbon economic development. The tighter the schedule for global carbon neutrality, the less carbon emission is permitted for the growth of developing countries. So it is essential for developed countries to lead the transition to net zero emissions, and also for developing nations to rapidly follow up for deep emission reduction, but the two would by no means become carbon neutral at the same time.

During the current implementation of the Paris Agreement, developed countries increasingly sought to downplay and forsake the CBDR principle by pushing the rhetoric of collective responsibilities and actions and more ambitious efforts by all parties, and ignoring their responsibility to assist developing countries in terms of funding, technology and capacity building in order to mitigate and adapt to climate change. The CBDR principle is the bedrock for a global climate governance system featuring fairness, rationality and win–win cooperation. It represents an institutional safeguard for achieving sustainable development by all countries and addressing the climate crisis in a coordinated and win–win manner. It is also the guiding principle that must be observed in global climate governance and international cooperation for a long time to come. China should adhere to the strategic positioning of developing countries, consolidate and build the strategic support for developing countries, maintain and uphold the basic principles enshrined in the Convention and the Paris Agreement, and promote the building of a global climate governance system and cooperation process featuring fairness, justice and win–win cooperation.

11.2.2 Positioning China as a Developing Country

The distinction between developed and developing countries in global climate governance is a political and historical concept of responsibility. Annex I to the Convention adopted by the United Nations Conference on Environment and Development (UNCED) in 1992 designated 41 countries undertaking the responsibilities and obligations of developed countries, including the countries of the former Soviet Union and the economies in transition in Eastern Europe, aside from the traditional post-industrialized economies. These countries enjoyed relatively developed economies with a higher per capita GDP. Their per capita greenhouse gas emissions have been much higher than those of developing countries. These countries have the historical and realistic responsibility for global climate change. In the context of global actions against climate change, these countries belong to developed nations, while non-Annex I countries fall under the category of developing countries. China is a non-Annex I country, which means that its status as a developing country is recognized under the Convention.

Both of the negotiations of the Kyoto Protocol adopted by the UN Climate Conference in 1997 and the Post-2007 Bali Roadmap insist on binding quantified absolute emission reduction obligations for developed countries, while developing countries are only required to take appropriate mitigation actions of their own in their sustainable development priority areas. As a developing country, China has not undertaken the obligation of quantitative emission reduction under the Kyoto Protocol, nor is it obliged to contribute to the "Green Climate Fund". Under the "Bali Roadmap", China set the NDC targets of a 40 ~ 45% drop in carbon dioxide emissions per unit of GDP by 2020 from the 2005 level, 15% of non-fossil energy, and forest stock growing to 1.3 billion cubic meters, etc., which are the voluntary emission reduction efforts in the priority areas of sustainable development such as energy transformation and ecological protection—in line with China's status as a developing country.

The Paris Agreement adopted by the UN Climate Conference at the end of 2015 established the post-2020 new global climate governance mechanism, which sought to keep global temperature rise below 2 °C above pre-industrial levels while pursuing efforts to limit the increase to 1.5 °C. It also created a new bottom-up structure where countries submit their own targets and action plans for NDCs—a new framework applicable to all Parties. Meanwhile, the mechanism also observed the principles of the Convention: equity, common but differentiated responsibilities and respective capabilities, taking into account the varied national conditions. The elements of climate change mitigation, adaptation, finance, technology, capacity building and transparency still demonstrate the flexibility of differentiated responsibilities and actions for developing and developed countries.

More than twenty years since the signing of the Convention, major changes have taken place in the economic development and greenhouse gas emissions of both developing and developed countries. In 1990, developed countries in Annex I accounted for only 22.2% of the world's population, 83% of the global GDP, nearly 70% of global greenhouse gas emissions, and 82% of global cumulative emissions since the industrial revolution. This confirmed the primary responsibility of developed countries for global climate change, and their due obligations were identified accordingly. By 2016, developed countries in Annex I saw their share of global GDP dropping to 63%, their emissions to 38%, and cumulative historical emissions to 68%. In contrast, emissions in developing countries were surging, with the per capita carbon dioxide emissions increasing from 12% to 34% of that in developed countries. This showed that, for one thing, with the economic development and emissions increase, developing countries should continuously scale up efforts to slash emissions and assume more responsibilities and obligations. For another, the historical and realistic responsibilities of developed and developing countries remained unchanged at the core, and the principle of common but differentiated responsibilities would be relevant for a long time to come.

In 1990, China only produced 10% of the world's CO_2 emissions, and the per capita emissions was less than half of the world average. Yet by 2016, China's emissions had made up 28% of the world's total emissions, more than the US and EU combined; China's per capita had reached 1.5 times of the world average, exceeding that of the EU; its share of cumulative global CO_2 emissions had risen from 5.2% in

2005 to 12.7%, which was still much lower than the US share of 25%, and the per capita cumulative emissions was only 68% of the world average. On the one hand, this means that the characteristics and status of China as a developing country had not been fundamentally reversed and would be maintained for a considerable period of time in the future; On the other hand, with the economic development, growing comprehensive national strength and rapidly increasing greenhouse gas emissions, China and other emerging developing countries must ramp up their actions in the global fight against climate change and enhance their NDC targets in order to make greater contributions.

In view of China's rising national strength and growing proportion of its greenhouse gas emissions in the world, there have been voices in the international community that questioned and challenged China's status as a developing country. U.S. climate negotiators even proposed to create a "graduation" for developing countries by delisting China, South Korea, Singapore and other emerging countries from non-Annex I so that they would undertake similar responsibilities and obligations as Annex I developed countries. However, this involves a modification of the Convention and a consensus by all parties, and therefore is hard to be put on the negotiation agenda. The division between Annex I developed countries and non-Annex I developing countries will not change and cannot be changed for a long time to come. Of course, flexibility should be shown in terms of the responsibilities and obligations in case of changed national circumstances in some countries.

The CBDR principle and flexibility embodied in the bottom-up NDC mechanism and all elements of the Paris Agreement are conducive to China's long-term strategic positioning as a developing country, providing space and the opportunities for China to safeguard national interests and play an active role in global climate governance. China's NDC targets by 2030 under the Paris Agreement—60% ~ 65% reduction in CO_2 emission per unit of GDP from the 2005 level, 20% non-fossil energy share in primary energy consumption, and peaking of CO_2 emissions around 2030 and making best efforts to peak earlier—were based on the principles of Paris Agreement and China's status as a developing country. These ambitious targets for emission reduction are aligned with its national conditions and development stage. China's NDC targets only cover energy-related CO_2 emissions and are relative emission reduction targets for cutting CO_2 emissions per unit of GDP, which is different from the quantified economy-wide absolute greenhouse gases emission reduction targets implemented by developed countries. With economic development and increasing overall national strength, China will also need to ramp up emission reduction efforts and take the initiative to assume more international responsibilities in the future. After peaking CO_2 emissions before 2030, China needs to strive for an economy-wide absolute reduction of all greenhouse gas emissions.

On May 20, 2020, the White House released the United States Strategic Approach to the People's Republic of China, further demonstrating its strategic intent to suppress and contain China. It strenuously denies China's status as a developing country, stressing that China is already a "mature economy" but still gains many benefits that are not accessible to the United States through the title of a developing nation. The document emphasizes that the United States views China as an equal rival, and will examine the benefits arising from China's being a developing country in international institutions and rules and ask for revisions to turn China into a "developed country". Upholding and cementing China's developing country status echoes China's national conditions and stage of development and safeguards China's legitimate rights and interests in international affairs. It is also conducive to expanding the strategic support of developing countries to defuse the plots and measures of the United States and other western countries to isolate and suppress China. Global climate change has always been an important area for developed and developing countries to compete and for big powers to contest. It will also be a major arena in which the United States and other western countries deny China's status as a developing country, disunite China and other developing countries, and apply pressure on and shift their responsibilities and obligations to China. In the wake of COVID-19, the United States will amplify its demands and position for China to take equal responsibilities and obligations in the field of climate change, and will continue to put pressure on China's emission reduction and funding commitments. Consolidating and maintaining China's strategic positioning as a developing country in global climate governance can strengthen China's solidarity with the vast number of developing countries, strive for a relatively loose carbon emission space for China to realize modernization, and set an example for China to maintain its developing country status in other fields.

11.2.3 The 1.5 °C Global Temperature Target

The objective of the Convention, adopted by the United Nations Conference on Environment and Development in 1992, is to stabilize "greenhouse gas concentrations in the atmosphere at a level that would prevent dangerous anthropogenic interference with the climate system", and "such a level should be achieved within a time frame sufficient to allow ecosystems to adapt naturally to climate change, to ensure that food production is not threatened and to enable economic development to proceed in a sustainable manner". Due to the lack of scientific research at that time, no specific control targets of temperature rise was proposed. With the progress of scientific research, the European Union first proposed to set a global climate target of keeping temperature rise below 2 °C above pre-industrial levels. The Copenhagen Climate Conference in 2009 failed to reach agreement on the second commitment period of the Kyoto Protocol, but managed to build consensus on the 2 °C goal. Later, the European Union, Small Island States and the Least Developed Countries went a step further by raising the target to 1.5 °C, and their voices were getting louder. Therefore, the Paris Agreement, after balancing various opinions, put forward "holding the increase in the global average temperature increase to well below 2 °C above pre-industrial levels and pursuing efforts to limit the temperature increase to 1.5 °C above pre-industrial levels", and requested IPCC to submit a technical document on the impact of global warming of 1.5 °C above pre-industrial levels and related pathways of global greenhouse gas emissions.

In October 2018, IPCC issued the Special Report on Global Warming of 1.5 °C, suggesting that the 1.5 °C target is technically feasible and could enable notable reduction in climate risks and negative impacts compared to 2 °C, and is generally conducive to achieving the 2030 Sustainable Development Goals (SDGs) of the United Nations [6]. However, the 1.5 °C target requires more urgent efforts in emission reduction. Specifically, net zero CO_2 emission and deep emission reduction of other greenhouse gases should be achieved by the middle of this century, which would incur a marginal cost 3 to 4 times larger than the 2 °C target. The release of the special report sparked worldwide repercussions. The United Kingdom has revised its Climate Change Act to introduce a carbon neutral target for 2050. The European Union updated its climate ambition for 2030 in the Green Deal, with a 50–55% cut in greenhouse gas emissions from the 1990 levels and a target of net zero emission (i.e., carbon neutral) by 2050, which will be protected by legislation. Some states in the United States, such as California and New York, also set targets for carbon neutrality by the middle of this century. Now, developed countries, including New Zealand, France, Canada, Spain, Sweden, Denmark and Norway, and developing countries such as Chile, Ethiopia, Costa Rica and Fiji have also outlined their own goals of achieving net zero emission by 2050 or before. COP26, which has been postponed to 2021 due to the pandemic, will continue to focus on strengthening emission reduction efforts of all countries. And it is expected that more countries will commit to a carbon neutral target with net zero emissions by 2050 in their low emission development strategies for the mid-century to be submitted in 2020.

The 1.5 °C target requires that global carbon emissions peak as soon as possible and begin to decline rapidly afterwards, so much so that CO_2 emissions are reduced by 45% by 2030 compared to 2010, global net zero emission is achieved by 2050, and achieving deep reductions in other greenhouse gases. According to the NDC targets of all countries under the Paris Agreement and the current trend, global greenhouse gas emissions will continue to rise by 2030, and there will be an annual emission reduction gap of about 11 ~ 21.5 billion tons to achieve the 2 °C target. Global warming is likely to reach 1.5 °C between 2030 and 2050 and to reach about 2.7 °C or even exceed 3 °C by the end of this century. If the current rate of global warming continues, the 2 °C target will not be achieved, not to mention the 1.5 °C target. Therefore, it is essential for all Parties to significantly enhance and update their NDC targets and actions and encourage all stakeholders to step up their ambitions.

As the all-round implementation of the Paris Agreement settles in, the top priority is to adhere to the goals and principles of the Paris Agreement, facilitate its comprehensive, balanced and effective enforcement, encourage all parties to ramp up their own efforts and boost joint international actions. COP26 should press for consensus on the market mechanism of Article 6 of the Paris Agreement, wrap up negotiations on the Paris Rulebook, and promote cooperation under the global market mechanism. It should review global emission reduction actions before 2020 and the implementation of the second commitment period of the Kyoto Protocol. It is also necessary to focus on how to ensure the delivery of US$100 billion each year committed by developed countries and other pressing issues such as climate disasters and adaptation in vulnerable areas of developing countries, and avoid empty talk on grand visions without concrete immediate action in order not to delay global joint response.

To keep global temperature rise under 1.5 °C, the world as a whole should become carbon neutral by the middle of this century. The historical and current high emissions of developed countries are self-evident, and the cumulative carbon emissions so far alone could push up the temperature by 1.2 °C. With increasing urgency of global long-term emission reduction, the pressure and risks for ensuring equitable access to sustainable development are mounting, as developing countries are left with smaller emission space in the process of modernization. To attain the 2 °C target, a maximum of 117 ~ 150 billion tons of carbon emissions are allowed in the future. To achieve the 1.5 °C target, the future emission space is 42 ~ 58 billion tons, which is 1/2 ~ 2/3 less than the 2 °C scenario, and primarily come from increased efforts by developing countries where more urgent and ambitious low-carbon transition is required for sustainable development. This points to the need for greater financial and technical support from developed countries. Under the 1.5 °C target, the world as a whole should achieve net zero CO_2 emissions by 2050. However, it is impossible for all countries to achieve net zero emissions simultaneously. Developed countries should take the lead in achieving net zero CO_2 emissions around 2035, and achieve a certain level of negative emissions of all greenhouse gases by 2050, so as to leave the developing world necessary carbon emissions space for their modernization efforts and an equitable access to sustainable development [7].

The establishment of the long-term goals of tacking climate change is a multilateral political decision under the Convention. The description of the long-term global temperature target in the Paris Agreement embodies the hard-won consensus by all Parties. It's built on the scientific assessment of the interplay between climate change and the economy, society and environment, accommodating the impact and loss of climate change, adaptation and mitigation, and cost and sacrifices for emission reduction. Due to varied national conditions and stages of development, different countries face different climate risks and have different priority areas of sustainable development. This underscores the importance of coordinating the diverse interests and requests of all parties and global cooperation, and promoting joint actions for win–win cooperation. China should continue to play its role as a responsible major country in global climate governance, actively leverage its influence as major developing country, boost exchanges and communication with other countries and international organizations, and lay the groundwork for building consensus on COP26 related issues.

China will still in the stage of building modernization before the middle of this century, and will face greater challenges for achieving carbon neutrality compared to developed countries. It entails faster and more drastic transformation of China's energy and economic systems. In the United States, for instance, the average age of coal-fired power plants is already about 45 years with a higher cost of generating electricity than that of natural gas and renewable energy. It's natural for them to be naturally decommissioned or replaced by clean electricity technologies. By comparison, the median age of coal-fired power plants in China is only 12 years. Most advanced supercritical and ultra-supercritical large coal-fired power plants were built in the past decade, and huge stranded infrastructure cost would be incurred should these plants are rapidly phased out. Furthermore, a large uptake of renewable power would place mounting strains on the grids, energy storage and peak shaving and stable operation, requiring groundbreaking technologies in the space of battery energy storage, hydrogen storage, smart grid, etc. It also involves faster and greater electrification penetration for energy end-users, which can only be made possible through innovations and industrialization of state-of-the-art technologies. The very frontier and hot spots of the world's technology competition, these technologies are a symbol for a country's technological innovation capabilities and technological competitiveness. This research found that China is currently not technologically prepared for carbon neutrality by 2050, and the cost of deep emission reduction will rise steeply in a non-linear fashion. Therefore, more in-depth research and analysis are required for the economic evaluation on the emission reduction pathways and technology options for carbon neutrality, study on the costs of urgent transformation, thus providing China's wisdom and solutions for a global validation of the long-term targets.

11.2.4 Carbon Pricing and Carbon Border Tax

Article 6 of the Paris Agreement calls for the creation of a global mechanism for carbon market cooperation to boost national emission reduction and international cooperation. Similar to the Clean Development Mechanism (CDM) under the Kyoto Protocol, it encourages collaborative emission reduction between developed and developing countries. While contributing to emission reduction of developed countries at a low cost, it helps promote low-carbon transition and sustainable development of developing nations. The cooperation of market mechanisms among nations under the Paris Agreement will spur the development of global carbon pricing mechanism, cement the Measurement, Reporting and Verification (MRV) system of all countries, foster technological innovation for emission reduction and international cooperation, and enable global collective emission reduction at a lower cost. The negotiations of Article 6 of the Paris Agreement on international market mechanism mainly involved measures for preventing carbon leakage and double-counting of emission reductions to ensure environmental integrity, and coordinating interests of parties with CDM. The international community also cherishes a desire for COP26 to reach agreement on the implementation rules for Article 6, so as to promote cooperation among countries and their own capacity building.

  The urgency and trend of global low-carbon transformation will also trigger changes and disputes in global economic and trade rules. To illustrate, some developed countries erect "green trade barriers" by raising the standards for energy efficiency or eco-friendliness of commodities, and suppress the growing trade competitiveness of developing countries through trade protectionism. The European Union (EU), the United States (US) and other developed countries have been planning to implement Border Carbon Adjustments (BCAs) such as carbon tariffs. Prior to

COP25 in November 2019, the EU launched the Green Deal, establishing the target of 50% to 55% of emission reduction by 2030 and net zero emission by 2050 through legislation, making Europe the first climate-neutral continent [8]. The EU policy package also included carbon border adjustment for selected industries before 2021 as a critical component of Green Deal on the diplomatic front, which would be the centerpiece for EU's future climate negotiations. The Republican and Democratic think tanks in the US are also contemplating the possibility and planning of carbon tariffs on imported products and carbon benefits at home in case US returns to the Paris Agreement in the future. Such unilateralism and protectionism tendencies in global climate governance needs a meticulous review and high vigilance.

The unilateral imposition of carbon tariffs by developed nations does not conform to the principles of the Convention. Article 3.5 (Principles) of the Convention clearly states: "the Parties should cooperate to promote a supportive and open international economic system that would lead to sustainable economic growth and development in all Parties, particularly developing country Parties, thus enabling them better to address the problems of climate change. Measures taken to combat climate change, including unilateral ones, should not constitute a means of arbitrary or unjustifiable discrimination or a disguised restriction on international trade." The carbon border tax levied by developed countries for the sake of domestic emission reduction policies and actions obviously contradicts this clause.

For emerging developing countries whose exports are mostly medium and low-end manufacturing products with high-energy consumption and low added value, carbon tariff will seriously erode their trade competitiveness [5]. The input–output method is employed to make the following calculations based on the latest foreign trade data from the Statistical Office of the European Union and China Trade and External Economic Statistical Yearbook: 1.53 billion tons of CO_2 emissions were embodied in China's exports in 2018, and 542 million tons for imported goods. CO_2 embodied in the net exports comprised roughly 10.5% of the total national emissions, of which 270 million tons were exported to the EU, accounting for 17.6%; 286 million tons went to the US, or 18.7%. Imported carbons from the EU and US were 31 million tons and 44 million tons respectively, making China the net exporter of embodied CO_2 emissions. The EU adopted carbon border tax on grounds of preventing carbon leakage, targeting imports with higher embodied CO_2 than domestic products that might push up global carbon emissions. The details of the EU carbon border adjustment mechanism and its target countries and its coverage of sectors and products are not yet clear. The measures might include a carbon tariff on specific industries or products, or imposing mandatory purchase of emission allowance under the EU carbon market on imported products, or a domestic carbon tax rebate or purchased allowance for exported products. In short, the purpose is to preserve the competitiveness of local enterprises in the EU and the international market, transfer the cost and pressure of emission reduction to countries that "have not made proper commitments", namely developing countries. If a carbon tax of 40 ~ 60 US dollars per ton is charged on China's exports to Europe and the United States, the tax would make up 2.6% ~ 4.0% of the total export, which would put the strain on China's exporters.

The trend of global economic integration has accelerated the adjustment and division of labor in the global industrial chain. While massively transferring high-energy-consuming raw material and low-end manufacturing industries to developing countries, developed countries have focused on the development of high-tech and modern service industries, thus occupying a top position in the international industrial value chain. The developing world, on the other hand, is pushed to the middle and low-end with low added value, high energy consumption and emission intensity, producing textiles, household appliances, motorcycles and other daily consumer and low-end manufacturing goods to be exported to developed countries. While these mid—and low-end products are produced in developing countries, they are consumed in developed nations. But under current rules, energy consumption and CO_2 emissions are calculated based on the origin of production. Developed countries consume these products, but are not accounted for emissions in the manufacturing process. That is, in fact, a shift of emission responsibilities and obligations in the form of carbon transfer or carbon leakage. Developed countries are obliged to assist developing countries to enhance energy efficiency and reduce CO_2 emissions during the production of these products, reduce the amount of emissions transferred to developing countries, and share the responsibility of cutting the transferred emissions.

In developed countries, the ambitious emission reduction targets and the implementation of carbon tax or carbon emission trading policy would increase the cost of emission reduction for enterprises. For example, imposing carbon border tax on imports from developing countries on grounds of protecting local enterprises is tantamount to a transfer of emission reduction cost to their developing counterparts. By relocating the high energy-intensive and low-end raw materials manufacturing industries to developing countries, developed countries have brought down their emissions and passed on domestic emission reduction costs to developing countries through imposing carbon tax on developing countries' products and using the collected tax proceeds to subsidize or encourage domestic enterprises' technology innovation and emission reduction. Currently, some developed countries are planning to implement a carbon tax rebate for their exports to developing countries, that is, to refund the increased cost of domestic enterprises under their own carbon tax or carbon market mechanism at the time of export so that they can stay competitive in the international market. This is a unilateral act of using carbon tax policy to safeguards own interests in international economic and trade competition.

As the world gears up for more urgent emission reduction, the unilateral implementation of carbon border tax by developed countries is, on the whole, an unfair act of trade protectionism through the transfer of responsibility and cost of emission reduction in the name of combating climate change. It not only harms global cooperation on climate change, technology innovation, emission cost reduction, and the related targets and pathways, but also gives rise to a new round of trade disputes and frictions, undermines WTO rules, and overshadows the prospect of a fair and just international climate governance mechanism for win–win cooperation as well as the process of global cooperation on climate change. It would only jeopardize its own image and undercut its influence and leadership in global climate governance. Under the mega trend of global economic and industrial convergence, developed countries should ramp up technology transfer and financial support to developing countries, promote technological innovation, energy conservation and efficiency improvement, and emission reduction in developing countries, thereby promoting carbon emissions reduction and fostering win–win cooperation among all countries in the world.

The pretext for EU and other developed countries to implement carbon tax is that the insufficient emission reduction efforts and the low emission reduction cost of companies in developing countries have created unfair advantages over companies in developed countries. In response to this concern, China should consolidate and speed up the development of the domestic carbon market and gradually include major high-emission sectors in the carbon market, as a clear signal of carbon pricing for energy-intensive products to write off or weaken the basis for developed countries to impose carbon border tax, and play a leading role in the development of the global carbon market and the carbon pricing mechanism in the future.

At present, most of China's export manufacturing products are at the middle and low-end of the international industrial chain with high energy consumption and low value rate. As a result, China has become a net exporter of embodied CO_2 emissions in foreign trade. While accelerating the transformation towards green, low-carbon and circular economy under the new development concept, China should also drive the transformation and upgrading of export-oriented enterprises, optimize the export mix, control and reduce the export of energy-intensive products, and move toward the higher end of the value chain to enhance quality and efficiency. At the same time, China should actively promote the export of new and high technologies and modern services, increase the weight of services in its export, effectively reduce the outward transfer of embodied carbon emissions, and gradually balance embodied emissions from imports and exports. This is the fundamental countermeasure for China to deal with the carbon border tax of developed countries in international trade and build the international competitiveness of its products and technologies. It is also a major step to reduce the total amount of domestic CO_2 emissions and should be a critical component of China's strategy of green and low-carbon economic transformation and high-quality development.

11.3 Intensive Engagement and Active Leadership in Global Climate Governance and International Cooperation

11.3.1 Xi Jinping's Thought on Ecological Civilization Is of Important Guiding Significance to Global Low-Carbon Development and Transformation

Upholding Xi Jinping's Thought on Ecological Civilization, promoting global ecological progress and achieving harmonious development between man and nature is the ultimate solution to climate change and an inevitable path toward global sustainable development. Xi's ideology of eco-civilization contains a range of expressions, including "the harmonious coexistence between man and nature", "green is gold", "good ecological environment represents the greatest well-being of the people", "mountains, waters, forests, fields, lakes and grassland are a community of life", "preserving ecological environment through the most stringent system and the rigorous rule of law", etc. This notion of eco-civilization is universally relevant for all countries, especially the developing nations. It provides directions for them to build ecological culture systems based on ecological values, coordinates economic development, social progress and environmental protection, facilitates low carbon transition of the energy and economic systems, and embarks on a path of climate-friendly, low carbon economic development. In his proposal for building global ecological civilization, Xi Jinping underlined the importance of "putting in place an ecological system with the respect for nature and green development at the heart, addressing the problems brought by industrial civilization, pursuing harmony between man and nature, and realizing sustainable development of the world and all-round development of man." The notion of global ecological civilization will also be a vital guiding ideology for global collaborative actions against ecological crisis, in particular climate change, prompting the transformation of human society from industrial civilization to ecological civilization, steering a paradigm shift of world economic and social development, and achieving the harmonious development between man and nature [9].

Xi Jinping's Thought on Eco-civilization advocates the establishment of a green, low carbon and circular economic system for sustainable development through the building of a clean, low-carbon, safe and efficient energy system, development of green finance and creation of a legal and policy framework for green production and consumption, etc. It contributes China's wisdom and expertise to the global cooperation on climate change and other ecological crisis of the planet, promotes the harmonious coexistence between man and nature and efforts to build a clean and beautiful world, thus driving the progress of global environmental governance and global ecological civilization.

Guided by Xi Jinping's Thought on Ecological Civilization, China is firmly committed to the path of civilized development featuring increased productivity, well-being of the people and sound ecology. It implements the strictest system for ecological and environmental protection in an effort to build a beautiful China. It adheres to the concept of green, low-carbon, circular and sustainable development in line with the path of climate-friendly, low-carbon economic development encouraged by the Paris Agreement. China seamlessly integrates climate change efforts with sustainable development at home, creating a win–win scenario for the economy, people's livelihood, energy, environment and CO_2 emission reduction, and making enormous strides in energy reform and CO_2 emission reduction. It has become an important contributor and torchbearer in advancing energy reform and low-carbon economic transformation in the world. China's successful experience in energy and economic transformation, new urbanization, industrial transformation and upgrading, environmental governance, as well as its policy framework of energy conservation, carbon reduction and ecological civilization can inform and inspire other developing countries. Its remarkable achievements and successful experience in ecological progress and environmental protection, as well as the concept and practice of building an economic, institutional and ecological safety system for ecological progress will shed a new light on the world ecological progress and sustainable development.

The report to the 19th National Congress of the Communist Party of China reaffirmed China's commitment to peaceful development in the new era, underscoring China's stance in building a community with a shared future for mankind, upholding the notion of extensive consultation, joint contribution and shared benefits in global governance, and actively participating in the reform and development of the global governance system, and constantly contributing China's wisdom and expertise. The report also highlighted China's pledge to environmental friendliness, cooperation in tackling climate change, protection of the planet, and contribution to global ecological safety. Xi Jinping's Thought on Global Ecological Civilization and the concept of building a community with a shared future for mankind embody Chinese wisdom and proposal for global environmental governance, and will play an important role in guiding the global efforts in building ecological civilization and moving toward a climate-friendly and low-carbon development path.

11.3.2 Actively Promote and Lead the Process of Global Climate Governance and Climate Negotiations

China will continue to assume responsibilities and obligations as a developing country for a long period of time. But in the long run, its strategic positioning shall be adjusted as its development stage changes. By the middle of this century when China becomes a great modern socialist country, it must gradually water down its status as a developing country, and proactively undertake the international responsibility for climate change, and actively promote and lead the process of global climate governance and climate negotiations in order to measure up to its rising status as a major country. As pointed out in the report to the 19th National Congress of the Communist Party of China, "taking a driving seat in international cooperation to respond to climate change, China has become an important participant, contributor and torchbearer in the global endeavor for ecological civilization". Xi Jinping also stressed that China should "implement a proactive strategy on climate change management, promote and guide the establishment of a fair, reasonable and win–win global governance system, showcase its image as a major responsible country, and build a community with a shared future for mankind".

The pioneering role in climate change negotiations is best seen in the capability to coordinate the varied positions and interests of parties, and to show impact, inspiration and leadership in building common ground of global targets and positions of all parties and seeking balance of varied interests. The ability to do so would facilitate consensus-building and unanimously accepted action plans, guide the direction and pace of global climate governance rule-making and cooperation process, thus taking the international moral high ground, improving international image and leadership, better safeguarding and expanding national interests, and demonstrating a country's soft power.

In the course of chronic climate negotiations, countries in the world are divided into three forces, namely, the "two camps"—developing countries and developed countries; the Umbrella Group headed by the EU and the United States; and the group of developing countries—"G77+ China" with the "Basic Four" as the core. China, in reality, has become the "backbone" of the "BASIC Countries" and the group of developing countries, and has maintained close communication and dialogues with major developed countries. It has actively bridged differences of all parties in climate negotiations and built mutual trust for cooperation, and has demonstrated its rising influence and leadership. China has become the "go-to" nation for coordinating differences for Paris climate conference and subsequent hosts of COP conferences and the secretariat of UNFCCC.

The China's leadership and pioneering role in global climate governance does not mean that it should contribute beyond its capabilities, given its current national condition and stage of development; nor should it share extra obligations that the United States has given up at greater cost. Rather, China should properly capture and lead the principles and trends of global climate governance, and set the stage for the reform and development of a fair and just international governance system. In the negotiations and full implementation of the Paris Agreement, China should adhere to the principles of the Convention and the Paris Agreement, consolidate the solidarity of developing countries, expand strategic support for developing countries, strengthen the coordination and communication with developed countries and promote the comprehensive, balanced and effective implementation of the Paris Agreement in terms of mitigation, adaptation, funding, technology, capacity building and transparency. China should also urge developed countries to fulfill their obligations to take the lead in emission reduction and provide financial and technical support to developing countries, so as to safeguard the legitimate rights and interests of China and other developing countries [3]. China advocates for a new concept of global governance featuring mutual respect, fairness, justice and win–win cooperation. China takes addressing climate change as an opportunity for sustainable development of all countries and promotes mutually beneficial cooperation for collaborative development. This is conducive to expanding the areas and space for voluntary cooperation among countries and building the shared interests of all parties for a shift of climate negotiations from a "zero-sum game" to win–win cooperation. The new type of leadership and pioneering role that China has demonstrated in climate governance concepts and cooperation approach—different from the United States and Europe—are gaining increasing recognition by the international community. Addressing climate change represents the shared interest of mankind and enables more common ground and potential for win–win cooperation than other global risks and regional hotspot issues in the political, economic and social arenas, promising broad prospects and political will for cooperation. Climate change provides a stage for China to actively engage in the reform and development of global governance. China can make efforts to make it a pioneering and successful example of building international relations featuring mutual respect, fairness, justice and win–win cooperation in the new era and building a community with a shared future for mankind.

11.3.3   Actively Promote International Cooperation on Climate Change and South-South Cooperation on Climate Change Under the Belt and Road Initiative

At present, the issue of how to strengthen low-carbon transition in the Belt and Road Initiative (BRI) projects to address climate change has drawn wide attention from foreign media and sparked concerns, doubts and critiques. To dispel the misgivings, China should take concrete actions and make actual progress to let the fact speak for itself. Therefore, the BRI cooperation should highlight the ideology of building global ecological civilization and the notion of green, low-carbon and circular development, strengthen the targets and policy incentives for tackling climate change, and create a bright spot of international pragmatic cooperation on climate change.

In facilitating "Belt and Road" cooperation, China is taking pragmatic actions to put into practice the new concept of global governance of building a community with a shared future for mankind, and to achieve win–win cooperation and common development. Over the course of "Belt and Road" cooperation, China should follow the guiding principle of ecological civilization and the notion of green development so that it can be aligned with the sustainable development strategies of the countries involved. China should improve the "connectivity" among countries and regions, and take green and low-carbon development as the cardinal principle for promoting the development of and cooperation on the Belt and Road Initiative.

"Green and low-carbon" is the key to addressing the negative impacts of disputes on the Belt and Road Initiative (such as investment in coal-fired power plants and the transfer of high-carbon and high-polluting capacities). In the future, China needs to summarize its best practices in climate change management and environment and climate governance, come up with tailored solutions of green low carbon technologies, project financing, business models based on the specific needs of countries involved in BRI, and assist BRI countries and regions to achieve their NDC targets and sustainable development goals. On the other hand, China can use a variety of instruments to expand climate financing channels to secure funding for the green BRI projects, for example, exploring innovative investment and financing models to attract more investors for the low-carbon development in BRI countries; harnessing the role of Asian Infrastructure Investment Bank, Silk Road Fund and South-South Cooperation Fund on Climate Change, etc. to create synergies, in conjunction with bilateral development aid, investment from multilateral development banks, private investment funds and philanthropic funds. Financial institutions should follow the Green Investment Principles and take low carbon and sustainable development as guiding principles for BRI investments. By vigorously tapping the potential of the PPP model of third-party cooperation and win–win cooperation, it will enable BRI projects to gain financial and credit support from local governments, strengthen the coordination between local governments and investors and operators of green and low-carbon projects, and enhance climate risk management and cash flow management.

China's South-South cooperation on climate change refers to China's cooperation with one or more developing countries, as well as with developed countries or UN agencies, in jointly supporting developing countries in enhancing their capacity to cope with climate change and advancing project cooperation. As a part of China's foreign aid, China's South-South cooperation on climate change follows the principle of non-interference in the internal affairs of recipient countries and aims at mutual benefit, "win–win" and "self-reliance". China's south-south cooperation, primarily through foreign aid, involves cooperation at bilateral, multilateral, regional and interregional levels. In the past few years, China has focused on some small and medium-sized projects in neighboring countries to meet their practical needs. For example, China assisted the construction of a residential housing project dubbed "safe island" in the Maldives to protect local residents from tsunamis and seawater erosion; China helped Bangladesh and the Maldives set up extreme weather warning systems to improve their ability to warn against climate and natural disasters; China State Oceanic Administration established the Framework Plan (2011–2015) for International Cooperation in the South China Sea and Adjacent Seas, which listed "ocean and climate change" and "marine disaster prevention and reduction" as the main funding areas; China worked with surrounding countries in conducting the research projects of "ocean–atmosphere interaction and observation of the tropical sea in southeast Indian Ocean" and "observation of the Indian monsoon outbursts". China has also strengthened scientific and technological cooperation with Africa, implementing more than 100 joint scientific and technological research demonstration projects. Since 2012, China has offered climate change foreign aid in conducting a wide array of projects, including research and capacity building for renewable energy utilization and marine disaster warning; development, popularization and application of LED lighting; technology demonstration on the comprehensive utilization of straw; research on utilizing and promoting wind-solar complementary power production system; experiment and demonstration on the efficient utilization of drip irrigation, etc., assisting developing countries in improving their ability to tackle climate change.

Developing countries that involved in key cooperation projects under the Belt and Road Initiative should be the target countries for South-South cooperation on climate change. During infrastructure construction and production capacity cooperation in these countries, Chinese companies should bear in mind the impacts of climate change and take into account the impacts of extreme natural disasters caused by climate change in the project design, and in the meantime, maximize the use of energy-saving and low-carbon technologies, adopt the same or higher standards for environment and energy efficiency, and promote the resource-saving and circular industrial development model across the whole industry chain, and jointly build low carbon industrial parks or demonstration zones. Chinese companies should select suitable projects and build buildings, infrastructure and production capacity that are green, low-carbon and recognizable. These projects should be included in the South-South flagship climate cooperation scheme for setting up 10 low-carbon demonstration zones, launching 100 mitigation and adaption programs, and providing 1,000 places on climate-change training programs to meet the needs of these countries for low carbon infrastructure under their national climate change strategies, improve their ability to respond to climate change, and enlarge the scope of their cooperation with China and explore more investment opportunities. Chinese companies should also assist host countries in planning and implementing adaptation, mitigation and capacity building projects within the framework of international cooperation on climate change, and seek international financing sources (e.g., the Green Development Fund) and funding from international institutions (e.g., the World Bank). In a word, China should make the joint response to climate change a crucial component, guiding principle and bright spot of the BRI strategy to enhance its international image and influence.

It is also necessary to strengthen South-South scientific and technological cooperation and technology transfer on climate change. Amidst low-carbon transition, the vast number of developing countries have huge demand for technology and the market potential is enormous. China has advanced, practical and low-cost low-carbon technologies that are urgently needed by developing countries, such as energy conservation and new energy. When conducting international production capacity cooperation in foreign markets, Chinese businesses should pay more attention to technology cooperation and transfer. China should actively promote joint R&D and technology transfer platforms or technology parks with the host country to address the key challenges of localization and industrialization in the process of advanced low-carbon technology transfer. This will not only improve innovation and industrialization of advanced technology in host countries, but also enhance the international competitiveness of China's leading technologies. China should play a pioneering role in exploring the model for technology transfer and the mechanism and pathway for mutual benefits within the framework of addressing climate change, and upgrade it to strategic cooperation at national level, which will certainly have a broad impact on the international community, drive and promote the development of financial and technical cooperation and related actions under the framework of climate change management.

Efforts should be made to quickly implement and make good use of the China South-South Climate Cooperation Fund. China has announced that it will invest 20 billion RMB to establish the China South-South Climate Cooperation Fund and has launched the flagship climate cooperation scheme for low-carbon demonstration, adaptation, mitigation and capacity building in developing countries, which were well-received by the international community and anticipated by developing countries. Now it is urgent to secure the funding sources and finalize the management mechanism to make use of the fund as soon as possible. The South-South Climate Cooperation Fund should not be operated or used in isolation. Rather, it calls for overall planning and inter-departmental coordination. It should be deemed as a seed fund or guiding fund to leverage and marshal other sources of funding, including funds from public and private sectors, and be channeled to projects via grant and commercial and mutually beneficial cooperation to maximize the role of the fund. Various channels for China's foreign aid and foreign cooperation can embed the element of climate cooperation, add and improve project selection and design to accommodate the policy incentives and technological pathway for mitigating natural disasters, saving energy and reducing carbon, so that these channels can be integrated into the framework of South-South climate cooperation to produce more visible and influential outcome on a large scale. Meanwhile, technical assistance and personnel training should be provided to developing countries to address the pressing issues pertaining to the implementation of the Paris Agreement, for example, greenhouse gas inventory development, MRV system building, and long-term low emissions development strategy preparation, etc. Developed countries have counted all channels of their financial support to developing nations as climate support, which greatly exaggerated the amount of their financial assistance. China also needs to sort through its financial support to developing countries under the Belt and Road Initiative, highlight policy incentives for tackling climate change, and integrate various financial sources to enable greater outcome and impact.

On one hand, China should continue to step up the cooperation with the European Union and other developed countries and regions, especially in the field of creating carbon emission trading market. By learning from the experience and lessons of international carbon markets, China can enhance the top-level planning of its carbon market and anticipate potential problems. On the other hand, the design and pathway of China's carbon market based on its institutional system at the current stage of development also provide important reference for other developing countries. So far, many developing nations have expressed their willingness to learn from China's experience in carbon market. With the steady progress of China's carbon market, its impact on the global carbon market will see a major boost, whether measured by its market size or pricing power. China should actively participate in international rule-making during the development of global carbon pricing mechanism and market mechanism under the Paris Agreement. Enhancing international cooperation on the carbon market will not only benefit the long-term development of China's carbon market, but also contribute to China's strategy of opening up and better align with China's efforts to build a community with a shared future for mankind.

The Paris Agreement calls for active participation and voluntary actions by all sectors of society of non-parties except those at national level. At present, various types of climate coalitions and associations formed by local governments, cities, industries, enterprises and civil society are cropping up, such as the World Alliance for Low Carbon Cities and the Oil and Gas Climate Initiative. These organizations not only put forward and develop common low carbon targets and action plans using a "bottom-up" approach, but also advocate code of conduct at the city, industry, corporate, financial investment and social levels. They recommend advanced technical standards, promote carbon labels and low-carbon certification, encourage green investment, strengthen self-discipline actions and cooperation, communicate and share their experience, becoming a vital force in advancing the implementation of Paris Agreement. Currently, with the active participation at all levels, China should make holistic planning and guide the directions for stakeholders to expand all-round influence, catalyze positive momentum in all fields, build its own competitive advantages, and contribute to the improvement of the comprehensive national competitiveness and international influence.

Its' crucial to reinforce exchanges and communication among international think tanks in the field of climate change and harness the role of public diplomacy. The exchanges and cooperation among think tanks contribute to mutual understanding, promote mutual trust, and help publicize China's climate governance concepts and domestic actions and results, disseminate China's values of ecological civilization and expand global influence. This is also an important area for China to enhance international influence and discourse power, and a symbol of the country's soft power. China should foster the growth of climate change think tanks and strengthen the coordination and cooperation among think tanks from different disciplines and with research focus. These think tanks are expected to study China issues under a global horizon and analyze global issues from a Chinese perspective. They should conduct comprehensive research and build discourse power in climate science, policies, technologies, etc., and showcase their influence and credibility in the field of science and policies in order to provide scientific and technical support to China's torchbearer role in global climate governance.

11.3.4 Actively Respond to the Post-Covid Global Governance

The raging COVID-19 represents a global public crisis that will profoundly affect and reshape the world political, economic and trade landscape, and intensify the turbulence of once-in-a-century transformation. Global climate change is a long-term ecological crisis for mankind. Enormous uncertainties exist for the post-Covid global climate cooperation, which will inevitably spark international concern and competition among major countries. The Paris Agreement mandates all parties to report by 2020 their efforts in implementing and enhancing NDC targets and action plans, and to submit long-term low-carbon emission development strategies for 2050. The international community has great expectations on major countries in the world, and China has particularly come under the spotlight. Now, as China is fighting against the pandemic and planning for economic recovery and development, it should take into account both domestic and international situations to make overall planning for domestic high quality development and economic security and global response to climate change and low carbon transition. It should spearhead global energy revolution and low carbon economic development trend, showing the spirit of a major country that is responsible for the common interests of mankind, thereby influencing the trend of international public opinion and dealing with a more complicated international situation after the pandemic.

The COVID-19 has complicated relations between major countries. It has become more prominent that the United States and other western countries have ramped up their long-term strategies and actions to contain China's peaceful rise. The pandemic has plunged the world into a severe economic recession, and the top priority of economic recovery for all governments is to safeguard people's livelihood and shore up the weakness in their industrial chains and enhance the safety and competitiveness of key industries, which inevitably undercuts the priority of the green and low-carbon transformation of energy and economy. Plus, the reduced supply of funding in the wake of the pandemic dampens the investment and actions needed to combat climate change. Frequent exchanges and dialogues are underway between major countries, think tanks, and international agencies on the coordination of the response to the pandemic and the long-term climate crisis. The UN Secretary-General António Guterres urged governments to place addressing climate change at the heart of economic recovery efforts as climate change is the deeper crisis of the world. Therefore, adhering to the post-pandemic policy orientation of "green economic recovery" and reinforcing NDC targets and actions under the Paris Agreement have increasingly become the consensus and shared focus of the international community.

At present, the EU and other developed countries and some international organizations are pushing for the emission reduction target in line with the 1.5 °C future in climate negotiations. In the "European Green Deal", the EU set out a strategy of carbon neutrality by 2050. The UK has amended its Climate Change Act to include carbon neutrality by 2050 into the legislation. The US Climate Alliance—covering California, New York and other states with over 65% of the US population and GDP—has been formed, with most of the states announcing carbon neutrality by 2050. Similar goals have been announced by some developing countries, such as Chile, Ethiopia and some small island states and least developed countries. Quite many countries and cities have declared to go 100% renewable in 2030–2050, and laid out timetable for phasing out coal and coal power as well as gasoline vehicles. The willingness and actions for deep decarbonization worldwide during and after the pandemic might be weakened in the short run, but it has become increasingly urgent for the medium- and long-term transitions towards low-carbon energy system and economy.

Climate change will become an ever-important arena for the post-pandemic competition between major countries. Despite the withdrawal of the Trump administration from the Paris Agreement, the US remains a Party to the UNFCCC. Adhering to the "America First" unilateralism during climate negotiations, the US has repeatedly asked China and other emerging developing countries to shoulder the same obligations as itself. President Joe Biden declared that the US must lead again on the global stage and will rejoin the Paris Agreement on his first day in office, followed by a summit of the world's major emitters to accelerate global progress. Massive emergency investments will be made to secure a clean energy system and economy with net zero emissions by 2050. He also publicly declared to form a united front with democratic allies of shared values and take tough actions against China through carbon border tax and stringent standards for investment and financial subsidies. In response to this, China should cooperate with India and other emerging developing countries and those who share similar positions, adhere to the principle of common but differentiated responsibilities in the light of different national circumstances, boost extensive and pragmatic cooperation and joint actions with countries of different types and stages of development as well as national organizations, strengthen multi-level exchanges and dialogues with developed countries, pursue cooperation on the shared interests of mankind, and defuse the growing tensions with major powers. On the other hand, China should double down on domestic strategies and urgent actions, formulate and announce the targets, strategy and implementation pathway of deep decarbonization by 2050 with the goal of keeping global temperature rise below 2 °C and striving to hold it under 1.5 °C, adapt to the current urgent situation of global emission reduction and showcase the posture of a major country that proactively assumes international responsibilities. It's essential to follow Xi Jinping's thought on ecological civilization and the concept of building a community with a shared future for mankind, and actively promote the development of global governance system and the international cooperation.

Tackling climate change is in the common interest of mankind. It represents the international moral high ground contested by major countries, and calls for close cooperation and concerted actions by all countries. The current global climate governance follows the multilateral mechanism of broad participation and equal consultation by all countries, which requires extensive communication and consultation between developed and developing countries. China has demonstrated growing influence and coordination capability in the field of climate change, and the international community also expects China to show further leadership and bridge more differences. Therefore, there is also room and need for China, the US and other developed countries to carry out dialogue, consultation and cooperation in the post-Covid-19 era. China should give full play to its diplomatic strength in global climate governance and adhere to multilateralism to respond to the unilateralism of the US and resolve the challenges and pressure.

References

1. UNFCCC (United Nations Framework Convention on Climate Change). Kyoto Protocol to the United Nations Framework Convention on Climate Change [2020-03-29]. https://unfccc.int/doc uments/2409
2. COP of UNFCCC. Paris Agreement. United Nations Climate Change, 2015-12-12 [2018-05-18]. https://unfccc.int/resource/docs/2015/cop21/chi/l09c.pdf
3. Jiankun He (2012) The situation of global climate governance after the Paris Agreement and China's leading role. China Environ Manag 10(1):9–14
4. Qimin C, Jiankun He (2013) Cognition, politics and comprehensive assessment of climate equity—How to comprehensively view the application of the "Common Zone" Principle in the Durban Platform. China Popul, Resour Environ 23(6):1–7
5. Jiankun He (2012) Global green, low-carbon development and equitable international system construction. China Popul, Resour Environ 22(5):15–21
6. IPCC (2018) Special report on global warming of 1.5°C.Cambridge University Press, Cambridge, UK
7. Jiankun He (2012) Global green and low-carbon development and fair international system construction. China Popul, Resour Environ 22(5):15–21
8. European Commission. A European green deal [2020-03-29]. https://ec.europa.eu/info/strategy/priorities-2019-2024/european-green-deal_en#documents

9. Jinping Xi (2017) Securing a decisive victory in building a moderately prosperous society in all respects and winning the great victory of socialism with Chinese characteristics for a New Era. People's Publishing House, Beijing

Chapter 12
Conclusions and Policy Recommendations

Based on the analysis from this research project, the following considerations and recommendations are presented for China's formulation of goals and pathways for its long-term low carbon development strategies.

12.1 China Should Formulate Domestic Long-Term Low-Carbon Development Strategy

Climate change is the greatest threat facing the sustainable development of human society. It has become a universal consensus internationally to protect the ecological safety of the earth and to address this challenge cooperatively. The core of climate actions is to mitigate anthropogenic GHG emissions – primarily CO_2 emissions from fossil fuel consumption. This has been the driver for the revolutionary reforms of the energy system and the low-carbon transformation of economic development patterns around the world. Developing advanced energy and deep decarbonization technologies are in the center stage for global technological innovation, thus the center of competition for every nation's advanced technological core competitiveness. Furthermore, a country's ability to develop advanced technologies in the energy and economic low-carbon transition would become a manifestation of a country's economic, trade, and technological competitiveness. Establishing and forming of a clean, low-carbon, safe, and efficient energy system and a green, low-carbon and circular economy have become a fundamental feature of modern energy and economic systems.

The 19th National Congress of the Communist Party of China set the goal and overall strategy of building China into a great modern socialist country by the middle of this century. Xi Jinping's thought on ecological civilization is an important guidance for realizing the harmonious development between human and nature and

© China Environment Publishing Group Co., Ltd. 2022
Institute of Climate Change and Sustainable Development of Tsinghua University et al.,
China's Long-Term Low-Carbon Development Strategies and Pathways,
https://doi.org/10.1007/978-981-16-2524-4_12

building a beautiful China. It's also crucial to embark on the path of green, low-carbon and circular development, promote coordinated governance and comprehensive strategies of the economy, society, environment and climate change, and accelerate low-carbon transformation of energy and economy, which is the ultimate solution to advancing ecological progress and sustainable development. China's policy and best practices in ecological civilization and low-carbon transition of the energy and economic systems will provide Chinese wisdom and experience to the world's energy and economic low carbon revolution, making new contributions to boosting global ecological progress, protecting eco-safety of the Earth, building a community with a shared future for mankind, and spurring new progress of humanity.

The Paris Agreement has set the global climate goal of limiting global temperature rise to well below 2°C and striving to keep it below 1.5°C. This is the political consensus and shared vision of all nations around the world based on the scientific assessment of the IPCC, and is the long-term goal of the global climate governance. Urgent actions on emission reduction are needed to achieve the goal, which involves a global net zero emission of GHGs – or global carbon neutrality - in the second half or even the middle of the century. The Paris Agreement also requires Parties to submit their long-term low greenhouse gas emission development strategies by 2020 in a joint effort to achieve long-term global emissions reduction targets. Currently, 121 countries have announced their goals and vision of achieving carbon neutrality by 2050. Many countries and cities have also vowed to go 100% renewable from 2030 to 2050, together with timetables for the phase-out of coal and coal power and the banning of gasoline vehicles.

The willingness and actions for deep decarbonization worldwide during and after the pandemic may be weakened in the short run, but the urgency of low-carbon energy and economic transition in the medium and long term will be ever more apparent. The international community places high expectations on China to announce ambitious long-term low-carbon development goals and strategies. Following President Xi Jinping's instruction on addressing climate change to "promote domestic development and build international reputation", China should pursue a pathway of long-term emission reduction to keep global temperature rise well below 2°C and ideally below 1.5°C compared to pre-industrial levels, coordinate domestic sustainable development and global climate governance, define the goals and measures for low carbon transformation by 2050, and incorporate them into the overall strategy and planning of China's socialist modernization by the middle of the century.

By the middle of the century, China will achieve the long-term goal of deep decarbonization, build a clean, low-carbon, safe and highly efficient energy system based on new and renewable energy, and largely phase out fossil energy. This will effectively curb the discharge of conventional pollutants from the source, and ultimately ensure the compliance with the PM2.5 targets in key polluting regions in China as the potential of end-of-pipe control diminishing. The clean energy system, primarily consisting of new and renewable energy by 2050, will ensure that most cities achieve the $PM_{2.5}$ concentration less than 15 $\mu g/m^3$, and more than 80% of cities achieve the high standard value of less than 10 $\mu g/m^3$. Meanwhile, the advent of smart renewable energy supply system will provide the fundamental safeguard to national energy security, and decouple economic growth and social development from resource use and environmental impacts.

The goals, scenarios and implementation pathways of China's medium- and long-term low-carbon development should, on the one hand, be grounded on the country's national conditions, capabilities, and modernization process with a outlook to ensuring the realization of China's goals for socialist modernization in the new era, and on the other hand, enable China to undertake its international responsibilities and contributions for ecological safeguard of the Earth and the common cause of humankind in in tandem with its growing comprehensive national strength and international influence. Its responsibilities and process should be different from that of developed nations. The goal and vision of China's low-carbon transition could be achieved through long-term efforts that would be widely recognized by the international community and would enhance China's competitiveness and influence.

12.2 The Long-Term Low-Carbon Development Strategy is a Goal-oriented Comprehensive Development Strategy

Development lies at the heart of the long-term low-carbon development strategy that prompts the transition towards green, low-carbon, and sustainable economic and social development and the harmonious co-existence between people and nature. Therefore, China's long-term low-carbon development strategy should not only ensure the realization of developing China into a great modern socialist country that is prosperous, strong, democratic, civilized, harmonious, and beautiful by the middle of the century, but also follow the pathway of deep decarbonization towards the goal of limiting the global warming well below 2°C and ideally within 1.5°C. In other words, the long-term low-carbon development strategy is oriented toward the both the goal of building a modern socialist country and the global climate target of long-term emission reduction. It is a crucial component of the overall goals and strategies of China's socialist modernization in the new era, and is incorporated into overall all development strategy. China's domestic and international imperatives can be coordinated to rejuvenate the Chinese nation and, in the meantime, enable China to contribute in new ways to global ecological civilization and to building a community with a shared future for humankind.

The long-term low-carbon development strategy should provide holistic coordination of the goals and visions of economic development, social equity and prosperity, resource conservation and efficient utilization, ecological and environmental protection and climate change. It should aim for collaborative and integrated governance to create a win–win situation for sustainable development, rather than a single strategy of emission reduction. The long-term low-carbon development strategy should not only strengthen energy and economic transformation and implement advanced technological innovation policies and measures, but also pay more attention to the implementation of nature-based solutions (NbS). By protecting, restoring, improving and enhancing the functionality of ecosystems, improving climate resilience, strengthening carbon storage of forests, grassland, wetlands, and agricultural land, reducing carbon sources and increasing carbon sinks, this would provide synergetic co-benefits in creating opportunities for economic growth and employment, improving the environmental quality of water, soil and air, ensuring food safety, and protecting biodiversity. Such efforts are guided by the notion of ecological civilization, which prioritizes ecology and promotes the harmonious co-existence between human and nature. NbS offers an important way to tackle climate change and promote coordinated governance and development of the economy, society, environment, and climate change, paving the way for building a beautiful China. Thus, it's vital to integrate policies of climate adaptation and mitigation with those of resource-saving and environmental protection to enable comprehensive management of ecosystems such as mountains, rivers, forests, farmland, lakes and grasslands. Vigorous steps should be taken to enhance afforestation, protect natural forests, return farmland to forests and grasslands, build forest shelterbelts, comprehensively control desertification, preserve water and soil and carry out other ecological projects to reduce carbon sources and increase sequestration. NbS is of great significance for achieving carbon neutrality in the future. In China, the annual increase of carbon sinks such as forests could reach around 700 million tCO_2 by the middle of this century. NbS will play an important role in offsetting the remaining greenhouse gas emissions from industrial production, land use, land use changes and other hard-to-abate sectors.

12.3 Long-Term Low-Carbon Development Strategy Should Demonstrate the Coordination of the "two Stages"

The characteristics of China's current industrialization and urbanization stages makes it more challenging for China to achieve deep decarbonization by the middle of this century compared with developed countries. The targets and pathways of low-carbon transformation should be formulated in coordination with the overall goals and tasks of the "two stages" of socialist modernization.

The 19th National Congress of the Communist Party of the People's Republic of China set forth goals and overall strategies for realizing socialist modernization in the new era. It comprehensively analyzed the international and domestic environment and the conditions for China's development, and developed arrangements and roadmaps for the two stages. In the first stage of 2020 to 2035, to achieve socialist modernization on the whole, China needs to reach the goal of a fundamental improvement in the domestic ecological environment, fulfill its international commitments on emission reduction, promote high-quality economic and social development, and solidify the technological, industrial, policy and market groundwork for achieving deep decarbonization by 2050. In the second stage of 2035 to 2050, on the path to building a great modern socialist country and a beautiful China, China needs to achieve deep decarbonization that is consistent with the goal of limiting global temperature rise well below 2°C and striving to keep it below 1.5°C and pave the way for achieving net zero emissions of all GHGs as soon as possible after 2050.

In the first stage (2020–2035), China's emission reduction targets and measures are mainly based on the internal demand and planning for domestic sustainable development, including resource conservation, environment protection and air quality improvement, in an effort to strengthen the policy lever of CO_2 emission reduction for maximized synergies. This strategy is formulated based on the integrated assessment of domestic and international circumstances. In the second stage (2035–2050), with China's advancing modernization, growing comprehensive national strength and the increasing urgent need to reduce global emissions, China should give more prominence to the 2°C pathway and proactively take on international responsibilities that echo China's growing comprehensive national strength and international influence. From the perspectives of enhancing international competitiveness, impact and torchbearer role in the process of low-carbon transformation, China should continue to ramp up its emission reduction efforts and achieve the deep decarbonization goal of near zero emission by 2050, which mirrors China's comprehensive national strength and international influence as a leading modern socialist country.

12.4 Goals and Transition Pathways for the 2050 Long-Term Low-Carbon Development Strategy Should Be Clarified

China's *Energy Production and Consumption Revolution Strategy (2016–2030)* published in 2016 unveiled the tentative target of non-fossil energy comprising over half of primary energy consumption in 2050, which is used as assumptions for the reinforced policy scenario in this study. Under the reinforced policy scenario, CO_2 emissions will amount to 6 ~ 6.5 billion tons by 2050, still far from achieving the 2°C target. China's long-term low-carbon strategy for the middle of the century should use the 2°C target as guidance for emission reduction, aligned with the current policy of energy conservation and carbon reduction and the 2030 NDC targets and implementation plan to peak emissions by 2030, after which China should continue to accelerate absolute CO_2 emission reduction, transit from the reinforced policy scenario to the 2°C scenario, promote the development of carbon capture and storage (CCS) technology and increase carbon sinks from agriculture and forestry to enable a deep decarbonization pathway that is in line with the 2°C target.

The deep emission reduction pathways for achieving the 2°C target by 2050 entails near-zero CO_2 emissions, implying that the net CO_2 emissions should be reduced to about 2 billion tons, which is equivalent to the projected world average of 1–1.5 tons of per capita emissions by 2050, equivalent to about 80% reduction from the peak of CO_2 emissions around 2030. Non-CO_2 GHG emissions should also peak around the same time with CO_2 emissions. Emission reduction efforts should be continuously strengthened so that by 2050, all GHG emissions will be reduced by 70% from the peak. The total energy consumption should also peak around 2035 at somewhere below 6.5 billion tce and, in 2050, should be reduced by more than 20% from the peak level. Non-fossil energy should account for more than 70% of primary energy consumption, and about 90% of the total electricity generation would come form non-fossil electricity, building a near-zero carbon emission energy system mainly comprising of new and renewable energy with the safety of energy supply fundamentally guaranteed and the emissions of conventional pollutants controlled at source. By 2050, China would see a ten-fold increase in GDP compared to 2005, an over 80% reduction in energy intensity per unit of GDP from that of 2005, and an over 95% drop in CO_2 intensity, and be able to decouple sustainable economic and social development from CO_2 emissions. To enable the deep decarbonization to achieve the 1.5°C target by 2050, China needs to further strengthen economy-wide emission reduction of all GHGs, peak CO_2 emissions around 2025, achieve net-zero emissions of CO_2 and reduce all GHG emissions by 90% by 2050. After 2050, China needs to ramp up efforts to strengthen emission reduction of non-CO_2 GHGs, increase carbon sinks, and use negative emissions technologies such as BECCS and CDR to achieve net-zero emissions of all GHGs and carbon neutrality at the earliest time possible. This would require more arduous efforts, larger investments and incur greater costs.

Establishing positive and clear long-term low-carbon goals and strategy not only provides top-level planning for China's long-term low-carbon development, but also conforms to and promotes the global climate governance process, and reflects the "common but differentiated responsibilities" between developing countries such as China and developed countries. Aiming at deep decarbonization with zero CO_2 emissions by 2050, China will fundamentally form a clean energy system with new and renewable energy as the major energy source and reduce conventional pollutants from the source which will help maintain the $PM_{2.5}$ concentration below the standard of 1 5 $\mu g/m^3$ in key polluting regions and even below 10 $\mu g/m^3$ in more than 80% of cities, contributing to the construction of a Beautiful China.

China's 2050 deep decarbonization goal will require large investment for the transformation of energy infrastructure and the large-scale grid integration of renewable energy that involves construction of energy storage facilities and smart grids. Specifically, the total investment required for the 2°C pathway is around 100 trillion RMB during the period from 2020 to 2050, equivalent to 1.5–2.0% of GDP annually; and the 1.5°C pathway requires around 138 trillion RMB investment for the energy system over the next 30 years, which is more than 2.5% of GDP annually. The rapid transformation of energy and power technology system would initially increase the cost of energy and electricity supply in the short-term, which would decline in the long-term. In addition, the rapid transformation would result in early retirement of traditional energy infrastructure including coal-fired power plants and incur economic losses from stranded assets. Wholistically speaking, the low-carbon transformation of energy and economy under the 2°C or even 1.5°C scenario is technically feasible and economically viable, but needs to overcome obstacles in areas such as technology diffusion, financing mechanism, and management system. The development of advance energy technologies and transformation of the energy

system will foster new drivers of economic growth and employment, accelerate electrification of end-user sectors such as industry, transportation and building, enhance
energy efficiency, promote the digital transformation of industries and society, and
enable China to spearhead global economic and social transformation.

China's long-term low-carbon emission strategy should also strengthen regulation
and control of non-CO_2 GHGs such as methane, nitrous oxide and fluorine gases,
so as to achieve economy-wide deep reduction of all GHGs emissions. At present,
non-CO_2 GHGs account for 16% of the total GHG emissions. As the costs of initial
emission reduction are relatively low, it is possible for non-CO_2 GHG emissions
to peak around the same time with CO_2 emissions. However, the marginal costs
for achieving deep emission reduction would increase steeply. Deep reduction of
non-CO_2 GHG emissions by 2050 would be more difficult than CO_2 reduction.
Therefore, it requires R&D of groundbreaking deep emission reduction technologies
and proliferation of GHG removal and other negative emissions technologies such as
CDR, in order to offset the remaining non-CO_2 emissions under the future scenario
of deep decarbonization.

12.5 China Should Enhance and Update its 2030 NDCs

China's NDCs under the Paris Agreement include: by 2030, the CO_2 emissions per
unit of GDP should fall by 60–65% from the 2005 level; the share of non-fossil fuels
in primary energy consumption should increase to around 20%; the forest stock
should increase by 4.5 billion m^3 on the 2005 level; and total CO_2 emissions should
peak in around 2030 and making best efforts to peak earlier. These targets were
formulated based on China's 2020 NDCs, which China pledged during the 2009
Copenhagen Climate Conference. With the efforts made during the 12th and 13th
Five-Year plan periods, by the end of 2019, China's CO_2 emissions per unit of GDP
fell by 48.1% from the 2005 level, which exceeded the target of 40–45% by 2020;
the forest stock increased by 1.3 billion m^3 compared to 2005, ahead of the 2020
timeframe; the proportion of non-fossil fuels reached 15.3% by 2019, exceeding the
target of 15% by 2020. These progress and achievements laid the groundwork for
the implementation of the 2030 NDCs, setting the stage for the further enhancing
and updating the 2030 NDCs.

The Paris Agreement requires all Parties to submit their reports on the implementation and update of NDCs and actions by the end of 2020. Currently, 114 countries
have announced that they will enhance and update their NDCs; the EU has pledged to
increase the 2030 emissions reduction target from 40% to 50–55%. The international
community is eyeing China for its enhanced and updated NDCs for 2030, and the
expectations are running high.

China adheres to the new development concept in a new era, promoting ecological
progress, enhancing policies for the establishment of a green, low-carbon, circular
industrial system as well as a clean, low-carbon, safe and efficient energy system.
Building on the targets and policy measures of strengthened energy conservation and

CO_2 emissions reduction during the 14th Five-Year Plan period, the 15th Five-Year Plan would keep and reinforce the trend and pace of the low carbon transition. By 2035, China would attain goal for the first stage of modernization, with a modern country taking shape. If we assume that China's GDP will double over the next 15 years and the per capita GDP is calculated at the current constant price of USD 20,000, then the average annual GDP growth rate is expected to be about 4.8% over the next 15 years. The average annual GDP growth during the 14th Five-Year Plan period might slow down compared with that of the 13th Five-Year Plan period, but is still expected to exceed 5%. From 2020 to 2030, China's annual GDP growth rate will remain at around 5% on average, and its energy intensity per unit of GDP will continue to decrease at relatively high rate of no less than 14%. By 2030, China would keep its total energy consumption under 6 billion tce, fulfilling the total energy consumption target set by the Energy Production and Consumption Revolution Strategy (2016–2030) while sustaining economic growth. New energy and renewable energy will continue to grow rapidly. While meeting the rising energy demand, the development potential of renewable energy would be further released, and its share would increase rapidly to about 25% by 2030, exceeding the NDC target of 20%. By then, the proportion of primary energy in power generation would increase from the current 45% to about 50%, which is in line with the China's target of generating 50% of the total electricity from non-fossil fuels by 2030 as stated in *the Energy Production and Consumption Revolution Strategy (2016–2030)*. Owning to the dual effect of energy conservation and substitution, by 2030, CO_2 emissions per unit of GDP would fall by 65–70% from the 2005 level. The installed power generation capacity of hydro, wind and solar energy would reach approximately 500 GW respectively, producing new sources of economic growth and creating new jobs. Under this circumstance, CO_2 emissions would enter a peak plateau by 2025 and reach a stable peak by 2030. According to forestry development plans, China's forest stock would amount to 21 billion m^3 by 2030, an increase of 5.5 billion m^3 more than that in 2005. In 2030, there is potential and possibility for all NDC targets to be over-achieved and achieved ahead of schedule.

Based on the above analysis, China should further enhance and update the 2030 NDC targets, which would promote low-carbon transformation and high-quality development of the energy system and the economy, respond to the expectations of the international community, and advance the global climate governance.

12.6   China's Goals and Policy Measures for Addressing Climate Change Should Be Strengthened in the 14th Five-Year Plan

China is the first country to have contained the Covid-19 pandemic and achieved economic recovery. The global spotlight is on how to embody the concept of green and low-carbon development and how to strengthen targets and policy guidance for energy conservation and carbon reduction in the 14th Five-Year Plan. They are also expected to become the "weathervane" for the world's realization of post-pandemic high-quality and sustainable green recovery. The 14th Five-Year plan period is a key stage for China to accelerate economic transformation and upgrading and promote high- quality development. The 14th Five-Year Plan should be aligned with the short- and- long-term efforts to enhance China's 2030 NDC targets as a whole. It's recommended to strengthen climate goals and measures in the 14th Five-Year Plan and to include a dedicated plan for addressing climate change in it to specify the goals, tasks, policies and actions.

The impact of Covid-19 pandemic on the global industrial chain as well as supply and demand will create uncertainties for the future economic development and energy transition. While securing steady progress of its economy and society, China needs to follow the new concept of development, accelerate industrial transformation and upgrading and the high-quality development, endeavor to develop the digital economy and high-tech industries technologies, and spur the transition of an economy driven by scale and speed to one that centers on quality and efficiency. During the 14th Five-Year Plan period, China should aim for a drop in energy intensity per unit of GDP of at least 14% and a cap of total energy consumption at 5.5 billion tce while securing an annual GDP growth of over 5%. As wind, solar and other renewable power become increasingly cost-competitive with coal-fired power generation, non-fossil energy during the 14th Five-Year Plan period could maintain an average annual growth of about 7%, the average annual growth rate during the 13th Five-Year Plan period. There is considerable potential for the development of corresponding energy storage systems and smart grids. By 2025, non-fossil energy would account for 20% of primary energy consumption. With the rapid development of new energy and renewable energy, new power demand during the 14th Five-Year Plan period can be mostly met by the increasing power supply from non-fossil fuels to effectively curb the rebound of coal power supply and coal consumption. During the 14th Five-Year Plan period, newly built coal-fired power plants should be strictly controlled, unless they are built for peak shaving, district heating and other essential needs. The dual effect of energy saving and energy mix improvement could reduce the CO_2 intensity per unit of GDP by 19–20%, and reduce the total CO_2 emissions to less than 10.5 billion tons at the end of the period.

It's important for the 14th Five-Year Plan to keep the targets of energy-saving and CO_2 emission reduction stated in the 12th and 13th Five-Year Plans. In particular, the binding target for the reduction of CO_2 intensity per unit of GDP should be kept and highlighted because it's a landmark for China's NDC commitment under the Paris Agreement. Based on this, China could implement and realize the phased NDC arrangements of cutting CO_2 emissions per unit of GDP by 60–65% from the 2005 level. It would showcase China's consistent climate policies and positive attitude of continuously reinforcing policy actions, and help China respond to the international community's concerns and expectations for China to continue to drive the global cooperation on climate change.

The 13th Five-Year Plan put forward the expected goal of holding total energy consumption under 5 billion tce at the end of the period, and this goal will be achieved.

During the 14th Five-Year Plan period, it is recommended to set total CO_2 emissions cap target to buttress China's efforts to implement and fulfill its commitment of peaking CO_2 emissions in around 2030. In fact, setting the targets for total energy consumption and non-fossil fuels as a percentage of primary energy consumption would be sufficient to control total CO_2 emissions. Adding CO_2 cap target would better underscore China's policy determination and new actions for tackling climate change. During the 14th Five-Year Plan period, CO_2 peaking activities should be conducted at the regional and sectoral level, and developed provinces and cities in the eastern coast and energy-intensive and high-carbon emission industries should be encouraged to set their peak targets first. Furthermore, the Ten-Year Action Plan for Peaking CO_2 Emissions should be formulated and executed in a way that seamlessly connects with the 14th Five-Year Plan and ensures regional coordination, both near and long term, thus providing the policy safeguard for an earlier peak of CO_2 emissions.

12.7 China Should Improve Policies and Institution Building for Addressing Climate Change

Institutional building for addressing climate change is a fundamental guarantee for China's long-term low-carbon development strategy, as well as a critical component of the institutional building for ecological civilization. China should establish a sound system of laws, regulations, and fiscal, taxation and financial policies and implementation mechanisms.

Climate legislation should be strengthened to secure the implementation of climate strategies, mechanisms and policy systems and the realization of long-term emission reduction targets. The enactment of dedicated laws is conducive to providing clear legal guidelines for nationwide climate actions, and to ensuring the realization of long-term strategic goals and measures against climate change. It's essential to strengthen and enhance the fiscal, financial and taxation policy framework to underpin the long-term low-carbon development, and governments should increase financial input and tax incentives to create an investment and financing mechanism that favors a low-carbon transition. It's also crucial to establish green investment guidelines, improve the green credit mechanism, issue green securitization products and green bonds, and put in place green credit incentives and guarantee mechanisms to encourage enterprises, financial institutions and private investors to invest in green growth.

Efforts should be made to accelerate the construction of a national carbon market and harness the fundamental role of market mechanisms in promoting low-carbon energy and economy transformation. The 13th Five-Year Plan period has witnessed notable progress in the carbon market pilots in five cities and two provinces, which has generated positive impact across the world and paved the way for a national carbon market. Now, it's vital to follow the direction of President Xi Jinping to launch a national carbon market, enhance policies and regulations, start the trading as early as possible, and expand its coverage from the power industry to the iron and steel, cement, aluminum, petrochemical, chemical, aviation and other energy-intensive industries as soon as possible, leverage the crucial role of market mechanisms in reducing GHGs emissions, and reduce economy-wide emission reduction costs. The market mechanisms should be combined with government regulatory measures to ensure the realization of national emission reduction targets. Market carbon price signals should be used to guide private investment and industrial technology upgrading so as to promote enterprises to conserve energy and reduce CO_2 emission reduction. At the same time, carbon allowances should be properly distributed to shape a system for capping total CO_2 emissions and a carbon emission monitoring, reporting and verification system at the corporate level. This would facilitate the institutional building for addressing climate change. A robust carbon market system with sound carbon pricing mechanism not only represents the fundamental institutions of China's long-term low-carbon development, but also is an effective institutional tool for hedging and countering the "carbon border adjustment" that the EU and the US have planned to adopt in international trade. It would also enable China to actively play an influential and guiding role in the establishment and development of a global carbon price mechanism.

It's important for China to deepen reforms and strengthen the leading role of the government in implementing the low-carbon transformation strategy. A target-oriented accountability for energy conservation and carbon reduction should be reinforced for governments at all levels, and targets for energy conservation and carbon reduction should be included in national and local five-year plans. It is also important to make innovations in the macro-control system for energy, establish and enhance energy legislation, and reform and improve the fiscal, taxation, and financial policy framework for promoting low-carbon development, the pricing mechanism for energy products, and the resource and environmental taxes and fees system. It's crucial to strengthen the reform of the energy market mechanism, establish a uniform and open market structure and system that promote just, fair and effective competition, and strengthen the standards for energy-efficient technologies and energy efficiency labeling and policies for industry access. Guided by the national medium- and long-term strategies and goals, government binding targets, mandatory standards, fiscal, tax, and financial policies should be reinforced and integrated with market mechanisms to improve the institutional safeguard for energy revolution and low-carbon development.

China needs to strengthen the institutional building and policy support systems for technological innovation and industrialization of cutting-edge technologies, plan ahead the R&D and industrialization of revolutionary deep decarbonization technologies, leverage China's huge market to support the technological upgrading of industries and the reform of the energy system via technological innovation, so as to sharpen the technological strength and market competitiveness of energy enterprises. Under the future 2°C scenario of deep decarbonization, the marginal cost of emission reduction in most cases would feature a steep rise in a nonlinear pattern, which requires revolutionary and strategic breakthroughs in frontier technologies. It's important to step up the R&D and penetration of deep CO_2 emission reduction technologies such as large-scale energy storage and smart grid technologies in the case of large-scale grid integration of renewable energy, BECCS for negative CO_2 emissions, technologies for the production, storage and use of hydrogen as a clean, zero-carbon and secondary energy source, and zero-carbon technologies for producing chemical, steel, cement, petrochemical and other raw materials. Moreover, it is necessary to ramp up R&D and demonstration of technologies for deep emission reduction of methane and other non-CO_2 GHGs, so as to enable deep emission reduction of all GHGs. China should spearhead the R&D and demonstration of these revolutionary technologies, achieve technological breakthroughs and industrialization as soon as possible, and make these technologies mature and affordable, so as to play a leading role in driving the low-carbon transformation of the global economy and society.

12.8 China Should Take a Proactive Role in Responding to the New Situation of Global Climate Governance in the Post-Covid-19 Era

The world economy has been hit hard by the Covid-19 pandemic. The US's attempts in reshaping the world economic order in the wake of the pandemic features a clear strategic intention to contain and isolate China across the board. Coping with climate change would be an important area of competition among major economies. China is also confronted with the arduous task of restoring the economy, securing people's livelihood and shoring up weakness in the industrial chain, with a grim situation with uncertainties both at home and abroad. Though the current and short-term scourge of Covid-19 might weaken the global response to climate change and low-carbon transition, the long-term trend would remain unchanged and it would become increasingly urgent for the low-carbon transition. In the post-Covid-19 era, it has become a consensus of the international community to adhere to the policy of "green recovery and low-carbon transformation" and to enhance the NDC targets and international cooperation under the Paris Agreement.

China's achievements in implementing its climate strategy and curbing CO_2 emissions are globally recognized. China has made a historical contribution to the adoption and enforcement of the Paris Agreement, and has become the backbone of the developing countries in the climate tussle between developed and developing nations. Tackling the global climate crisis represents the shared interest of the entire humanity and the international moral high ground for major power competition. At present, global climate governance adheres to a multilateral mechanism featuring broad participation and consultation on an equal footing, and requires extensive communication and consultation between developed and developing countries. China has already exhibited increasing influence and coordination capability on climate change issues, and the international community also expects Chinese to further assume leadership and play a "bridging" role. Despite the plummeting relations between China and the US, there are still space and opportunities for the two sides to consult, negotiate and cooperate on climate change. Active participation in and promotion of global climate governance would help China build up the strategic support of developing countries. China should continue along the path of multilateralism in respond to unilateralism of the US and should expand its diplomatic advantage in the field of climate change, striving to resolve any challenges or pressure.

With the complex and hugely uncertain international circumstances in the post-Covid era, China must actively promote global climate governance and cooperation and should foster the institution building of a fair, just and win–win global climate governance system. China should uphold the principles of equity, common but differentiated responsibilities and respective capabilities, strengthen exchanges and communications with countries and nation groups with varied national conditions and in different stages of development, and promote comprehensive, balanced and effective implementation of the elements and provisions pertaining to mitigation, adaptation, funding, technology, capacity building and transparency in the Paris Agreement. China should strive to promote a conclusion in the negotiations on Article 6 of the Paris Agreement and an all-round adoption of the Paris Rulebook at COP26 and promote pragmatic cooperation of the international community. China should endeavor to make addressing climate change a pioneering and successful example of Xi Jinping's thoughts on building a global ecological civilization and a community with a shared future for mankind.

Actively promoting South-South cooperation in addressing climate change is a key step for China to expand its strategic support in climate field and live up to its responsibility as a major country. In the green BRI development and project cooperation, China should adhere to green and low-carbon guidelines and standards, and strictly rein in the export of coal-fired power plants and other high-carbon projects. It's important to fulfill President Xi Jinping's pledge at the Paris Climate Conference to set up a 20 billion RMB South-South Climate Cooperation Fund to support and assist developing countries in capacity building, and ease the pressure from the international community for China's funding contribution.

To sum up, amid the complex circumstances at home and abroad, China should maintain its strategic focus, ramping up the planning and actions on climate change at home. Follow the guidance of building socialism in the new era, China needs to coordinate short-term and long-term aspects, and domestic and international interests. China should specify the targets, policies and measures for energy conservation and CO_2 emission reduction in the two documents which are due to be submitted to the UNFCCC Secretariat by 2020, namely, the report on the implementation and update of 2030 NDCs and the low-carbon emission strategy by the middle of the century, and in China's 14th Five-Year Plan. On the domestic front, China should promote industrial transformation and upgrading and high-quality development, and foster a coordinated governance and win–win situation for economic development, energy security, environmental protection and addressing climate change. On the international stage, China should respond to the general expectations of the international community, resolve such pressures as emissions reduction and financing, thereby building up a reputation of a major developing country taking responsibility for a shared vision of the world and constantly improving its international competitiveness and influence.

Appendices

Tables

See Tables A.1, A.2, A.3, A.4

© China Environment Publishing Group Co., Ltd. 2022

Institute of Climate Change and Sustainable Development of Tsinghua University et al.,
China's Long-Term Low-Carbon Development Strategies and Pathways,
https://doi.org/10.1007/978-981-16-2524-4

Table A.1 Data of energy consumption & CO$_2$ emissions in 2030 & 2050 under the policy scenario

2030

		Power consumption		Primary energy direct consumption(million tce)					End use energy consumption(million tce)	CO$_2$ direct emissions(Gt)
		Power(trillion kWh)	electrothermal equivalent(million tce)	Coal	Oil	Gas	Non-fossil	Subtotal		
End users	Industrial	5.7	700	1,180	530	200	70	1,970	2,670	4.54
	Building	2.6	320	150	0	350	40	540	860	0.97
	Transport	0.3	40	0	540	20	0	560	600	1.16
	Others	0.6	80	100	50	30	50	230	310	0.41
	Total	9.2	1,130	1,430	1,120	590	160	3,300	4,430	7.07
Power supply		9.2		1,450	0	110	1,190	2,750		4.01
Total primary energy consumption				2,880	1,120	700	1,350	6,060		11.08
Primary energy consumption mix(%)				4,760	1,840	1,160	2,240	0		

2050

		Power consumption		Primary energy direct consumption(million tce)					End use energy consumption (million tce)	CO$_2$ direct emissions(Gt)
		Power(trillion kWh)	electrothermal equivalent(million tce)	Coal	Oil	Gas	Non-fossil	Subtotal		
End users	Industrial	6.2	760	900	390	320	80	1,680	2440	3.69
	Building	4.1	500	90	0	370	40	500	990	0.83
	Transport	0.3	40	0	520	20	10	550	590	1.11
	Others	0.8	100	30	20	20	30	100	190	0.15
	Total	11.4	1,400	1,020	920	720	160	2,820	4220	5.79
Power supply		11.4		1,160	0	150	2,110	3,410		3.29
Total primary energy consumption				2,180	920	870	2,260	6,230		9.08
Primary energy consumption mix(%)				3,490	1,480	1,400	3,630			

Table A.2 Data of energy consumption & CO_2 emissions in 2030 & 2050 under the reinforced policy scenario

2030

		Power consumption		Primary energy direct consumption(million tce)					End use energy consumption(million tce)	CO_2 direct emissions(Gt)
		Power(trillion kWh)	electrothermal equivalent(million tce)	Coal	Oil	Gas	Non-fossil	Subtotal		
End users	Industrial	5.7	700	1,180	530	200	70	1,970	2,670	4.54
	Building	2.6	320	150	0	350	40	540	860	0.97
	Transport	0.3	40	0	540	20	0	560	600	1.16
	Others	0.6	80	100	50	30	50	230	310	0.41
	Total	9.2	1,130	1,430	1,120	590	160	3,300	4,430	7.07
Power supply		9.2		1,450	0	110	1,190	2,750		4.01
Total primary energy consumption				2,880	1,120	700	1,350	6,060		11.08
Primary energy consumption mix(%)				4,760	1,840	1,160	2,240	0		

2050

		Power consumption		Primary energy direct consumption(million tce)					End use energy consumption (million tce)	CO_2 direct emissions(Gt)
		Power(trillion kWh)	electrothermal equivalent(million tce)	Coal	Oil	Gas	Non-fossil	Subtotal		
End users	Industrial	6.2	760	900	390	320	80	1,680	2440	3.69
	Building	4.1	500	90	0	370	40	500	990	0.83
	Transport	0.3	40	0	520	20	10	550	590	1.11
	Others	0.8	100	30	20	20	30	100	190	0.15
	Total	11.4	1,400	1,020	920	720	160	2,820	4220	5.79
Power supply		11.4		1,160	0	150	2,110	3,410		3.29
Total primary energy consumption				2,180	920	870	2,260	6,230		9.08
Primary energy consumption mix(%)				3,490	1,480	1,400	3,630			

Table A.3 Data of energy consumption & CO$_2$ emissions in 2030 & 2050 under the 2 °C scenario

2030

		Power consumption		Primary energy direct consumption(million tce)					End use energy consumption(million tce)	CO$_2$ direct emissions(Gt)
		Power (trillion kWh)	electrothermal equivalent(million tce)	Coal	Oil	Gas	Non-fossil	Subtotal		
End users	Industrial	6.1	740	980	370	290	90	1730	2470	3.82
	Building	2.5	310	80	0	270	60	410	720	0.65
	Transport	0.4	50	0	500	20	10	530	580	1.07
	Others	0.6	80	40	10	10	30	90	170	0.14
	Total	9.6	1180	1100	870	590	190	2760	3940	5.69
Power supply				1340	0	120	1430	2890		3.73
Total primary energy consumption				2440	870	710	1620	5640		9.42
Primary energy consumption mix(%)				4320	1550	1260	2860	0		

2050

		Power consumption		Primary energy direct consumption(million tce)					End use energy consumption(million tce)	CO$_2$ direct emissions(Gt)
		Power(trillion kWh)	electrothermal equivalent(million tce)	Coal	Oil	Gas	Non-fossil	Subtotal		
End users	Industrial	7.8	960	240	140	170	140	690	1650	1.2
	Building	3.7	450	10	0	180	80	260	710	0.31
	Transport	0.8	100	0	250	20	40	300	400	0.55
	Others	0.9	110	10	10	10	10	30	140	0.04
	Total	13.1	1610	250	400	370	270	1290	2900	2.1
Power supply		13.1		220	0	150	3540	3910		0.83
Total primary energy consumption				470	400	520	3810	5200		2.92
Primary energy consumption mix(%)				910	770	1000	7320			

Table A.4 Data of energy consumption & CO_2 emissions in 2030 & 2050 under the 1.5°C scenario

2030

		Power consumption/production		Primary energy direct consumption(million tce)					End use energy consumption (million tce)	CO_2 direct emissions(Gt)
		Power(trillion kWh)	electrothermal equivalent(million tce)	Coal	Oil	Gas	Non-fossil	Subtotal		
End users	Industrial	6.3	770	730	200	260	120	1,300	2,070	2.76
	Building	2.6	320	50	0	270	60	370	690	0.56
	Transport	0.6	70	0	480	20	10	510	580	1.04
	Others	0.6	80	20	20	10	10	50	130	0.09
	Total	10	1,230	800	700	550	200	2,240	3,480	4.45
Power supply		10		1,060	0	120	1,840	3,010		2.99
Total primary energy consumption				1,860	700	660	2,030	5,250		7.44
Primary energy consumption mix(%)				3,540	1,320	1,260	3,870			

2050

		Power consumption/production		Primary energy direct consumption(million tce)					End use energy consumption (million tce)	CO_2 direct emissions (Gt)
		Power(trillion kWh)	electrothermal equivalent(million tce)	Coal	Oil	Gas	Non-fossil	Subtotal		
End users	Industrial	8.0	980	80	70	60	220	430	1,410	0.46
	Building	3.9	480	0	0	40	90	140	620	0.08
	Transport	1.6	200	0	70	10	60	150	350	0.17
	Others	0.8	90	10	10	10	20	40	140	0.04
	Total	14.3	1,750	90	150	130	400	770	2,520	0.75
Power supply		14.3		180	0	150	3,900	4,230		0.72
Total primary energy consumption				270	150	280	4,310	5,000		1.47
Primary energy consumption mix(%)				540	300	550	8,610			

Annex

Annex I, II and non-annex in Chapters 6 and 11 are referring to three main groups under the UNFCCC (United Nations Framework Convention on Climate Change). Annex I Parties include the industrialized countries that were members of the OECD (Organisation for Economic Co-operation and Development) in 1992, plus countries with economies in transition (the EIT Parties), including the Russian Federation, the Baltic States, and several Central and Eastern European States. Annex II Parties consist of the OECD members of Annex I, but not the EIT Parties. Non-Annex I Parties are mostly developing countries. For more detail, please find in the link: https://unfccc.int/parties-observers.

© China Environment Publishing Group Co., Ltd. 2022
333
Institute of Climate Change and Sustainable Development of Tsinghua University et al.,
China's Long-Term Low-Carbon Development Strategies and Pathways,
https://doi.org/10.1007/978-981-16-2524-4

Printed in the United States
by Baker & Taylor Publisher Services